Ergonomics and Human Factors
for a Sustainable Future

Andrew Thatcher
Paul H. P. Yeow
Editors

Ergonomics and Human Factors for a Sustainable Future

Current Research and Future Possibilities

Editors
Andrew Thatcher
Psychology Department
University of the Witwatersrand
Johannesburg, South Africa

Paul H. P. Yeow
School of Business
Monash University Malaysia
Bandar Sunway, Malaysia

ISBN 978-981-10-8071-5   ISBN 978-981-10-8072-2   (eBook)
https://doi.org/10.1007/978-981-10-8072-2

Library of Congress Control Number: 2018933381

This Palgrave Macmillan imprint is published by the registered company Springer Nature Singapore Pte Ltd. part of Springer Nature.
The registered company address is: 152 Beach Road, #21-01/04 Gateway East, Singapore 189721, Singapore

CONTENTS

1   **Introduction**   1
Andrew Thatcher and Paul H. P. Yeow

2   **A Sustainable System-of-Systems Approach: Identifying the Important Boundaries for a Target System in Human Factors and Ergonomics**   23
Andrew Thatcher and Paul H. P. Yeow

3   **Defining Sustainable and "Decent" Work for Human Factors and Ergonomics**   47
Knut Inge Fostervold, Peter Christian Koren, and Odd Viggo Nilsen

4   **Human Factors Issues in Responsible Computer Consumption**   77
Paul H. P. Yeow, Wee Hong Loo, and Uchenna Cyril Eze

5   **Human Factors and Ergonomics in Interactions with Sustainable Appliances and Devices**   111
Kirsten M. A. Revell and Neville A. Stanton

6 Human Factors and Ergonomics in the Individual
Adoption and Use of Electric Vehicles 135
Thomas Franke, Franziska Schmalfuß, and Nadine Rauh

7 HFE in Biophilic Design: Human Connections
with Nature 161
Ryan Lumber, Miles Richardson, and Jo-Anne Albertsen

8 Building Sustainable Organisations: Contributions
of Activity-Centred Ergonomics and the Psychodynamics
of Work 191
Claudio Marcelo Brunoro, Ivan Bolis,
and Laerte Idal Sznelwar

9 Green Buildings: The Role of HFE 211
Erminia Attaianese

10 Human Factors and Ergonomics: Contribution
to Sustainability and Decent Work in Global Supply
Chains 243
Klaus J. Zink and Klaus Fischer

11 Natural Resource Use, Institutions, and Green
Ergonomics 271
Ashutosh Sarker, Wai-Ching Poon, and Gamini Herath

12 Examining the Challenges of Responsible Consumption
in an Emerging Market 299
Fandy Tjiptono

13 Promoting Green Technology Financing: Political Will
and Information Asymmetries 329
Jothee Sinnakkannu and Ananda Samudhram

14 **Lives We Have Reason to Value**   355
    Dave Moore

15 **Ergonomics and Human Factors for a Sustainable Future:**
    Suggestions for a Way Forward   373
    Andrew Thatcher and Paul H. P. Yeow

Index   391

Notes on Contributors

Jo-Anne Albertsen is a graduate student in the College of Life and Natural Sciences at the University of Derby, UK.

Erminia Attaianese is an Assistant Professor at the University of Naples Federico II, Italy. She is an architect and a registered European Ergonomist. Her PhD looked at rehabilitation in the built environment.

Ivan Bolis completed his doctorate at the University of São Paulo, Brazil, where he is doing post-doctoral work in Production Engineering Department. His doctorate was on creating sustainability in organization.

Claudio Marcelo Brunoro is a Professor at the University of São Paulo, Brazil. His doctoral thesis was on the work and sustainability theme— Work and Sustainability: contributions of ergonomics of work activity and psychodynamics. He has experience in the area of production engineering work, technology and organization, activity ergonomics, work psychodynamics, and sustainability.

Uchenna Cyril Eze is an Associate Professor at the International College, Beijing Normal University-Hong Kong Baptist University. He is currently acting Associate Dean.

Klaus Fischer was recently the Deputy Scientific Director at the Institute for Technology and Work at the University of Kaiserslautern, Germany. Klaus Fischer studied industrial engineering with specialization in chemistry at the Technical University of Kaiserslautern. After studying, he worked as a research assistant at the University of Karlsruhe (TH) in projects on sustainable energy system development in the countries of the ASEAN states.

Knut Inge Fostervold is a Psychologist and an Associate Professor of Psychology at the University of Oslo with emphasis on quantitative research methodology. He has been a board member of the Norwegian Association for Ergonomics and Human Factors for several years. He has been head of the Ergonomics and council member of the International Ergonomics Association (IEA). Knut was recently made a fellow of the International Ergonomics Association.

Thomas Franke is a Professor of Engineering Psychology and Cognitive Ergonomics at the University of Luebeck, Germany. His research interests are in sustainability, energy efficiency, health, mobility, and automation. He focuses on aspects of user acceptance, user experience, users-system interaction, and interface design.

Gamini Herath is Professor of Economics at the School of Business, Monash University Malaysia. He is the Leader of the Sustainable Development cluster of Global Asia in the 21st Century (GA21) multidisciplinary research platform of Monash University Malaysia. He is a Senior Fellow of the Jeffrey Sachs Centre for Sustainable Development, Sunway University, Malaysia.

Peter Christian Koren is an Associate Professor in the Department of Media, Culture and Social Studies at the University of Stavanger, Norway.

Wee Hong Loo is a PhD graduate in the School of Business, Monash University Malaysia.

Ryan Lumber is a doctoral student and a lecturer at De Montfort University, UK. Ryan is working on nature connectedness (the subjective sensation of belonging to a wider natural community), specifically on the pathways to nature connection and its associated benefits to wellbeing and pro-environmentalism.

Dave Moore is a Senior Lecturer at the Auckland University of Technology, New Zealand. Dave was past co-chair of the Human Factors and Sustainable Development Technical Committee of the International Ergonomics Association. He is a past president of Human Factors and Ergonomics Society, New Zealand (HFESNZ).

Odd Viggo Nilsen works at the Akershus County Council in Oslo, Norway, where he is a human resources consultant with experience in health safety and the environment, management development, and organizational surveys.

Wai-Ching Poon is Associate Professor of Economics at the School of Business, Monash University Malaysia.

Nadine Rauh is a graduate student in Psychology at the Chemnitz University of Technology, Germany, where she specializes in work psychology.

Kirsten M. A. Revell is a Research Fellow within Engineering and the Environment at the University of Southampton, UK. She recently worked on the Future Flight Deck project focusing on evaluating how technology could enable changes not only to the design of the flight deck cockpit but also the tasks and functions undertaken by pilots and configurations of pilot crewing (e.g. single pilot operations). She is currently looking at the design of human interactions with autonomous vehicles, as part of the Hi:DAV project, funded by Jaguar Land Rover and EPSRC.

Miles Richardson is Head of Psychology and Deputy Head of Life Sciences at the University of Derby, UK. He is an associate editor of the journal *Ergonomics*, and he also leads the Nature Connectedness Research Group.

Ananda Rao Samudhram is a Lecturer in the Department of Accounting at the School of Business, Monash University Malaysia.

Ashutosh Sarker is Senior Lecturer of Economics in the Department of Economics at Monash University Malaysia.

Franziska Schmalfuß is a graduate student in Psychology at the Chemnitz University of Technology, Germany, where she specializes in electromobility, usability, and human factors.

Jothee Sinnakkannu is an Associate Professor in the Banking and Finance at the School of Business, Monash University Malaysia.

Neville Stanton is Professor of Human Factors in Transport within Engineering and the Environment at the University of Southampton, UK. He is a Fellow of the British Psychological Society and the Chartered Institute of Ergonomics and Human Factors. The Institute of Ergonomics and Human Factors awarded him the Otto Edholm Medal in 2001 for his contribution to basic and applied ergonomics research, the President's Medal in 2008 to the HFI-DTC, and the Sir Frederic Bartlett Medal in 2012 for a lifetime contribution to ergonomics research. In 2007 the Royal Aeronautical Society awarded him the Hodgson Medal and Bronze Award with colleagues for their work on flight deck safety.

Laerte Idal Sznelwar is a Professor in the Department of Production Engineering at the Polytechnic School of the University of São Paulo, Brazil. He is a member of the team of the Laboratoire de Psychologie du Travail et de l' Action at CNAM.

Andrew Thatcher is Chair of Industrial/Organisational psychology in the Psychology Department at the University of the Witwatersrand, South Africa. He is Chair of the Human Factors and Sustainable Development Technical Committee of the International Ergonomics Association. He is one of the editors of the journal *Ergonomics* and is the ergonomics expert on the World Green Building Council's technical committee on wellbeing, health, and productivity in green buildings. He is the immediate past president of the Ergonomics Society of South Africa.

Fandy Tjiptono is currently an Associate Professor at the School of Business, Monash University Malaysia.

Paul H. P. Yeow is an Associate Professor at the School of Business, Monash University Malaysia. He was the former Head of Discipline in E-Business/Information Systems and Marketing Department at Monash University. He was a recipient of Monash University's Pro-Vice Chancellor's Research Award. He has been a guest editor and an editorial board member for the journal of *Applied Ergonomics*.

Klaus J. Zink is a Professor at the Institute for Technology and Work at the University of Kaiserslautern, Germany. He was President of the Gesellschaft für Arbeitswissenschaft from 1997 to 1999. He was also a member of the Council of the International Ergonomics Association from 1995 to 2000 and 2004 to 2009, and a member of the Executive Committee of the International Ergonomics Association from 2000 to 2003. In 2000 he became a Fellow of the International Ergonomics Association.

LIST OF FIGURES

Fig. 1.1   Sustainable system-of-systems (SSoS) model for HFE   11

Fig. 1.2   Organisation of the book according to the SSoS model   14

Fig. 2.1   HFE nested hierarchy of sustainable system-of-systems   26

Fig. 2.2   An HFE sustainable system-of-systems using a green building as the target system   29

Fig. 2.3   A simplified link analysis only considering some of the important technical aspects   37

Fig. 4.1   HFE issues influencing responsible computer consumption behaviour (RCCB) (Note: Model 1 consists of only the Theory of Planned Behaviour [TPB] antecedents; Model 2 consists of the TPB and new contextual antecedents; + refers to a positive relationship; H refers to the hypotheses tested in this research; *RCCI* responsible computer consumption intention, *HFE* human factors and ergonomics)   82

Fig. 5.1   A simplified diagram exploring the link between device design, mental models, interaction patterns, and goal achievement. Arrows denote hypothesised influence that has directed the authors' body of research   117

Fig. 5.2   Boost redesign at the device level, to promote an appropriate mental model of device function that communicates a predefined time period of operation, using the analogy of a clock   127

Fig. 5.3   Boost redesign at the control system level, to promote an appropriate mental model of design function that communicates the 'conditional rule' for boiler operation using an electrical circuit analogy   128

Fig. 8.1   A complex system-of-systems approach
            (Thatcher & Yeow, 2016)   192
Fig. 10.1   Ergonomic interventions in different phases of global value
            creation (Adapted from Kubek, Fischer, & Zink, 2015, p. 74)   257
Fig. 11.1   Schematic representation of the integrated model   278
Fig. 11.2   Polycentric governance to manage river water quality
            (Sources: Adapted from Ostrom (1990, 2005))   282
Fig. 12.1   A basic marketing system perspective   302
Fig. 12.2   Responsible consumption segments   312
Fig. 12.3   Responsible consumption segments in Indonesia
            (*Attitude * Intention to Buy Green Products*)
            (**Notes:** X^2 = 48.847, ρ = 0.000)   314
Fig. 12.4   Responsible consumption segments in Indonesia
            (*Attitude * Past Purchase of Green Products*)
            (**Notes:** X^2 = 10.398, ρ = 0.001)   314
Fig. 13.1 CO_2 emissions: 2000–2010. Malaysia, compared with
            averages of the world, OECD and upper middle-income
            nations (Data Source: World Bank DataBank)   333
Fig. 13.2   Information asymmetries and expectations gaps   344
Fig. 13.3   Analysis of reasons for rejection of GTFS-based loan
            applications   347
Fig. 14.1   What success on board looks like—elements identified
            by the different levels of staff on a commercial fishing vessel   362
Fig. 14.2   Matrix of the values identified as being of highest personal
            priority by each crew level   363

LIST OF TABLES

Table 1.1   Global, complex, and interlinked asymmetries threatening
            sustainability   3
Table 1.2   The chronological development of HFE and sustainability   8
Table 4.1   Demographic information and types of computer ownership   90
Table 4.2   Factor analysis for the attributes measuring
            the independent variables   91
Table 4.3   Factor analysis for attributes measuring responsible computer
            consumption intention (RCCI)   93
Table 4.4   Factor analysis for attributes measuring responsible computer
            consumption behaviour (RCCB)   93
Table 4.5   Discriminant validity analysis   95
Table 4.6   Hierarchical regression analysis for responsible computer
            consumption intention (RCCI)   96
Table 4.7   Hierarchical regression analysis for responsible computer
            consumption behaviour (RCCB)   97
Table 7.1   The nine values of biophilia based on Kellert
            and Wilson (1993)   164
Table 8.1   Definitions of (social) corporate sustainability   195
Table 8.2   Number and position of respondents in each
            area of the companies   201
Table 11.1 Connections among rules and levels of analysis   281
Table 11.2 Understanding of environmental awareness and health issues   287
Table 11.3 Daily water usage routine   289
Table 12.1 Consumer social responsibility measures   309

Table 12.2   Mean differences between social responsibility domains   310
Table 13.1   GTFS-based loan eligibility and general criteria   335
Table 13.2   Key findings from interviews   343
Table 13.3   Summary statistics: paid-up capital, prior experience,
             costs and amount of loan applied for in RM (millions)
             and as percentage of project cost   345
Table 13.4   Tests of significance: means of approved verses rejected
             applications   346
Table 13.5 χ^2-tests: impact of energy producer and technology source
             categorisations on GTFS loan approvals   346
Table 13.6   Summary of findings   348

CHAPTER 1

Introduction

Andrew Thatcher and Paul H. P. Yeow

INTRODUCTION: DEFINING THE PROBLEM

For most of human existence our species has had a fairly negligible impact on the planet. However, this changed roughly 250 years ago with the advent of the industrial revolution. The development of technologies powered by fossil fuels, rather than human or animal power, saw radical improvements in working conditions and living standards for the majority of people in what became known as the developed world. Further technological revolutions such as the development of electricity and significant advancements in medical care have seen further improvements in human wellbeing (Hecht et al., 2012). Once again, the people most likely to benefit from these innovations have been in the developed world. However, recently it has become evident that these developments also have significant negative consequences. These consequences can be sometimes artificially separated into environmental issues and social issues even though they are clearly interrelated. From an environmental perspective, the massive increase in the burning of fossil fuels has had a concomitant

A. Thatcher (✉)
Psychology Department, University of the Witwatersrand, Johannesburg, South Africa

P. H. P. Yeow
School of Business, Monash University Malaysia, Bandar Sunway, Malaysia

© The Author(s) 2018
A. Thatcher, P. H. P. Yeow (eds.), *Ergonomics and Human Factors for a Sustainable Future*, https://doi.org/10.1007/978-981-10-8072-2_1

1

increase in the amount of greenhouse gases (GHGs) being released into our environment. The effect of increased GHGs, particularly into the atmosphere, has been a steady increase in global temperatures known as global climate change (Incropera, 2016). Scientists predict that an increase in the average global temperature of 2 °C will be catastrophic for human habitation of the planet (Hansen et al., 2016). From a social perspective, these developments have seen an explosion in the human population as people live longer, healthier lives. However, the benefits of these developments have not been consistently realised by people, leading to widespread social inequalities, especially based on access to financial and physical resources (Piketty, 2014; Stiglitz, 2015; Vitousek, Mooney, Lubchenco, & Melillo, 1997).

An increase in benefits for humans has led to environmental degradation as the carrying capacities of several key ecosystems have battled to maintain equilibrium with the rapid increase in the size of the population (Vitousek et al., 1997). What we are experiencing now as a planet is a complex web of interacting systems known as wicked problems (Incropera, 2016) or even super-wicked problems (Levin, Cashore, Bernstein, & Auld, 2012). Wicked and super-wicked problems are characterised by being difficult to define, where the leading causes of the problems (i.e. humans) are also expected to find the solutions, with no obvious right or wrong answers, and with rapidly approaching deadlines (Levin et al., 2012; Rittel & Webber, 1973). In fact, the influence of humans on the planet's ecosystems is now so great that scientists have declared that we have entered a new geological epoch driven by human-led changes to geochemical cycles, the Anthropocene age (Steffen, Grinevald, Crutzen, & McNeill, 2011).

Thatcher and Yeow (2016a) characterised these interrelated problems as a set of three global asymmetries: resource asymmetries, asymmetries in the accumulation and distribution of waste, and legislative asymmetries (see Table 1.1). As far as *resource asymmetries* go, there are significant global imbalances in access to food, shelter, basic sanitation, healthcare, jobs, energy, clean water, education, consumer goods, productive land, and communication infrastructure (Hecht et al., 2012). This has led to poverty, famine, disease, and ultimately war as people fight over basic resources (Kelley, Mohtadi, Cane, Seager, & Kushnir, 2015). These asymmetries are also not good for the environment as people attempt to exploit increasingly marginal resources, leading them ever-closer to collapse (e.g.

Table 1.1 Global, complex, and interlinked asymmetries threatening sustainability

Problem	Human consequences	Environmental consequences
1. Resources asymmetries:		
Water, food, land, sanitation, energy, housing, education, jobs, healthcare, cultural expression	Poverty, hunger, disease, cultural subjugation and intolerance, exploitative labour practices; health	Land degradation, drought, deforestation, water pollution, monocultures, GMO, atmospheric pollution
2. Asymmetries in accumulation and distribution of waste:		
CO_2, CO, O_3 depletion, VOC, heavy metals, e-waste	Food security, health, disease spread	Global climate change, desertification, oceanic and land dead zones, ocean garbage patches, species extinction
3. Legislative asymmetries:		
Worker protections, technology transfer, labour broking, operational relocation, environmental protections	Exploitative labour practices (e.g. child labour, modern slave labour, unequal compensation for work), culturally and anthropometrically inappropriate technology, social conflict, health	Land degradation, freshwater depletion, unequal global distribution of waste, species threatened and extinction

deforestation, polluted water, and land degradation). Similarly, *waste accumulation and distribution* has occurred asymmetrically. For example, until recently carbon dioxide and other greenhouse gases were primarily released from technology in developed countries, but the effects will be felt worldwide and those least able to cope will be affected the worst (Samir & Lutz, 2017). The accumulations of carbon dioxide, heavy metal, plastic, nitrogen, and other wastes have already had detrimental effects on human health (Patz, Frumkin, Holloway, Vimont, & Haines, 2014; Pimentel et al., 2007) including respiratory problems, cancers, immune system defects, and birth defects. In addition, climate change accelerates these impacts through the spread of diseases to previously unknown geographical regions (Martens, 2013). Similarly, asymmetries in waste result in significant damage to ecosystems including nitrogen eutrophication of land and water (Rabotyagov, Kling, Gassman, Rabalais, & Turner, 2014) and coral reef degradation (Pratchett, Hoey, & Wilson, 2014). *Legislative asymmetries* refer to the fact that there is little global consistency in legal protections for human and environmental rights. Some geographical regions have robust legislation and effective enforcement, other regions have good legislation but poor enforcement, and other regions have neither good legislation nor effective enforcement. This situation is made more complex by changing political regimes that use worker and environmental protections as political tools (see Sarfaty, 2017). The consequences for people are exploitative labour practices, social conflict, and poor health. The environmental consequences include land degradation, species extinction, and unequal waste distribution.

As implied before, these issues are complex and highly interconnected. For example, land resource availability (a resource asymmetry) may also be influenced by legislation which gives some people more land rights than others (a legislative asymmetry) or by waste if the land had previously been a toxic waste dump (an asymmetry in the accumulation and distribution of waste). Land degradation is also exacerbated by other resource asymmetries. For example, financial resource limitations are likely to result in the over-exploitation of land resources through overgrazing, deforestation for agricultural land and fuel, the degradation of limited arable land through poor agricultural methods, and the use of outdated industrial machinery that is inefficient (Thatcher, 2013). The relationships between "problems" and "consequences" are also multi-directional. Deforestation contributes to climate change, which exacerbates land asymmetries. Deforestation for fuel results in atmospheric pollution, which exacerbates the asymmetries in the accumulation and distribution of waste, which leads to respiratory health problems that are exacerbated by asymmetries in the provision of healthcare systems. Table 1.1 summarises the asymmetries and some of their likely human and environmental consequences.

Scientifically there is little doubt that human activities have been primarily responsible for these devastating human and environmental consequences (Cook et al., 2013; Incropera, 2016; Maibach, Myers, & Leiserowitz, 2014; Oreskes, 2004). In fact, the latest synthesis of the assessment report of the Intergovernmental Panel on Climate Change (2015) suggests that the human and environmental consequences are worsening. The important questions that arise are whether the positive developments initiated by the industrial revolution can be sustained into the future, whether the gains can be equitably distributed to all people, and whether this can be achieved without destroying our life-supporting ecosystems.

Sustainability or Sustainable Development?

In their editorial to the special issue on sustainability in the journal *Ergonomics*, Haslam and Waterson (2013) noted how the meaning of the term "sustainability" had changed over time. The first use of the term "sustainability" is attributed to von Carlowitz (1713) who was writing about forestry and the need to provide a continuous supply of wood for human requirements (i.e. building materials, fuel for heating and cooking, physical supports for mining operations, and raw materials for the manufacturing of products). A dictionary definition of sustainability refers to the ability of something to endure either at a certain rate or level, or over some indeterminate "long-term" period. More recently, sustainability has become synonymous with the need to ensure that we (humans) need to preserve our natural environment. However, Johnston, Everard, Santillo, and Robèrt (2007) estimated that there were literally hundreds of definitions of sustainability (and that was more than ten years ago). In essence, sustainability is an issue of resource scarcity or damage, either currently or potentially at some time in the future (Johnston et al., 2007). Johnston et al.'s (2007) definition of sustainability has four components: (1) reduce our extraction of raw materials from the earth's crust; (2) reduce materials produced by society; (3) prevent the degradation of nature; and (4) remove barriers that prevent people from meeting their needs. Despite the inherent contradictions in this definition, it is important to note that sustainability refers not to ecosystems but to human social systems. When we raise the issue of sustainability, we are actually talking about sustained human survival within a resource-constrained environment.

Sustainable development then refers to sustained improvements in human social structures. The popularisation of the term is often attributed to the World Commission on Environment and Development's (WCED, 1987) definition of "development that meets the needs of the present without compromising the ability of future generations to meet their own needs". The WCED definition, also known as the Brundlandt definition, emphasises sustainable human social development that needs to be considered over an inter-generational time frame. However, others have considered the term sustainable development to be an oxymoron (Redclift, 2005). This is because development is often equated with growth (i.e. economic growth, population growth, consumption growth, etc.). As Meadows, Meadows, Randers, and Behrens (1972) noted more than four decades ago, and reiterated three decades later (Meadows, Randers, &

Meadows, 2004), there are very real ecosystemic carrying capacity limits to growth. In fact, Meadows et al. (2004) have suggested that human growth has already exceeded the carrying capacity of earth's ecosystems and what is actually needed is degrowth (Fournier, 2008). Redclift (2005) noted that too much human development inevitably means environmental degradation. This argument rests on the fact that while economic and social developments seemingly have no limits, there are very real limits to the carrying capacities of the supporting ecological systems that provide nutrients, water, air, and materials for our survival. As we discussed in the introduction to this chapter when discussing asymmetries, the question becomes how do we achieve equitable human development within the carrying capacity of earth's ecosystems?

Johnston et al. (2007) noted that sustainability and sustainable development are frequently used interchangeably and yet we have shown that they have subtly different meanings. Both terms refer to the threat to continued human existence, at least within acceptable levels of health and wellbeing. We prefer to use the term sustainability when referring to general issues related to the continuance of a particular human system and sustainable development when referring to a need to address an issue of a perceived social inequity that requires corrective action. In essence though, both terms refer to human (as a species) survivability under resource constraints.

Historically, What Has Been the Human Factors and Ergonomics (HFE) Response?: 1980s–2005

Can HFE address these sustainability and sustainable development issues? If we accept that humans are the primary agents responsible for these problems, then this opens up the possibility that concerted human effort can help produce the solutions. Since HFE is a discipline focused on humans in the system (Wilson, 2014), then we are ideally placed to make a difference. Indeed, this point has been emphasised several times in the last two and a half decades (Drury, 2014; Hanson, 2013; Helander, 1997; Moray, 1995; Nickerson, 1992; Nickerson & Moray, 1995; Steimle & Zink, 2006; Thatcher, 2013; Vicente, 1998) with various different recommendations being made. These recommendations have included the design of low-resource technologies so that people will interact with them in ways that facilitate low-resource use, the design of tasks and jobs in various sectors related to sustainability, the design of interventions that change people's behaviour to reduce resource use, and helping facilitate behaviour change at the organisational level and even across organisations in a value chain.

Until recently though, the response from the HFE discipline has been sporadic and weak. The earliest consideration of the sustainability challenges in the HFE literature was Wisner's (1985) work on anthropotechnology. While anthropotechnology is not, strictly speaking, a treatise on HFE's response to sustainability challenges, Wisner did highlight the dangers of wholesale transfer of technology without due consideration of the context in which it would be used. In this context, Wisner's (1985) work was the forerunner for considering the sustainable transfer of technology. However, it wasn't until the early 1990s before the HFE community was drawn into the challenges related to our sustained survival as a species. Nickerson (1992, 1993), in his Human Factors and Ergonomics Society address and subsequent book, was the first to draw our attention to the problems and the possible solutions where HFE could contribute. This work was expanded on by Moray (1995) and Nickerson and Moray (1995) where a clear set of guidelines for the HFE discipline were laid out. The guidelines included (1) the need to consider the HFE discipline's value structure in relation to environmental and social problems, (2) embrace complexity by taking a multidisciplinary approach, (3) focus on the underlying needs of people and not only their wants, and (4) consider cross-cultural factors, especially where technology transfer might have to occur.

On Moray's (1995) first point, little happened until Hancock and Drury (2011) and Moore and Barnard (2012) reiterated the need to reconsider the value-set for the HFE discipline. To date, only Lange-Morales, Thatcher, and García-Acosta (2014) have made an attempt to define what these values would look like for HFE, with little commentary or reflection from the general HFE discipline. Little much happened between 1995 and 2006. Garcia-Acosta (1996) wrote a Masters' thesis and introduced the term "ergoecology" to demonstrate the gap in HFE theory about human-ecological relationships. However, this work appeared in Spanish and garnered little attention beyond a small number of Spanish-speaking colleagues. Charytonowicz (1998) used the term "ecological ergonomics" at an Organisational Design and Management conference to show the relationship between the built environment, humans, and the natural environment, but this term also gained little traction. Vicente (1998) reiterated Moray's point about the need to embrace complex systems in addressing the global sustainability challenges and provided some examples of the HFE approach to finding solutions. Further, in a series of research papers, Sauer and his colleagues investigated how HFE considerations in the design of the interfaces of basic household products (i.e. a domestic kettle and a vacuum cleaner) could reduce resource use. Sauer, Wiese, and Rüttinger (2002) found that the simple placement of controls reduced electricity use. Sauer, Wiese, and Rüttinger (2004) found that automating certain functions also reduced electricity use. Sauer, Wiese, and Rüttinger (2002, 2003) and Wiese, Sauer, and Rüttinger (2004) found that improving the labelling of displays and controls improved resource use (i.e. reduced energy and water consumption), while Sauer and Rüttinger (2004) found that the size of a product (in this case a domestic kettle) affected resource use. The chronological development of HFE and sustainability issues is outlined in Table 1.2.

Table 1.2   The chronological development of HFE and sustainability

1980s	Anthropotechnology term introduced (Wisner, 1985)
1990s	HFE needs to address global challenges *HFES Annual Meeting* address (Nickerson, 1992)
	Keynote address to *IEA Congress* in 1993 on need to address global challenges (Moray, 1995)
	Ergoecology term first introduced in Spanish (Garcia-Acosta, 1996)
	Systems approach required for ergonomics to solve global problems (Vicente, 1998)
	Eco-ergonomics term first used (Charytonowicz, 1998)
2000s	Sustainable development and human factors introduced (Steimle & Zink, 2006)
	Eco-ergonomics term revived (Brown, 2007)
	Human factors and sustainable development introduced (Zink, 2008a, 2008b)
	Green ergonomics term first used (Hedge, 2008)
	Formation of the Human Factors and Sustainable Development Technical Committee (HFSD TC) of the IEA in 2008
2010s	Special Issue on ergonomics and sustainability in *Ergonomics* (Haslam & Waterson, 2013)
	Green ergonomics defined in *Ergonomics* (Hanson, 2013; Thatcher, 2013)
	Ergoecology term revived and defined in English in *Theoretical Issues in Ergonomics Science* (Garcia-Acosta, Pinilla, Larrahondo, & Morales, 2014)
	Sustainable work systems (including supply chain ergonomics and lifecycle ergonomics) defined in *Applied Ergonomics* (Zink, 2014)
	Two Special Issues on human factors and climate change in *Ergonomics in Design* (Nemire, 2014a, 2014b)
	Sustainable system-of-systems model for HFE introduced in *Ergonomics* (Thatcher & Yeow, 2016a)
	Special Issue on ergonomics and sustainability in *Applied Ergonomics* (Thatcher & Yeow, 2016b)

Recent HFE Responses: 2006 to Present

The work that appeared in the HFE literature for the decade after Moray's and Nickerson's call could hardly be classified as coordinated or prolific. Still, there was some evidence that researchers in the HFE community were thinking about these problems. The growth in interest from the HFE discipline started with Steimle and Zink's (2006) book chapter in the "International encyclopedia of ergonomics and human factors" where they introduced the term "sustainable development and human factors". Their book chapter was the first in the HFE discipline to draw attention to the relationships between societal problems, economic problems, and environmental problems and the role that HFE could play in addressing this nexus. Zink (2008a) edited a book with 15 chapters looking at the relationship between HFE and the various issues of sustainable development. While the book focused on organisational and social issues, it laid the groundwork for what was to follow. Zink (2008b) went on to establish the Human Factors and Sustainable Development (HFSD) Technical Committee of the International Ergonomics Association (IEA), which was officially launched in 2009 at the IEA's Congress in Beijing. We must acknowledge that this book was produced with the support and commitment of members of the HFSD Technical Committee who constituted the bulk of the authors and reviewers of the chapters.

The response since the formation of the HFSD Technical Committee has been far more encouraging. There were 25 papers at the 2009 IEA Congress in Beijing, China, which referred to sustainability or sustainable development, including three sessions that were specifically hosted by the new Technical Committee. At the 2012 IEA Congress in Recife, Brazil, there were 35 papers specifically mentioning sustainability or sustainable development and three dedicated sessions hosted by the Technical Committee. There were also three sessions hosted by the Technical Committee at the 2015 IEA Congress in Melbourne, Australia, but only 15 papers that addressed issues of sustainability and sustainable development.

The reviews of the work published from 1992 to 2012 were not particularly positive. Martin Legg, and Brown's (2013) review covered the period 1995–2012. They specifically focused on peer-reviewed journal articles that looked at both issues of sustainability and HFE. Of the 1934 journal articles that met the initial inclusion criteria, only seven articles met the final criteria for inclusion in their review. They concluded that the link between HFE and sustainability was limited and required greater attention. Thatcher (2012), reviewing work largely published at HFE conferences from 2008 to 2011, found that the majority of this work was theoretical in nature and concentrated largely on economic and social sustainability. Thatcher (2012) found that an astonishing 44% of the papers provided no definition of sustainability or sustainable development. Radjiyev, Qiu, Xiong, and Nam (2015) review covered the period 1992–2011. They concluded that HFE had remained relatively absent from the debates over design and sustainability during the review period. Since these reviews were published, much has happened in the HFE domain. This book is an attempt to bring some of this work together to demonstrate to the broader HFE discipline what work has been taking place.

Similarly, the number of scholarly articles appearing in HFE journals has been increasing. In 2013 *Ergonomics* published a special issue entitled "ergonomics and sustainability" (Haslam & Waterson, 2013). The practitioner journal *Ergonomics in Design* published two special issues in 2014 on "combating climate change: the role of human factors/ergonomics" (Nemire, 2014a, 2014b). In November 2004, the journal *WORK* published a special issue on "green ergonomics" (Dorsey, Hedge, & Miller, 2014). *Human Factors*, in a section guest edited by Kermit Davis, published papers based on their 2015 human factors prize for excellence in HFE research with the theme of sustainability and resilience. The winners of the *Human Factors* award are represented by a chapter in this book. Most recently, *Applied Ergonomics* published a special issue on human factors for a sustainable future (Thatcher & Yeow, 2016b). This book emerged from the work published in the *Applied Ergonomics* special issue.

There is also a great deal of work that is being produced outside of these special issues. This work has included investigations on HFE issues in the adoption of different sustainable devices and systems such as home energy-saving devices (Katzeff, Nyblom, Tunheden, & Torstensson, 2012), home heating systems (Revell & Stanton, 2017), water-saving devices (Fang & Sun, 2016), electric vehicles (Cocron et al., 2013; Young, Birrell, & Stanton, 2011), clothing design (Robinette & Veitch, 2016), sustainable work systems (Bolis, Brunoro, & Sznelwar, 2014), and public transport (Aceves-González, May, & Cook, 2016). There is also work that has looked at developing the theoretical frameworks linking HFE to issues of sustainability such as ergoecology (García-Acosta, Pinilla, Larrahondo, & Morales, 2014), supply chain ergonomics and lifecycle ergonomics (Zink, 2014), and a sustainable system-of-systems model for HFE (Thatcher & Yeow, 2016a). Richardson et al. (2017) have researched extensively on the positive benefits of human connections with nature and what this means for HFE design. Meanwhile, Demirel, Zhang, and Duffy (2016) have looked at the role of digital human modelling in meeting sustainability objectives, and Siemieniuch, Sinclair, and Henshaw (2015) have looked at how applying HFE to the design of manufacturing systems might address sustainability problems. A summary of the major milestones in the emergence of sustainability and HFE is given in Table 1.2.

Organisation of This Book

The chapters in this book are organised according to Thatcher and Yeow's (2016a) sustainable system-of-systems (SSoS) model for HFE. The SSoS model conceptualises HFE systems as a nested hierarchy of sociotechnical systems. The concept of a nested hierarchy is derived from Costanza and Patten's (1995) observation that natural systems are ordered in this way based on size, complexity, longevity, and geographical reach. For HFE, the smaller systems might be characterised as micro-ergonomics, the intermediate systems as either meso-ergonomics (Karsh, Waterson, & Holden, 2014) or macro-ergonomics, and the large, global systems as mega-ergonomics (Samaras & Samaras, 2010). This nested hierarchy is represented in Fig. 1.1. At the individual micro-ergonomics level, we would consider HFE work that is interested in designing sustainable work systems that avoid fatigue and burnout and encourage decent,

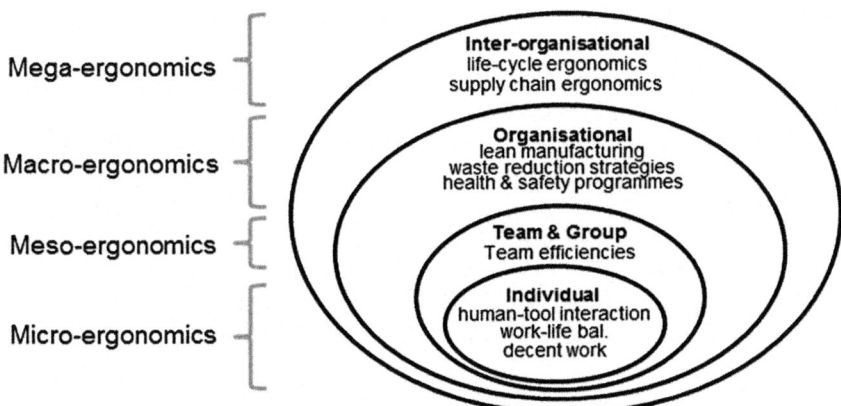

Fig. 1.1   Sustainable system-of-systems (SSoS) model for HFE

meaningful work and work-life balance (e.g. Docherty, Forslin, & Shani, 2002). Other micro-ergonomics interventions might include the design of human-tool interactions that facilitate and encourage sustainable behaviour (e.g. Adams & David, 2007; Revell & Stanton, 2016). At the meso-ergonomics level, it is important to consider resource-efficient work for teams and groups. Similarly, to the micro-ergonomics level, this would include sustainable teamwork systems and systems that facilitate sustainable behaviour for teams/groups (e.g. Torres, Teixeira, & Merino, 2009). Sustainable systems at the macro-ergonomics level involve issues with entire organisations such as the design of lean manufacturing systems (e.g. Genaidy, Sequeira, Rinder, & A-Rehim, 2009). At the mega-ergonomics level, HFE would be involved with systems that stretch across organisational and/or national boundaries. This might include approaches such as lifecycle and supply chain ergonomics (e.g. Zink, 2014). These hierarchical levels are necessary simplifications since it is also possible to have hierarchies of complexity, longevity, size, and geographical reach within a level (e.g. within micro-ergonomics).

Wilson (2014) uses the terms target system, sibling systems, parent systems, and child systems to describe the hierarchical relationships between the different system levels in the HFE hierarchy. The target system is the specific system of interest for the HFE investigation. Sibling systems are the systems at the same relative level of complexity, size, longevity, and geographical reach as the target system. Parent systems are the systems that encompass the smaller, underlying systems and child systems are those systems encompassed by the broader, parent system. In this parent-sibling-child SSoS model of HFE, the target system interacts with the surrounding systems and can have impacts on (and can be impacted upon by) systems seemingly distant (in geographical space and time). The sustainability of the target system will depend on the strength of the inter-relationships (multiple, strong bonds build system resilience), the ability of parent systems to provide support and system "memory", and the necessity of child systems to create opportunities to adapt to changes (Thatcher, 2016).

The chapters in this book are organised in a similar manner. Chapter 2 extends on the rather brief explanation of the SSoS for HFE given in this chapter. In Chap. 2, Thatcher and Yeow explore different methods for practitioners and researchers to use in order to find systems of relevance and importance for an HFE investigation or intervention. Chapters 3, 4, 5, and 6 each look at different components of the micro-ergonomics approach to sustainability in HFE. In Chap. 3, Fostervold et al. look at defining sustainable and decent work in the context of HFE. Yeow et al. (Chap. 4) look at how we might design computer consumption systems to encourage environmentally sustainable behaviours. Revell and Stanton (Chap. 5) look at the various attempts in the HFE field to design various (electrical) appliances such as home heating systems and other domestic appliances that facilitate sustainable behaviours. Finally from a micro-ergonomics perspective, Franke et al. (Chap. 6) look at the various HFE aspects of designing battery electric vehicles to encourage low-carbon transport. Lumber et al.'s chapter (Chap. 7) is the only chapter that examines aspects of meso-ergonomics. Their chapter looks more specifically at HFE implications for designing systems that incorporate nature. This chapter looks at both individual interactions (i.e. micro-ergonomics) and group interactions (i.e. meso-ergonomics) with nature.

Chapter 8 is the only chapter that looks at the role of HFE at the organisational level (i.e. macro-ergonomics). Brunoro et al. adopt a psychodynamic and activity theory perspective to understand how HFE can assist organisations to become more sustainable and encourage more sustainable behaviour from their employees. Chapters 9, 10, 11, 12, and 13 each look at different components of the mega-ergonomics approach to sustainability in HFE. In Chap. 9, Attaianese takes a lifecycle approach to the design of green buildings to demonstrate the different places where HFE might be involved in ensuring integration between the built environment and human interactions. This approach also demonstrates how the different levels in the SSoS hierarchy interact. While the target system for Attaianese is a green building, different human systems engage with the building at different points in its life cycle (e.g. the health and safety of workers at the construction stage and encourage appropriate sustainable behaviour during the building's usage stage). Chapters 10 and 11 look at HFE contributions at the inter-organisational level. Zink and Fischer (Chap. 10) examine the role of HFE in ensuring decent work in global supply chains, while Sarker et al. (Chap. 11) look at the role that HFE might have in getting different organisations to cooperate on joint sustainability initiatives. Chapters 12 and 13 look at the role of politics in facilitating or disrupting HFE interventions. Tjiptono (Chap. 12) explores the role of government policies in providing a supporting infrastructure for responsible consumption of resources. Sinnakkannu et al. (Chap. 13) look at political will and its influence on the large-scale adoption (or not) of sustainable technologies.

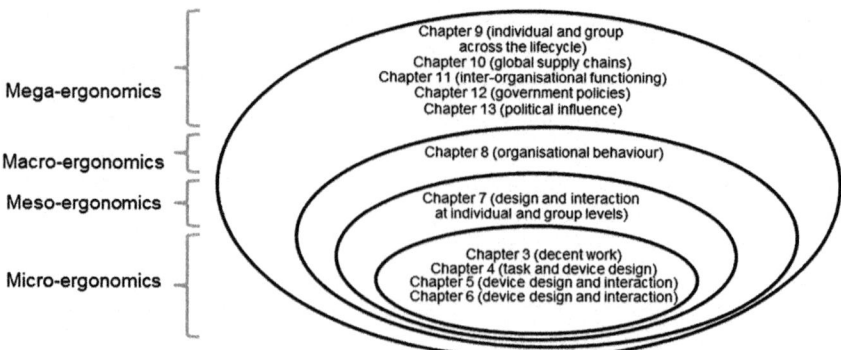

Fig. 1.2   Organisation of the book according to the SSoS model

The book concludes with two chapters that give a critical review of the work completed so far in the HFE field and attempt to lay out an agenda for future work. Moore (Chap. 14) encourages us to ensure that HFE foregrounds human wellbeing and meaning in its deliberations about sustainability. After all, if we are to escape our current predicament, we are going to need complete commitment from the vast majority of earth's human inhabitants. In Chap. 15, Thatcher and Yeow provide a set of priorities required of the HFE field in addressing the challenges of the twenty-first century. The organisation of the book in relation to the SSoS is given in Fig. 1.2.

AIMS OF THIS BOOK

The book was intended to build on all the theoretical and hypothetical cases that dominated the early work in this highly important area. We have attempted to showcase the work that has already been accomplished (and to show the many gaps/opportunities). The chapters in this book are essentially of two types. First, there are chapters that review the existing theoretical and empirical work that has already been published. These chapters are intended to provide a critical overview of this work culminating in suggestions for future research work. We believe that these chapters will be particularly useful for researchers and students wishing to understand and expand on the existing work. Chapter 3 on sustainable work systems; Chap. 5 on HFE interactions with sustainable appliances and devices;

Chap. 6 on the adoption of electric vehicles; Chap. 7 on biophilic design; Chap. 9 on green buildings; and Chap. 10 on supply chain ergonomics are examples. Second, there are chapters that report original research work. These chapters represent new areas of study that are still emerging. We believe that these chapters will be useful for readers by showing that there are a number of areas where significant contributions are still needed. Chapter 2 on identifying system boundaries in the sustainable system-of-systems model; Chap. 4 on responsible computer use; Chap. 8 on applying psychodynamic theory and activity theory to understanding organisational sustainability; Chap. 11 on institutional roles in sustainability; Chap. 12 on responsible consumption in emerging markets; and Chap. 13 on political influences in ensuring sustainability are examples. Chapter 14 is a reflective piece looking across all the previous chapters to identify gaps where new work is still required. At the outset, we wish to acknowledge that this book is not a complete representation of all the work being done in this emerging subfield. For example, we have not been able to include a chapter on lifecycle ergonomics or on the significant amount of work being done on lean manufacturing or in the design of recycling centres. We would also have liked to include a chapter reflecting on the values and ethics of the HFE discipline as well as a chapter on the importance of a participatory approach to ensure social justice and equity.

We believe that this book would also be useful as an introductory text for courses in HFE covering issues on sustainability. This book provides a systematic learning approach to this topic. First, this book summarises a significant proportion of the empirical work that has been done within the HFE discipline to address the issue of sustainability. Second, the book provides a set of possible tools that would enable HFE researchers and practitioners to systematically investigate how HFE might tackle sustainable issues. Third, the book provides a number of examples and case studies at the micro-, meso-, macro-, and mega-ergonomics levels. Finally, the book identifies gaps and provides directions for future research endeavours. Since the concept for this book arose from a Special Issue of the journal *Applied Ergonomics* (see Thatcher & Yeow, 2016b), interested readers are also referred to there for further case studies including HFE in the design of a water management system and a home heating system, the role of range stress in battery electric vehicles, and the design of recycling centres, the sustainability of organisations.

The terms "ergonomics" and "human factors" are often used interchangeably in the literature, although sometimes with slightly different meanings. In this book the term "human factors and ergonomics" is preferred (abbreviated to HFE throughout the chapter) when referring to the disciplinary field in general. According to the International Ergonomics Association's website, the official definition of "ergonomics (or human factors) is the scientific discipline concerned with the understanding of the interactions among humans and other elements of a system, and the profession that applies theoretical principles, data and methods to design in order to optimize human well being and overall system performance" (http://www.iea.cc). The official definition recognises the interchangeability of these two terms, and we have treated them as such throughout the book.

In this book we have listened to Moray's (1995) advice that addressing the challenges of the future requires embracing complexity through adopting a multidisciplinary approach. This book includes chapters from authors specialising in engineering, engineering management, business management, architecture, industrial design, environmental science, social science, psychology, economics, marketing, and human resource management in an attempt to show how are various disciplines share common ground in addressing the concerns of sustainability. There are very few HFE publications that attempt to integrate thinking from such a wide array of disciplines. Such an approach is required if we are to find sustainable solutions to current global problems.

References

Aceves-González, C., May, A., & Cook, S. (2016). An observational comparison of the older and younger bus passenger experience in a developing world city. *Ergonomics, 59*, 840–850.

Adams, P., & David, G. C. (2007). Light vehicle fuelling errors in the UK: The nature of the problem, its consequences and prevention. *Applied Ergonomics, 38*, 499–511.

Bolis, I., Brunoro, C. M., & Sznelwar, L. I. (2014). Mapping the relationships between work and sustainability and the opportunities for ergonomic action. *Applied Ergonomics, 45*, 1225–1239.

Brown, C. (2007, November 7–9). *Eco-ergonomics.* In Proceedings of the New Zealand Ergonomics Society Conference, Waiheke Island.

Charytonowicz, J. (1998). Ergonomics in architecture. In P. Vink, E. A. P. Konigsveld, & S. Dhondt (Eds.), *Human factors in organizational design & management VI.* Oxford, UK: Elsevier.

Cocron, P., Bühler, F., Franke, T., Neumann, I., Dielmann, B., & Krems, J. F. (2013). Energy recapture through deceleration – Regenerative braking in electric vehicles from a user perspective. *Ergonomics, 56*, 1203–1215.

Cook, J., Nuccitelli, D., Green, S. A., Richardson, M., Winkler, B., Painting, R., et al. (2013). Quantifying the consensus on anthropogenic global warming in the scientific literature. *Environmental Research Letters, 8*(2), 024024.

Costanza, R., & Patten, B. C. (1995). Defining and predicting sustainability. *Ecological Economics, 15*, 193–196.

Demirel, H. O., Zhang, L., & Duffy, V. G. (2016). Opportunities for meeting sustainability objectives. *International Journal of Industrial Ergonomics, 51*, 73–81.

Docherty, P., Forslin, J., & Shani, A. B. (Eds.). (2002). *Creating sustainable work systems: Emerging perspectives and practice.* London: Routledge.

Dorsey, J., Hedge, A., & Miller, L. (2014). Green ergonomics. *WORK: A Journal of Prevention, Assessment and Rehabilitation, 49*, 345–346.

Drury, C. G. (2014). Can HF/E professionals contribute to global climate change solutions? *Ergonomics in Design, 22*, 30–33.

Fang, Y. M., & Sun, M. S. (2016). Applying eco-visualisations of different interface formats to evoke sustainable behaviours towards household water saving. *Behaviour & Information Technology, 35*, 748–757.

Fournier, V. (2008). Escaping from the economy: The politics of degrowth. *International Journal of Sociology and Social Policy, 18*(11/12), 528–545.

García-Acosta, G. (1996). *Modelos de explicación sistémica de la ergonomía* [Models of systemic explanation of ergonomics]. Unpublished Master's thesis. Mexico City, México: Universidad Nacional Autónoma de México.

García-Acosta, G., Pinilla, M. H. S., Larrahondo, P. A. R., & Morales, K. L. (2014). Ergoecology: Fundamentals of a new multidisciplinary field. *Theoretical Issues in Ergonomics Science, 15*, 111–133.

Genaidy, A. M., Sequeira, R., Rinder, M. M., & A-Rehim, A. D. (2009). Determinants of business sustainability: An ergonomics perspective. *Ergonomics, 52*, 273–301.

Hancock, P. A., & Drury, C. G. (2011). Does human factors/ergonomics contribute to the quality of life? *Theoretical Issues in Ergonomics Science, 12*(5), 416–426.

Hansen, J., Sato, M., Hearty, P., Ruedy, R., Kelley, M., Masson-Delmotte, V., et al. (2016). Ice melt, sea level rise and superstorms: Evidence from paleoclimate data, climate modeling, and modern observations that 2 C global warming could be dangerous. *Atmospheric Chemistry and Physics, 16*, 3761–3812.

Hanson, M. A. (2013). Green ergonomics: Challenges and opportunities. *Ergonomics, 56*(3), 399–408.

Haslam, R., & Waterson, P. (2013). Ergonomics and sustainability. *Ergonomics, 56*(3), 343–347.

Hecht, A. D., Fiksel, J., Fulton, S. C., Yosie, T. F., Hawkins, N. C., Leuenberger, H., et al. (2012). Creating the future we want. *Sustainability: Science, Practice & Policy, 8*, 62–75.

Hedge, A. (2008). The sprouting of "green" ergonomics. *HFES Bulletin, 51,* 1–3.

Helander, M. G. (1997). Forty years of IEA: Some reflections on the evolution of ergonomics. *Ergonomics, 40,* 952–961.

Incropera, F. P. (2016). *Climate change: A wicked problem.* New York: Cambridge University Press.

Intergovernmental Panel on Climate Change. (2015). *Climate change 2014: Synthesis report.* Contribution of working groups I, II and III to the fifth assessment report of the Intergovernmental Panel on Climate Change. Geneva, Switzerland, IPCC.

Johnston, P., Everard, M., Santillo, D., & Robèrt, K. (2007). Reclaiming the definition of sustainability. *Environmental Science Pollution Research, 14,* 60–66.

Karsh, B. T., Waterson, P., & Holden, R. J. (2014). Crossing levels in systems ergonomics: A framework to support 'mesoergonomic' inquiry. *Applied Ergonomics, 45,* 45–54.

Katzeff, C., Nyblom, Å., Tunheden, S., & Torstensson, C. (2012). User-centred design and evaluation of EnergyCoach – An interactive energy service for households. *Behaviour & Information Technology, 31,* 305–324.

Kelley, C. P., Mohtadi, S., Cane, M. A., Seager, R., & Kushnir, Y. (2015). Climate change in the Fertile Crescent and implications of the recent Syrian drought. *Proceedings of the National Academy of Sciences, 112,* 3241–3246.

Lange-Morales, K., Thatcher, A., & García-Acosta, G. (2014). Towards a sustainable world through human factors and ergonomics: It is all about values. *Ergonomics, 57,* 1603–1615.

Levin, K., Cashore, B., Bernstein, S., & Auld, G. (2012). Overcoming the tragedy of super wicked problems: Constraining our future selves to ameliorate global climate change. *Policy Sciences, 45,* 123–152.

Maibach, E., Myers, T., & Leiserowitz, A. (2014). Climate scientists need to set the record straight: There is a scientific consensus that human-caused climate change is happening. *Earth's Future, 2,* 295–298.

Martens, P. (2013). *Health and climate change: Modelling the impacts of global warming and ozone depletion.* Oxford, UK: Taylor & Francis.

Martin, K. K., Legg, S. S., & Brown, C. C. (2013). Designing for sustainability: Ergonomics – Carpe diem. *Ergonomics, 56,* 365–388.

Meadows, D., Randers, J., & Meadows, D. (2004). *Limits to growth: The 30 year update.* London: Earthscan.

Meadows, D. H., Meadows, D. L., Randers, J., & Behrens, W. W. (1972). *The limits to growth.* A report for The Club of Rome's project on the predicament of mankind. Washington, DC, Potomac Books.

Moore, D., & Barnard, T. (2012). With eloquence and humanity? Human factors/ergonomics in sustainable human development. *Human Factors, 54,* 940–951.

Moray, N. (1995). Ergonomics and the global problems of the twenty-first century. *Ergonomics, 38*, 1691–1707.

Nemire, K. (2014a). Combating climate change: The role of human factors/ergonomics, Part 1 – Special Issue. *Ergonomics in Design, 22*, 3–3.

Nemire, K. (2014b). Introduction to the special issue on combating climate change: The role of human factors/ergonomics, Part 2. *Ergonomics in Design, 22*, 3.

Nickerson, R. S. (1992). What does human factors research have to do with environmental management? *Proceedings of the Human Factors and Ergonomics Society Annual Meeting, 36*, 636–639.

Nickerson, R. S. (1993). *Looking ahead: Human factors challenges in a changing world.* Hillsdale, NJ: Lawrence Erlbaum Associates.

Nickerson, R. S., & Moray, N. (1995). Environmental change. In R. S. Nickerson (Ed.), *Emerging needs and opportunities for human factors research* (pp. 158–176). Washington, DC: National Academy Press.

Oreskes, N. (2004). The scientific consensus on climate change. *Science, 306*, 1686.

Patz, J. A., Frumkin, H., Holloway, T., Vimont, D. J., & Haines, A. (2014). Climate change: Challenges and opportunities for global health. *Journal of the American Medical Association, 312*, 1565–1580.

Piketty, T. (2014). *Capital in the twenty-first century.* Cambridge, MA: Belknap Press.

Pimentel, D., Cooperstein, S., Randell, H., Filiberto, D., Sorrentino, S., Kaye, B., et al. (2007). Ecology of increasing diseases: Population growth and environmental degradation. *Human Ecology, 35*, 653–668.

Pratchett, M. S., Hoey, A. S., & Wilson, S. K. (2014). Reef degradation and the loss of critical ecosystem goods and services provided by coral reef fishes. *Current Opinion in Environmental Sustainability, 7*, 37–43.

Rabotyagov, S. S., Kling, C. L., Gassman, P. W., Rabalais, N. N., & Turner, R. E. (2014). The economics of dead zones: Causes, impacts, policy challenges, and a model of the Gulf of Mexico hypoxic zone. *Review of Environmental Economics and Policy, 8*, 58–79.

Radjiyev, A., Qiu, H., Xiong, S., & Nam, K. (2015). Ergonomics and sustainable development in the past two decades (1992–2011): Research trends and how ergonomics can contribute to sustainable development. *Applied Ergonomics, 46*, 67–75.

Redclift, M. (2005). Sustainable development (1987–2005): An oxymoron comes of age. *Sustainable Development, 13*, 212–227.

Revell, K. M., & Stanton, N. A. (2016). Mind the gap – Deriving a compatible user mental model of the home heating system to encourage sustainable behaviour. *Applied Ergonomics, 57*, 48–61.

Revell, K. M., & Stanton, N. A. (2017). *Mental models: Design of user interaction and interfaces for domestic energy systems.* London: CRC Press.

Richardson, M., Maspero, M., Golightly, D., Sheffield, D., Staples, V., & Lumber, R. (2017). Nature: A new paradigm for well-being and ergonomics. *Ergonomics, 60*, 292–305.

Rittel, H. W., & Webber, M. M. (1973). Dilemmas in a general theory of planning. *Policy Sciences, 4*, 155–169.

Robinette, K. M., & Veitch, D. (2016). Sustainable sizing. *Human Factors, 58*, 657–664.

Samaras, E. A., & Samaras, G. M. (2010). Using human-centered systems engineering to reduce nurse stakeholder dissonance. AAMI HF horizons 2010; *Biomedical Instrumentation & Technology, 44*, 25–32.

Samir, K. C., & Lutz, W. (2017). The human core of the shared socioeconomic pathways: Population scenarios by age, sex and level of education for all countries to 2100. *Global Environmental Change, 42*, 181–192.

Sarfaty, M. (2017). High risk US policy on climate change. *British Medical Journal* (Online), *357*, j1735.

Sauer, J., & Rüttinger, B. (2004). Environmental conservation in the domestic domain: The influence of technical design features and person-based factors. *Ergonomics, 47*, 1053–1072.

Sauer, J., Wiese, B. S., & Rüttinger, B. (2002). Improving ecological performance of electrical consumer products: The role of design-based measures and user variables. *Applied Ergonomics, 33*, 297–307.

Sauer, J., Wiese, B. S., & Rüttinger, B. (2003). Designing low-complexity electrical consumer products for ecological use. *Applied Ergonomics, 34*, 521–531.

Sauer, J., Wiese, B. S., & Rüttinger, B. (2004). Ecological performance of electrical consumer products: The influence of automation and information-based measures. *Applied Ergonomics, 35*, 37–47.

Siemieniuch, C. E., Sinclair, M. A., & deC Henshaw, M. J. (2015). Global drivers, sustainable manufacturing and systems ergonomics. *Applied Ergonomics, 51*, 104–119.

Steffen, W., Grinevald, J., Crutzen, P., & McNeill, J. (2011). The Anthropocene: Conceptual and historical perspectives. *Philosophical Transactions of the Royal Society of London A: Mathematical, Physical and Engineering Sciences, 369*(1938), 842–867.

Steimle, U., & Zink, K. J. (2006). Sustainable development and human factors. In W. Karwowski (Ed.), *International encyclopedia of ergonomics and human factors* (2nd ed.). London: Taylor & Francis.

Stiglitz, J. E. (2015). *The great divide: Unequal societies and what we can do about them.* New York: W.W. Norton & Company.

Thatcher, A. (2012). Early variability in the conceptualisation of 'sustainable development and human factors'. *WORK: A Journal of Prevention, Assessment and Rehabilitation, 41*, 3892–3899.

Thatcher, A. (2013). Green ergonomics: Definition and scope. *Ergonomics, 56,* 389–398.

Thatcher, A. (2016). *Longevity in a sustainable human factors and ergonomics system-of-systems.* 22nd Semana de Salud Ocupacional in Medellin, Colombia.

Thatcher, A., & Yeow, P. H. (2016a). A sustainable system of systems approach: A new HFE paradigm. *Ergonomics, 59,* 167–178.

Thatcher, A., & Yeow, P. H. (2016b). Human factors for a sustainable future. *Applied Ergonomics, 57,* 94–104.

Torres, M. K. L., Teixeira C. S., & Merino, E. A. D. (2009, August 9–14). *Ergonomics and sustainable development in mussel cultivation.* International Ergonomics Association 17th Triennial Congress [CD-Rom], Beijing, China.

Vicente, K. J. (1998). Human factors and global problems: A systems approach. *Systems Engineering, 1,* 57–69.

Vitousek, P. M., Mooney, H. A., Lubchenco, J., & Melillo, J. M. (1997). Human domination of Earth's ecosystems. *Science, 277,* 494–499.

von Carlowitz, H. C. (1713). *Sylvicultura oeconomica, oder haußwirthliche nachricht und naturmäßige anweisung zur wilden baum-zucht [Economic news and instructions for the natural growing of wild trees].* Leipzig, Germany: Braun.

Wiese, B. S., Sauer, J., & Rüttinger, B. (2004). Consumers' use of written product information. *Ergonomics, 47,* 1180–1194.

Wilson, J. R. (2014). Fundamentals of systems ergonomics/human factors. *Applied Ergonomics, 45,* 5–13.

Wisner, A. (1985). Ergonomics in industrially developing countries. *Ergonomics, 28,* 1213–1224.

World Commission on Environmental Development. (1987). *Our common future.* Report of the World Commission on Environment and Development. Oxford, UK: Oxford University Press.

Young, M. S., Birrell, S. A., & Stanton, N. A. (2011). Safe driving in a green world: A review of driver performance benchmarks and technologies to support 'smart' driving. *Applied Ergonomics, 42,* 533–539.

Zink, K. J. (Ed.). (2008a). *Corporate sustainability as a challenge for comprehensive management.* Heidelberg, Germany: Physica-Verlag.

Zink, K. J. (2008b). New IEA Human Factors and Sustainable Development Technical Committee. *HFES Bulletin, 51,* 3–4.

Zink, K. J. (2014). Designing sustainable work systems: The need for a systems approach. *Applied Ergonomics, 45,* 126–132.

CHAPTER 2

A Sustainable System-of-Systems Approach: Identifying the Important Boundaries for a Target System in Human Factors and Ergonomics

Andrew Thatcher and Paul H. P. Yeow

INTRODUCTION

The issues of global climate change and sustainability have been described as "wicked" problems (Murphy, 2012). This is not because the issues and responses are necessarily "evil", but because wicked problems are difficult to define, there are no right or wrong answers, and the solution pool constantly changes (Rittel & Webber, 1973). Levin, Cashore, Bernstein, and Auld (2012) have even argued that sustainability is actually a "super-wicked" problem because of four additional features: (a) time is now running out as we start to approach critical ecospheric "tipping points"; (b) the same entity that is creating the problems (i.e. us as humans) is also the

A. Thatcher (✉)
Psychology Department, University of the Witwatersrand, Johannesburg, South Africa

P. H. P. Yeow
School of Business, Monash University Malaysia, Bandar Sunway, Malaysia

© The Author(s) 2018
A. Thatcher, P. H. P. Yeow (eds.), *Ergonomics and Human Factors for a Sustainable Future*, https://doi.org/10.1007/978-981-10-8072-2_2

entity trying to resolve the problems; (c) there is no (or at best a weak) central authority to coordinate approaches; and (d) temporal psychological discounting continues to push needed action further into the future. According to Walker, Stanton, Salmon, Jenkins, and Rafferty (2010), complexity is characterised by three features: there is a multiplicity of interconnections, they are dynamic and nonlinear, and the boundary conditions are "fuzzy" boundaries. Our interconnected and complex world is characterised by the corporate globalisation agenda, the Internet Era (which eliminates time and distance boundaries and information asymmetry), and the growth of international travel networks. This interconnectedness results in unique challenges when considering sustainability. Hecht et al. (2012) summarises the complex sustainability challenge succinctly as the need to "meet the needs of the growing population in a way that restores and maintains the Earth's natural resources while promoting economic prosperity" (p. 63). Within human factors and ergonomics (HFE), Lange-Morales, Thatcher, and García-Acosta (2014) ideas encapsulate this by recommending that one of the HFE values for a sustainable world is through the appreciation of complexity.

Thatcher and Yeow (2016a) defined a number of negative consequences that characterise the problems that emerge from concerns about sustainability: resource asymmetries (e.g. unequal access to water, food, energy, hygiene, healthcare, education, etc.), waste accumulation and distribution asymmetries (e.g. atmospheric carbon dioxide, heavy metals, nitrogen oxides, sulphur oxides, volatile organic compounds, etc.), and regulatory asymmetries (e.g. international boundaries, access to markets, protection of labour, etc.). Of course there are also positive consequences of an interconnected world including greater sharing and distribution of knowledge and expertise, enormous diversity in possible solutions and opinions, and the ability to rapidly distribute essential resources (Hakansson & Ford, 2002). The problem with these increasingly complex systems though is that it is difficult to accurately understand the impacts the further one moves from the target system. For HFE (human factors and ergonomics), our global interconnected society requires us to think beyond simple human-tool interaction to consider how the raw materials were extracted from the ground or the way our information was gathered, the working conditions in the factories and sweatshops where the items were assembled, the sophistication of the transport networks that brought the items and services to a place of convenience, and our methods of disposal or reuse. In representing the complexity inherent in an

HFE response to sustainability issues, Thatcher and Yeow (2016a, 2016b) proposed a sustainable system-of-systems model to characterise these complex interrelationships. In this chapter we explore possible ways of coping with the fuzzy boundary problem in HFE.

SUSTAINABLE SYSTEM-OF-SYSTEMS FOR HFE

A system describes a number of interacting components confined within a boundary that defines a set of interrelated functions or purposes with a common aim and with links in a many-to-many mapping format (Wilson, 2014). Of interest to HFE though are systems that incorporate interactions with humans. There is growing recognition that smaller systems interact and become integrated to form much larger "super-systems" in what is known as a system-of-systems. According to Maier (1998) a system-of-systems describes an emergent class of systems that reflects connections between independent systems in their own right. A system-of-systems describes a set of geographically dispersed systems, where each component system can and does operate independently, and where the set of systems produces new and emergent features and side effects. Each component system in a system-of-systems may have its own goals and purposes and should be able to operate independently from the existence of a system-of-systems. The basic concepts of a sustainable system-of-systems approach to HFE (Thatcher & Yeow, 2016a, 2016b) were drawn from Wilson's (2014) work on complex systems and Costanza and Patten's (1995) work on sustainability. This work builds on an understanding of ecological systems, and therefore the sustainable system-of-systems for HFE approach falls within the green ergonomics domain (Thatcher, 2013), understanding the reciprocal relationships between human systems and natural systems.

In previous publications (Thatcher & Yeow, 2016a, 2016b), we have introduced the basic concepts of a sustainable system-of-systems for HFE and used this as a framework to show how the different approaches to understanding sustainability within HFE might be conceptualised as a coherent hierarchical model. This resolution is similar to Genaidy, Sequeira, Rinder, and A-Rehim (2009) recognition that sustainable work systems require a merging of micro- and macro-ergonomic approaches and Karsh, Waterson, and Holden (2014) work on bridging the gap between micro- and macro-ergonomic approaches through mesoergonomics. An example of Thatcher and Yeow's (2016a, 2016b) framework is given in Fig. 2.1.

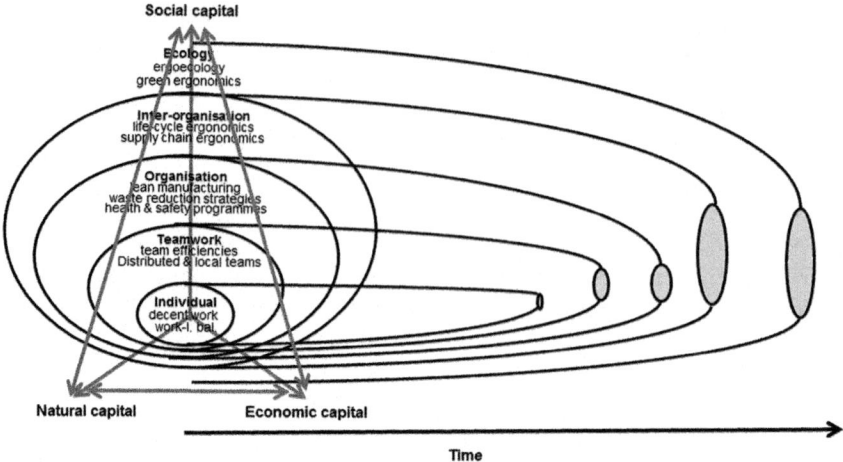

Fig. 2.1   HFE nested hierarchy of sustainable system-of-systems

Essentially, a sustainable system-of-systems contains three components: (1) a nested hierarchy of systems, (2) a time dimension for a system to be sustainable, and (3) an acknowledgement of multiple goals.

First, Costanza and Patten (1995) noted that natural systems are organised as a nested hierarchy of systems based on complexity and geographical distribution through space. Wilson (2014) used a similar concept called a parent-sibling-child system to denote encompassing systems (i.e. parent systems), systems that interact directly at the same hierarchical level (i.e. sibling systems), and smaller encompassed systems (i.e. child systems). Figure 2.1 shows how each broader scope of consideration encapsulates a system of smaller scope. Within HFE it is therefore important to understand how the sustainability of the target system (i.e. the system including humans that is the primary source of an HFE investigation or intervention) is also embedded in the sustainability of its parent, sibling, and child systems. If the target system is human interactions with the new workplace layout in a green building, then one needs to also consider the various sibling systems that interact with the workplace layout such as the work functioning of the team using the space or the availability of appropriate ergonomic furniture. The target and sibling systems will also be impacted by the various parent systems such as the lighting, heating, ventilation, and cooling systems or, at an even larger level, the geographical location (which determines how building occupants can interact with transport networks and other amenities). The target system (i.e. workplace layout) will also be influenced by various smaller systems (i.e. child systems) such as the types of work that need to be carried out. The concept of nested hierarchies of systems was also used in HFE by Karsh et al. (2014) as part of the meso-ergonomic approach to linking microergonomics to macroergonomics.

Second, the sustainable system-of-systems recognises that each system will only exist for a period of time but will inevitably have a termination point. Costanza and Patten (1995) noted that all systems have a natural lifespan (i.e. termination point) and systems that last too long will become brittle and unable to cope with changes imposed on it by the external environment (i.e. by parent systems). Similarly, systems that terminate prematurely could cause instability in the system-of-systems, especially systems that depend on their supporting infrastructure. It is argued that the natural time period over which a component system should be deemed sustainable is dependent on their relative position within the parent-sibling-child nested hierarchy. More complex and geographically dispersed systems will have longer natural lifespans than less complex, less dispersed systems.

For example, one would expect the green building workplace layout to have a natural lifespan roughly equivalent to the needs of the work team (i.e. sibling systems). The workplace layout would be expected to have a longer natural lifespan than its child systems (i.e. the work tasks). In fact, the workplace layout might actually constrain the ability to perform certain newly evolving work tasks. Larger, more complex systems would be expected to last longer than the office layout such as the lighting, heating, ventilation, and cooling systems. More complex systems might last even longer, such as the public transport network. Work systems (i.e. child systems) that last beyond their natural lifespan result in work practices and workplace layouts that become defunct and unable to cope with the changing nature of work (i.e. result in brittle systems). Similarly, public transport networks (i.e. parent systems) that terminate or change too quickly result in instability as the building occupants could struggle to get to work on time.

The ovals at the termination points in Fig. 2.1 are also important. They represent another aspect of the sustainable system-of-systems for HFE known as complex adaptive systems. According to Gunderson and Holling (2002), natural systems pass through a series of phases from exploitation, to consolidation, to destruction, and finally to reorganisation. It is beyond the scope of this chapter to discuss these aspects of adaptive change in detail and interested readers are directed to Thatcher (2016). For the purposes of this discussion, it is important to note that systems (including HFE systems) do not always permanently terminate but undergo a process of creative destruction and reorganisation that should be compatible with their supporting parent systems.

Third, a sustainable system-of-systems approach cannot be considered sustainable unless it recognises multiple goals. These goals also need to be considered across the different levels of the hierarchy due to the interconnected nature of a system-of-systems. A common (but certainly not the only) model of articulating multiple goals is Elkington's (1998) Triple Bottom Line (TBL) perspective of social, economic, and natural capital. By definition all organisations require human and natural capital to operate. Organisations are collections of people (i.e. social capital) that either manipulate natural capital (i.e. raw materials or information) or use natural capital to manipulate knowledge and products in order to create or maintain economic capital. Other multiple goal perspectives place a greater emphasis on human capital such as Hawkes' (2001) social, economic, natural, and cultural capital goals or Scerri and James' (2010) economic, ecological, political, and cultural goals. It is also important to note that the different goals are interlinked and that there needs to be a fair degree of balance between the different goals or the system-of-systems will collapse leading to non-sustainability. For example, a mistreatment of social capital within a green building (e.g. through poor indoor environmental quality) will result in disgruntled, poor performing employees, a poor organisational climate, and ultimately an organisation (or building) whose existence is in jeopardy.

Ultimately though, the sustainability of the system-of-systems rests on the availability, curatorship, and enrichment of natural capital. Without natural capital there will be no fresh air to breath, land and nutrients to grow food, or raw materials with which to construct our products or shelters. It is important to note that often the goals conflict with one another. For example, the need for employment under decent work conditions (a social goal) may conflict with the goal of the organisation to maximise profits (an economic goal). Conflicts can occur within a system as well as between the different hierarchical levels of the system-of-systems. Part of the challenge is to find sufficient balance with the goals to facilitate sustainability. The work of Carayannis, Bojicic-Dzelilovic, Olin, Rigterink, and Schomerus (2014) on conflict resolution suggests that goals and resolutions have become increasingly complex. In particular, transnationality has made it difficult to identify responsible authorities, actors, policies, and political entities. For example, the carbon dioxide produced by some countries in producing electricity to power industry (amongst many other human activities) is widely distributed around the globe and affects climate change not only in the country producing this greenhouse gas but also countries who played a negligible role in these increases. Finally, it bears mentioning that these models do not represent the only goals of a system. The goals referred to in this section should be viewed as high-level categories of goals for the system-of-systems, rather than specific system goals.

The example of a green building's workplace layout that has been used to illustrate the concepts contained in the sustainable system-of-systems for HFE is given in Fig. 2.2.

The sustainable systems-of-systems approach for HFE demonstrates that HFE researchers and practitioners need an appreciation of the complex relationships between various systems of interest. These interactions and the possible side effects of making changes to a target system require further consideration given the complexities of our modern world. In their mesoergonomic inquiry approach, Karsh et al. (2014) have already started to look at how nested systems interact across individual system

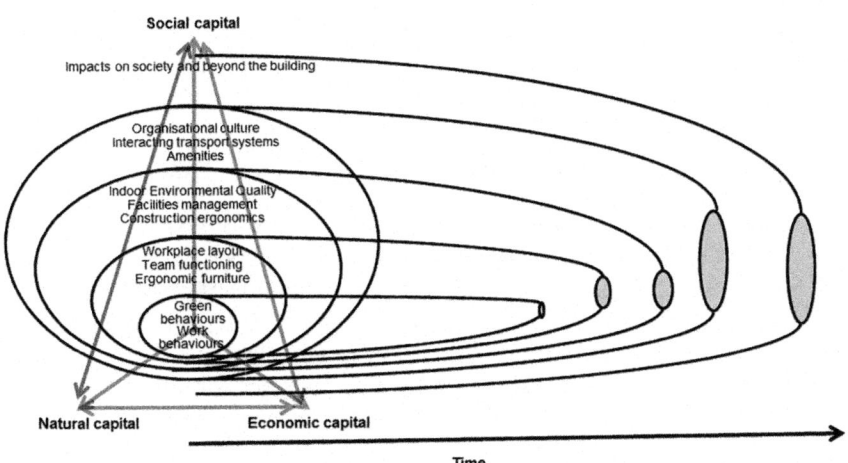

Fig. 2.2 An HFE sustainable system-of-systems using a green building as the target system

boundaries. While their work has started to look at cross-boundary interactions amongst systems up to the level of organisations, they acknowledged that further work was required to understand the interactions across boundaries of ever-larger hierarchies of systems such as political systems and socio-ecological systems.

According to Dekker, Hancock, and Wilkin (2013), while "a complex system is held together only by its local relationships" (p. 359), it may be increasingly difficult (perhaps even arbitrary) to determine where one system ends and another system begins. Complex systems are multi-valent (i.e. there are many-to-many critical relationships) and the complexity of relationships rapidly increases the further one moves away from the primary system of interest. The types of relationships also vary, with some relationships being competitive (e.g. access to a limited resource), others being cooperative (e.g. supply chains), and others being both competitive and cooperative (e.g. co-opetition). For HFE, some of these interactions might be distributed across geographical space as in supply chain ergonomics (Zink, 2014), while other interactions might be spread across time such as in lifecycle ergonomics (Zink, 2014). Wilson (2000) also emphasised that it was necessary to examine the complex interrelationships between HFE systems and argued that we mustn't only look at the systems but also place our attention specifically on understanding the interactions between systems. Wilson (2000) referred to this process as "complex interacting systems of interest" (p. 564), but unfortunately did not go further to try and explain how HFE researchers and practitioners would actually identify these "systems of interest".

What Is the Boundary of the System-of-Systems?

One of the limitations of the HFE sustainable system-of-systems model, as it has been articulated in our work thus far (Thatcher & Yeow, 2016a, 2016b), is that there are potentially limitless side effects and interactions from parent, sibling, and child systems, while HFE practitioners and researchers have limited resources to consider these limitless boundaries. In this chapter we argue that HFE will need to consider ways for researchers and practitioners to navigate this limitless boundary problem and identify the important or relevant parent, sibling, and child systems. Only in this instance will a target system within an HFE sustainable system-of-systems be addressable by current methods. While focusing largely on accident investigations, Salmon, Walker, Read, Goode, and Stanton (2017) acknowledge that the complexity of systems that HFE practitioners face currently outstrips the tools that the discipline currently has to locate system boundaries and analyse these complex situations in order to provide solutions. In the sustainable system-of-systems for HFE model, what is determined to be a parent, sibling, or child system will depend on the particular target system under investigation. A particular system (e.g. an organisation) might be a parent system (e.g. a lean manufacturing system), or a child system (e.g. a team efficiency system since it is at a level lower than the organisation), or a sibling system (e.g. a waste recycling system that interacts with the lean manufacturing system) in different contexts. In this chapter we consider the Pareto Principle, the Stakeholder Salience Theory, and the Network Theory as possible theoretical frameworks that might be applied to help determine appropriate boundaries for an HFE approach.

Pareto Principle

Pareto was an Italian economist who introduced the law of the vital few, or the 80/20 rule, where a few people make up the majority of the numbers (Amoroso, 1938). Pareto first proposed this principle through his observation that 80% of the land in Italy was owned by 20% of the population. Applied to the HFE sustainable system-of-systems model, this could mean identifying if the system-of-system boundary encompasses the majority. To put this in terms of the sustainable system-of-systems model would mean determining where the largest side effects on the target system (or by the target system) are located (i.e. from/onto parent, sibling, and child systems). In our green building example, the workplace layout could have numerous side effects on various child, sibling, and parent systems. Applying the Pareto Principle to determine which systems are of interest would mean focusing the analysis on the obvious HFE impacts, that is, the work system, the indoor environment, and the organisational culture. However, in this instance, one might ignore the impacts of various other systems that might also exert a significant influence on the workplace layout such as work habits, the changing nature of work, and the changing nature of technology needed to do that work. There are, of course, a number of limitations with the Pareto Principle approach. It hinges on whether one has the cooperation of the system owners/actors to do something about the side effects. It assumes that the systems which have the largest side effects are known a priori and that these do not change over time.

Stakeholder Salience Theory

One attempt to help the investigator identify key side effects is the Stakeholder Salience Theory. This theory invites the investigator to only consider the systems that address the key target system stakeholder concerns. According to Mitchell, Agle, and Wood (1997), stakeholders are "any group or individual who can affect or is affected by the achievement of the organisation's objectives" (p. 869). In the present context, the target system might affect/is affected by other systems in the overall system-of-systems that are owned by various stakeholders. In defining the boundary, HFE managers may include those systems belonging to stakeholders who have a higher degree of claims (or stakeholder salience). The degree of claim is dependent on three factors: power, legitimacy, and urgency. Power is defined as "a relationship among social actors in which one social actor, A, can get another social actor, B, to do something that B would not have otherwise done" (Mitchell et al., 1997; p. 869). Legitimacy is defined as the right for an entity to exist, in other words, "a generalized perception or assumption that the actions of an entity are desirable, proper, or appropriate within some socially constructed system of norms, values, beliefs, definitions" (Mitchell et al., 1997, p. 869). Lastly, urgency is defined as "the degree to which stakeholder claims call for immediate attention" (Mitchell et al., 1997, p. 869). Mitchell et al. (1997) suggested that those stakeholders that have more than one factor involved should be given a higher priority. Putting this into the context of the model presented here, we could consider the side effects of the target system on the other systems owned by the stakeholders with higher salience because of the urgent claim they have on the target system, the power they could exert on the target system if we do not address their claim, and where legitimacy to operate might be lost if the problem is not addressed.

Of course, in our green building workplace layout example, the Stakeholder Salience Theory immediately poses ethical problems with identifying salience. As HFE practitioners, one might argue that the most important stakeholders (certainly in terms of legitimacy and urgency) are the employees who have to work in a specific layout. These employees, as key stakeholders, are certainly more important than employees who only occasionally have to work in that space. But what about the organisation's management, who exert salience through their power and may suggest that any workplace layout would cease to exist unless the organisation itself is sustainable? What about those workers who only occasionally operate in that workplace layout but are highly influential when they do (e.g. maintenance workers looking after the lighting or air conditioning systems)? For example, maintenance workers are key to ensuring that the indoor environmental quality technologies (e.g. lighting systems, ventilation, and heating and cooling systems) are operating in a way that ensures optimal health, wellbeing, and effectiveness. However, they are only in the specific workspace temporarily, and their work could be negated by changes to workplace layout (e.g. moving desks to sit right beneath vents or furnishings that obscure light sources) or human intervention with the control mechanisms (e.g. changing the ambient temperature level on a control panel).

One of the problems with the Stakeholder Salience Theory is that the only way to determine salience is through the rather subjective lens of the people/person doing the investigating (in this example, the HFE practitioner, usually commissioned by management). Salience is socially constructed through the investigators and the people commissioning the investigation. Investigators (or commissioners) frequently predetermine who the "salient" stakeholders will be. Also, by just focusing on the salient stakeholders, the perceived "unimportant" stakeholders will be left out. As with our earlier example, by focusing attention on the primary building occupants of a specific workspace, temporary occupants (e.g. maintenance workers) may be left out, thereby producing workplace designs that don't optimise building energy-saving systems, for example. The Network Theory attempts to address this issue.

Soft Systems Methodology

Another alternative is to look at methods familiar to the information sciences. A popular method to identify and understand complex real-world problems is called Soft Systems Methodology, developed by Checkland (1972). The Soft Systems Methodology is often typified by the Clients, Actors, Transformation, Weltanschauung, Owner, and Environmental (CATWOE) analysis. To use the sustainable system-of-systems nomenclature we have developed, *Clients* refer to the important users of the *target* system. *Actors* refer to the various stakeholders involved in the implementation of the target system. Stakeholders could be drawn from the related sibling systems and first-order parent and child systems. Depending on the specific target system, stakeholders might include subordinates, suppliers, maintenance people, service providers, business partners, competitors, professional bodies, communities, and government officials. From an HFE perspective, stakeholders would also include the people who actually implement interventions and therefore may importantly also include the HFE researchers/practitioners themselves. For the purposes of this chapter, *Transformation* refers to the process of identifying the various inputs and outputs of the target system as well as the expected changes that they will undergo. In essence, this is the activity of Customers, Actors, and Owners identifying the relevant parent, sibling, and child systems as well as the important interactions between these systems.

Weltanschauung, also known as "worldviews", is a unique component of CATWOE and describes the different worldviews of all the various participants (i.e. the Customers, Actors, and Owners). In the CATWOE approach it is understood that the different participants will hold different worldviews, which could have an impact on the implementation of the target system. Since worldviews are a central component of CATWOE, further explanation of this concept is required. Bergvall-Kåreborn, Mirijamdotter, and Basden (2004) noted that CATWOE has been criticised for its poor definition of what is meant by worldviews. Checkland and Davies (1986) attempted to define three types of worldviews (W1, W2, and W3) relevant for CATWOE. W1 worldviews are assumptions about the target system's meaning, purpose, and functioning. As such, W1 worldviews are analogous to participants' conceptual models of the target system. Conceptual models refer to participants' pre-existing, holistic, internal representations of how the target system functions, including its purpose and structure (Richardson & Ball, 2009). W2 worldviews refer to the assumptions about how to make improvements to a problematic situation in the target system and apply to an understanding of the possible solution set and the relative advantages and disadvantages of the options in that solution set. W3 worldviews are the assumptions that we have about how reality is constructed. In terms of understanding systems in the sustainable system-of-systems model, W3 worldviews refer to how participants consider the world to be structured in relation to the target system. W3 worldviews would therefore be the personal assumptions related to an understanding of the relationships between the multiple goals of the system (e.g. ecocentric, econocentric, or sociocentric). For the purposes of this chapter, W1 and W3 worldviews are probably the most relevant. W2 worldviews refer specifically to situations where the target system is the focus of an intended change intervention (one of the main purposes of a CATWOE analysis). W2 worldviews therefore refer to the range of worldviews concerning possible change interventions amongst the CATWOE participants. In the sustainable system-of-systems model for HFE, we are trying to determine the relevant parent, child, and sibling systems and not necessarily trying to implement a change intervention for the target system. W2 worldviews would therefore only be relevant if the target system has been identified as being in need of change.

The *Owners* refer to the people who have decision making authority over the target system. Finally, the *Environmental* constraints describe the various external (to the target system) limitations such as legal requirements, resource constraints, and time limitations. In essence, these are aspects of the higher-order parent systems for most HFE target systems.

While not explicitly a tool to identify the relevant HFE systems of interest, CATWOE can be used to identify those people that need to be consulted in the system identification process. Specifically, these people would be the Customers, the Actors, and the Owners as well as their related worldviews. It would be necessary to draw knowledge from these people in order to identify the relevant systems (i.e. Transformations) of interest. In other words, it is not the role of the HFE expert acting alone to identify the related systems and system boundaries but to work in tandem with the other relevant stakeholders. Checkland (2000), in reviewing the Soft Systems Methodology, has noted CATWOE's successful implementation in understanding a variety of complex interrelated systems such as the redesign of the UK's National Health Service and various levels of complexity including the National Health Service itself, as well as government, service providers, patients, and other stakeholders. In addition, the Soft Systems Methodology has been applied to understand systems-of-systems (Jackson & Keys, 1984) as well as a number of sustainability problems (Bell & Morse, 2004; Paliwal, 2005; Zhang, Calvo-Amodio, & Haapala, 2015).

Network Theory

Stakeholder Salience Theory suggests that one needs to focus on the systems owned by salient stakeholders that one could have collaboration/control over, to exert influence. As it is, the world is a playground without a teacher when it comes to sustainability issues. There is no overseer of global sustainable issues such as natural resource depletion, waste, and poor working conditions. Instead several researchers have proposed that the world is made up of networks of organisations/governments that share resources and activities and have ties/relationships (Newman, 2003; Öberg, Huge-Brodin, & Björklund, 2012; Snehota & Hakansson, 1995).

Network Theory emerges from the comparison of random connectivity graphs (generated mathematically) with real-world networks (derived from a careful examination of naturally occurring networks). There are three properties of networks that are important for our discussion on interactions and side effects in systems-of-systems.

The first property, as Milgram (1967) demonstrated, is that simple, short paths can often be followed through a complex real-world network. This is a property known as the "small-world effect". There are actually two aspects to the small-world effect identified by Easley and Kleinberg (2010) that are important for this discussion: (1) while networks are complex, the multiplicity of connections naturally (and mathematically) leads to the emergence of short paths through the complexity; and (2) the natural qualities of human pattern recognition abilities make people unusually adept at finding these short paths. Within social networks this property has become popularised as the "six degrees of separation" phenomenon (i.e. that there are only six people/connectors that link an individual with some other arbitrary person). This property of a network means that it may be possible to identify the connected systems (and their paths—i.e. parent, sibling, and child systems) that might enable interventions to spread rapidly throughout the entire network. This implies that a careful analysis of the network of systems might provide important information on how to diffuse the impact of an intervention across the whole network. Easley and Kleinberg (2010) suggest that people are particularly good at identifying those paths. Of course, HFE systems are a combination of human and technical components (i.e. socio-technical systems) making network analysis more difficult. The simplified link analysis based on the network of workplace design features is illustrated in Fig. 2.3. In a full link analysis, the size of the nodes, the placement of the nodes in relation to other nodes, and the thickness of the connecting lines may also be important considerations. This information could be added after a systematic analysis of the network.

The second property is that effects tend to cluster within a network (Newman, 2003). This is because in real-world networks the relationships between entities are non-random (i.e. a relatively small number of systems within a system-of-systems will naturally be highly correlated with one another). This means that in a network there tends to be a large number of interrelationships between these entities. For the purposes of applying this property to identifying the side effects for a target system, the investigator is faced with the task of identifying the related systems with the highest

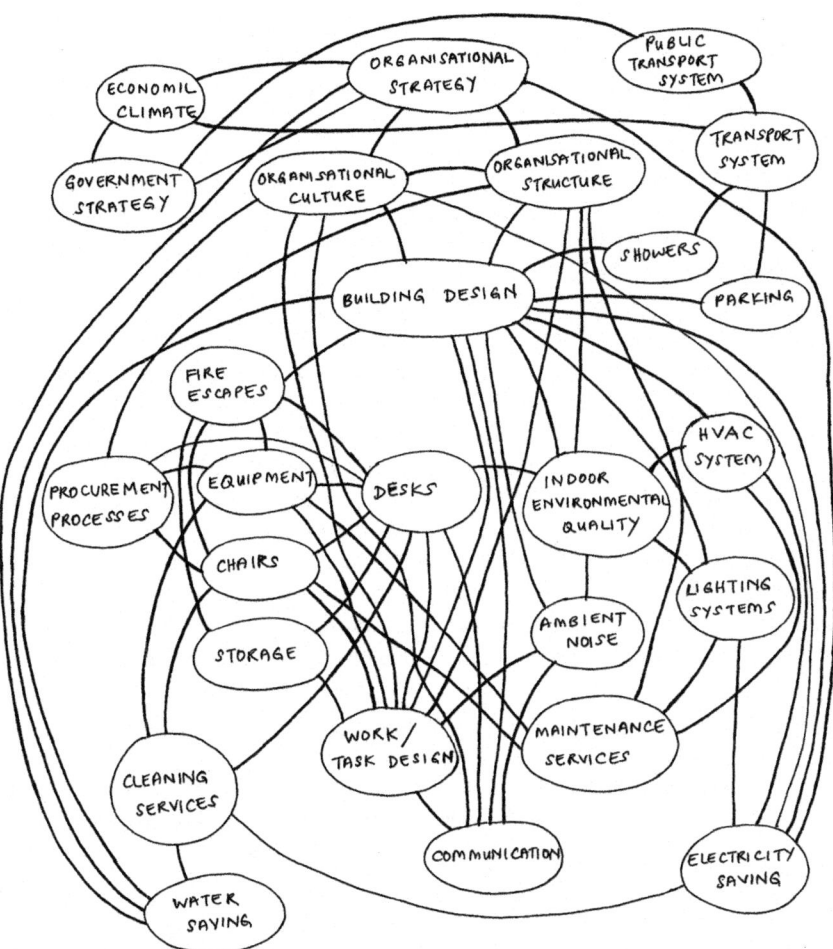

Fig. 2.3 A simplified link analysis only considering some of the important technical aspects

number of connections to the target system. This property might appear similar to the Stakeholder Salience Theory except with the important difference that it is the system relationships that are salient (depending on the type and number of connections) not the stakeholders/organisations/nodes themselves. Using Network Theory, it should be possible to carefully analyse the network connections to identify the key nodes (i.e. the key parent, sibling, and child systems). Since people are naturally adept at finding these connections, HFE methods and principles can therefore be applied to provide support to these identification processes. There are also multiple link analysis (Philip, Han, & Faloutsos, 2010) and social network analysis tools (Scott, 2012) available that can aid in conducting these types of investigations. In fact, a link analysis has already been used as a tool for an ergonomic workplace design (Brooks, 1998), although the scope of the links was fairly narrowly defined.

The third property is that networks are relatively resistant to random interventions/disruptions (Newman, 2003). This means that if an HFE intervention is attempted at random, there is a relatively high likelihood that there will be no significant impact on the system-of-systems. However, it has also been shown that the network will be rapidly disrupted when interventions target key components in the network. This brings to mind Meadows' (1999) concept of "leverage points"—strategic points in the system where small changes have big overall effects. Gunderson and Holling (2002) make a similar point in their complex adaptive cycle model but suggest that the "leverage points" may also be points in time rather than only points in space. According to Gunderson and Holling (2002), larger, more complex systems can be encouraged to enter a state of change if the system is already experiencing brittleness and a smaller, less complex system is already in a state of change. This suggests that the timing of change processes is also important to consider. If an intervention is carefully considered to target key components/leverage points in the system-of-systems, there is a high likelihood that this could have a broad significant impact across levels of the hierarchy. Obviously when making changes to a system as complex as the ecosphere, it is not easy to identify the key nodes. A reading of Meadows (1999) gives some rather abstract, but useful, suggestions for identifying leverage points (e.g. that negative and positive feedback loops are powerful contributors to change; information and transport "loops" are critical to making influential changes; and the ability to transcend paradigms opens up new opportunities for change). It should be noted though that Network Theory itself does not identify the actual leverage points. Network analysis can help identify potential intervention points. In social network analysis these leverage points are known as "influencers". There are possibilities to look at the work in the social network analysis sphere on calculating the relative extent of the influence through measures such as the Klout Index (Edwards, Spence, Gentile, Edwards, &

Edwards, 2013). There is certainly scope to see if similar influence measures could be developed within other types of link and network analyses.

The main focus of the network is continuous improvement for the overall good/interest of the network, since sustainability is not an end state but a process, similar to Kaizen's continuous improvement concept (Imai, 1986). Additionally, none of the actors (e.g. individuals, investigators, organisations, governments) have control over all the others in the network or can dominate the whole network (Hakansson & Ford, 2002). However, the actors are interdependent with each other, particularly in addressing sustainability issues (Öberg et al., 2012). They co-evolve with each other as they have reciprocal relationships analogous to Darwin's co-evolution theory where two species' genetic compositions evolve together because of their reciprocal relationships (Hakansson & Ford, 2002; Öberg et al., 2012).

OTHER CONSIDERATIONS

In order to facilitate cooperation, it is argued that one should consider the dominant social paradigm (Barnhart & Mish, 2017) of the societies in which the system-of-systems exists while addressing the side effects of the target system. In general one has to work within the constraint of the dominant social paradigm in introducing policies, laws, and regulations (e.g. carbon tax, incentives, etc.) to reduce the side effects on the related systems. This idea is incorporated in Lange-Morales et al.'s (2014) set of HFE values for a sustainable world through the value of respecting diversity. According to Barnhart and Mish (2017), US society is brought up with a materialistic/consumerism culture, where people work hard to earn income with the objective of buying more things/increasing their consumption in order to attain happiness. Expecting them to practise voluntary simplicity in life will probably not work as this behaviour is out of the norm of the society's values (i.e. consumerism). Social engineering, legal systems, or system-of-systems interventions that attempt to address sustainability issues have to be presented from the angle of the existing dominant social paradigm.

For example, according to a report by the Food and Agriculture Organization (FAO) of the United Nations, cattle account for 18% of greenhouse gas emission, more than the entire transport industry combined (Steinfeld et al. 2006). However, it could be argued that a worldwide ban on beef would not be a good solution in this context as beef is much liked and is a staple diet of consumers in many countries. Social compliance to such a ban would likely be poor and would spell political suicide for governments trying to implement such a ban (despite the obvious environmental benefits). However, having an ethical consumption tax to discourage the consumption of beef and also providing other acceptable alternatives to beef consumption, such as chicken and turkey (without tax) or even laboratory-manufactured beef (also called cultured meat, synthetic meat, or clean meat), might be a better solution. In fact, such a solution has been overwhelmingly supported in the UK (in 2015) and Denmark (in 2016).

When we demarcate the boundary of the system-of-systems to tackle a sustainability issue, we have to consider the law of unintended consequences (Merton, 1936). This "law" states that when we have a purposive social action, there may be an unexpected drawback or perverse/opposite results from our interventions. For example, as in our earlier paper (Thatcher & Yeow, 2016a), when we try to address the energy reduction issue and the CO_2 emission issue of consuming petroleum, countries like Malaysia consider biofuel that may have better burning potential and emits less pollution and which can have a relatively sustainable supply (compared to a source like petroleum which is naturally depleting and cannot be "grown"). However, as expected from the law of unintended consequences, the rapid expansion of palm estates to meet the growing demand of biofuels is causing extinction of species and a considerable haze through burning around the Southeast Asia region. Recently, people have become sick and schools are closing down, as the ecosystem is severely affected by the high air pollutant index (mostly caused by the burning of palm estate waste in Indonesia and Malaysia). The burning has also led to an unintended forest fire in Indonesia and Malaysia (Kodas, 2014). Thus, the overall extent of pollution is arguably much worse than when using petroleum.

The issue of unintended consequences is similar to the concept of emergence, a quality of complex systems, including system-of-systems (Maier, 1998). According to Dekker et al. (2013), emergence refers to properties of a system that produce outcomes or behaviours that cannot be predicted a priori. This is usually because our models of systems are necessarily simplified. Dekker et al. (2013) noted that emergence was particularly important in understanding notions of sustainability and consequences far (in network terms) from the source of an ergonomics intervention. The law of unintended consequences and the concept of emergence were carefully considered by Lange-Morales et al. (2014) in their value of respecting transparency and openness.

CONCLUSION

At this stage, the Network Theory initially appears to be a good candidate theory to assist researchers and practitioners in identifying the important child, sibling, and parent systems to focus on within a sustainable system-of-systems model. In particular, it is the key nodes within strong networks that matter in the effort to continuously improve the system-of-systems. Returning to our example of the atmospheric haze problem in Southeast Asia, what is required is a strong network of governments working in concert to address the haze pollution problem. This requires coherent interactions and the sharing of resources and activities such as identifying hot spots (i.e. key nodes), firefighting, cloud seeding, and planning haze preventive actions. From this example, it should also be obvious that identifying key nodes and finding solutions will require a multidisciplinary approach with a great deal of transdisciplinary knowledge sharing. As Manuaba (2007) has already noted, sustainability requires a systemic, holistic, interdisciplinary, and participatory (i.e. SHIP) approach. On this point, Wisner (1985) called for more engagement between ergonomics and anthropology, and Boudeau, Wilkin, and Dekker (2014) called for greater engagement between ergonomics and politics. Another way of accessing multidisciplinary input is through a method already familiar to HFE, the participatory ergonomics approach (Haines, Wilson, Vink, & Koningsveld, 2002). However, it is acknowledged that even with a multidisciplinary approach, the key nodes may be difficult to identify, in part because of personal agendas, hidden networks, and clandestine political motives (Carayannis et al., 2014).

It should also be noted that we don't believe that Network Theory offers a complete solution. This approach only indicates the theoretical possibility that nodes of influence might be identified, not that the nodes can be easily identified by current HFE methods. One possibility would be to combine a CATWOE analysis with a network analysis. The CATWOE analysis might prove a helpful technique in identifying the relevant customers, actors, and owners. These people would be able to provide valuable insights about worldviews that shape the relationships and transformations between systems within the network. As with Salmon et al. (2017), we suggest that new methods might have to be explored within the HFE discipline. Salmon et al. (2017) suggest that Event Analysis of Systemic Teamwork (Walker et al., 2006), Functional Resonance Analysis Method (Hollnagel, 2012), and Accimap (Svedung & Rasmussen, 2002) may be good candidates from the HFE field that could be expanded to deal with increased complexity. In addition, there are also a number of emergent graphical techniques such as mess mapping and resolution mapping (Horn & Weber, 2007), dialogue mapping (Conklin, 2006), cognitive redirective mapping (Schultz & Barnett, 2015), the socio-ecological matrix (Ali, 2004), and multicriteria analysis from sustainomics (Munasinghe, 2009) that require further exploration. While we have attempted to provide some theoretical solutions to identifying target system boundaries, it should be noted that wicked problems are managed, debated, and constantly renegotiated rather than solved, especially considering that the system-of-systems keeps evolving, with changing sustainability challenges.

References

Ali, S. H. (2004). A socio-ecological autopsy of the E. coli O157: H7 outbreak in Walkerton, Ontario, Canada. *Social Science & Medicine, 58,* 2601–2612.

Amoroso, L. (1938). Vilfredo Pareto. *Econometrica: Journal of the Econometric Society, 6,* 1–21.

Barnhart, M., & Mish, J. (2017). Hippies, hummer owners, and people like me: stereotyping as a means of reconciling ethical consumption values with the DSP. *Journal of Macromarketing, 37,* 57–71.

Bell, S., & Morse, S. (2004). Experiences with sustainability indicators and stakeholder participation: A case study relating to a 'Blue Plan' project in Malta. *Sustainable Development, 12,* 1–14.

Bergvall-Kåreborn, B., Mirijamdotter, A., & Basden, A. (2004). Basic principles of SSM modeling: An examination of CATWOE from a soft perspective. *Systemic Practice and Action Research, 17,* 55–73.

Boudeau, C., Wilkin, P., & Dekker, S. W. (2014). Ergonomics as authoritarian or libertarian: Learning from Colin Ward's politics of design. *The Design Journal, 17,* 91–114.

Brooks, A. (1998). Ergonomic approaches to office layout and space planning. *Facilities, 16,* 73–78.

Carayannis, T., Bojicic-Dzelilovic, V., Olin, N., Rigterink, A., & Schomerus, M. (2014). *Practice without evidence: Interrogating conflict resolution approaches and assumptions.* Justice and Security Research Programme, International Development Department, London School of Economics.

Checkland, P. B. (1972). Towards a systems-based methodology for real-world problem solving. *Journal of Systems Engineering, 3,* 87–116.

Checkland, P. B. (2000). Soft systems methodology: A thirty year retrospective. *Systems Research and Behavioral Science, 17,* S11–S58.

Checkland, P. B., & Davies, L. (1986). The use of the term 'Weltanschauung' in soft systems methodology. *Journal of Applied Systems Analysis, 13*, 109–115.

Conklin, J. (2006). *Dialogue mapping: Building shared understanding of wicked problems*. Chichester, UK: Wiley.

Costanza, R., & Patten, B. C. (1995). Defining and predicting sustainability. *Ecological Economics, 15*, 193–196.

Dekker, S. W., Hancock, P. A., & Wilkin, P. (2013). Ergonomics and sustainability: Towards an embrace of complexity and emergence. *Ergonomics, 56*, 357–364.

Easley, D., & Kleinberg, J. (2010). *Networks, crowds, and markets: Reasoning about a highly connected world*. New York: Cambridge University Press.

Edwards, C., Spence, P. R., Gentile, C. J., Edwards, A., & Edwards, A. (2013). How much Klout do you have... A test of system generated cues on source credibility. *Computers in Human Behavior, 29*, A12–A16.

Elkington, J. (1998). *Cannibals with forks: The triple bottom line of 21st century business*. Oxford, UK: Capstone.

Genaidy, A. M., Sequeira, R., Rinder, M. M., & A-Rehim, A. D. (2009). Determinants of business sustainability: An ergonomics perspective. *Ergonomics, 52*, 273–301.

Gunderson, L. H., & Holling, C. S. (2002). *Panarchy: Understanding transformation in human and natural systems*. Washington, DC: Island Press.

Haines, H., Wilson, J. R., Vink, P., & Koningsveld, A. E. (2002). Validating a framework for participatory ergonomics (the PEF). *Ergonomics, 45*, 309–327.

Hakansson, H., & Ford, D. (2002). How should companies interact in business networks? *Journal of Business Research, 55*, 133–139.

Hawkes, J. (2001). *The fourth pillar of sustainability: Culture's essential role in public planning*. Melbourne, Australia: Common Ground.

Hecht, A. D., Fiksel, J., Fulton, S. C., Yosie, T. F., Hawkins, N. C., Leuenberger, H., et al. (2012). Creating the future we want. *Sustainability: Science, Practice, & Policy, 8*, 62–75.

Hollnagel, E. (2012). *FRAM: The functional resonance analysis method: Modelling complex socio-technical systems*. Aldershot, UK: Ashgate.

Horn, R. E., & Weber, R. P. (2007). *New tools for resolving wicked problems: Mess mapping and resolution mapping processes*. Watertown, MA: Strategy Kinetics LLC.

Imai, M. (1986). *Kaizen – The key to Japan's competitive success*. New York: Random House.

Jackson, M. C., & Keys, P. (1984). Towards a system of systems methodologies. *Journal of the Operational Research Society, 35*, 473–486.

Karsh, B. T., Waterson, P., & Holden, R. J. (2014). Crossing levels in systems ergonomics: A framework to support 'mesoergonomic' inquiry. *Applied Ergonomics, 45*, 45–54.

Kodas, M. (2014). *How did palm oil become such a problem – And what can we do about it?* ENASIA, November 3, 2014. http://ensia.com/features/how-did-palm-oil-become-such-a-problem-and-what-can-we-do-about-it/. Accessed 13 July 2016.

Lange-Morales, K., Thatcher, A., & García-Acosta, G. (2014). Towards a sustainable world through human factors and ergonomics: It is all about values. *Ergonomics, 57,* 1603–1615.

Levin, K., Cashore, B., Bernstein, S., & Auld, G. (2012). Overcoming the tragedy of super wicked problems: Constraining our future selves to ameliorate global climate change. *Policy Sciences, 45,* 123–152.

Maier, M. W. (1998). Architecturing principles for systems of systems. *Systems Engineering, 1,* 267–284.

Manuaba, A. (2007). A total approach in ergonomics is a must to attain humane, competitive and sustainable work systems and products. *Journal of Human Ergology, 36,* 23–30.

Meadows, D. (1999). *Leverage points: Places to intervene in a system.* Hartland, VT: The Sustainability Institute.

Merton, R. K. (1936). The unanticipated consequences of purposive social action. *American Sociological Review, 1,* 894–904.

Milgram, S. (1967). The small world problem. *Psychology Today, 2,* 60–67.

Mitchell, R. K., Agle, B. R., & Wood, D. J. (1997). Toward a theory of stakeholder identification and salience: Defining the principle of who and what really counts. *Academy of Management Review, 22,* 853–886.

Munasinghe, M. (2009). *Sustainable development in practice: Sustainomics methodology and applications.* Cambridge, UK: Cambridge University Press.

Murphy, R. (2012). Sustainability: A wicked problem. *Sociologica, 6,* 1–23.

Newman, M. E. (2003). The structure and function of complex networks. *SIAM Review, 45,* 167–256.

Öberg, C., Huge-Brodin, M., & Björklund, M. (2012). Applying a network level in environmental impact assessments. *Journal of Business Research, 65,* 247–255.

Paliwal, P. (2005). Sustainable development and systems thinking: A case study of a heritage city. *The International Journal of Sustainable Development & World Ecology, 12,* 213–220.

Philip, S. Y., Han, J., & Faloutsos, C. (2010). *Link mining: Models, algorithms, and applications.* Berlin, Germany: Springer.

Richardson, M., & Ball, L. J. (2009). Internal representations, external representations and ergonomics: Towards a theoretical integration. *Theoretical Issues in Ergonomics Science, 10,* 335–376.

Rittel, H. W., & Webber, M. M. (1973). Dilemmas in a general theory of planning. *Policy Sciences, 4,* 155–169.

Salmon, P. M., Walker, G. H., Read, G. J. M., Goode, N., & Stanton, N. A. (2017). Fitting methods to paradigms: Are ergonomics methods fit for systems thinking? *Ergonomics, 60,* 194–205.

Scerri, A., & James, P. (2010). Communities of citizens and 'indicators' of sustainability. *Community Development Journal, 45*, 219–236.

Schultz, E., & Barnett, B. (2015). *Cognitive directive mapping: Designing futures that challenge anthropocentrism.* Nordes, 6. http://www.nordes.org/opj/index.php/n13/article/view/398/376

Scott, J. (2012). *Social network analysis.* London: Sage Publications.

Snehota, I., & Hakansson, H. (Eds.). (1995). *Developing relationships in business networks.* London: Routledge.

Steinfeld, H., Gerber, P., Wassenaar, T., Castel, V., Rosales, M., & De Haan, C. (2006). *Livestock's long shadow.* Rome: FAO.

Svedung, I., & Rasmussen, J. (2002). Graphic representation of accident scenarios: Mapping system structure and the causation of accidents. *Safety Science, 40*, 397–417.

Thatcher, A. (2013). Green ergonomics: Definition and scope. *Ergonomics, 56*, 389–398.

Thatcher, A. (2016). *Longevity in a sustainable human factors and ergonomics system-of-systems.* Plenary address at the 22nd Semana de Salud Ocupacional in Medellin Bogota.

Thatcher, A., & Yeow, P. H. P. (2016a). A sustainable system of systems approach: A new HFE paradigm. *Ergonomics, 59*, 167–178.

Thatcher, A., & Yeow, P. H. P. (2016b). Human factors for a sustainable future. *Applied Ergonomics, 57*, 1–7.

Walker, G. H., Gibson, H., Stanton, N. A., Baber, C., Salmon, P., & Green, D. (2006). Event analysis of systemic teamwork (EAST): A novel integration of ergonomics methods to analyse C4i activity. *Ergonomics, 49*, 1345–1369.

Walker, G. H., Stanton, N. A., Salmon, P. M., Jenkins, D. P., & Rafferty, L. (2010). Translating concepts of complexity to the field of ergonomics. *Ergonomics, 53*, 1175–1186.

Wilson, J. R. (2000). Fundamentals of ergonomics in theory and practice. *Applied Ergonomics, 31*, 557–567.

Wilson, J. R. (2014). Fundamentals of systems ergonomics/human factors. *Applied Ergonomics, 45*, 5–13.

Wisner, A. (1985). Ergonomics in industrially developing countries. *Ergonomics, 28*, 1213–1224.

Zhang, H., Calvo-Amodio, J., & Haapala, K. R. (2015). Establishing foundational concepts for sustainable manufacturing systems assessment through systems thinking. *International Journal of Strategic Engineering Asset Management, 2*, 249–269.

Zink, K. J. (2014). Designing sustainable work systems: The need for a systems approach. *Applied Ergonomics, 45*, 126–132.

CHAPTER 3

Defining Sustainable and "Decent" Work for Human Factors and Ergonomics

Knut Inge Fostervold, Peter Christian Koren, and Odd Viggo Nilsen

INTRODUCTION

How to ensure a sustainable planet Earth may be the major existential question faced by man today. Questions pertaining to sustainability and sustainable development (SD) have consequently engendered nascent attention also in the field of human factors and ergonomics (HFE). Despite its infancy, the discussion has been markedly diversified, ranging from general theoretical considerations to how HFE affects (or should affect) sustainable development within more restricted areas. An example of the former is the discussion of shared values ingrained in the fields of ergoecology and green ergonomics and how the acceptance of such values may contribute to the

K. I. Fostervold (✉)
Department of Psychology, University of Oslo, Oslo, Norway

P. C. Koren
Department of Media and Social Sciences, University of Stavanger, Stavanger, Norway

O. V. Nilsen
Akershus County Council, Oslo, Norway

© The Author(s) 2018
A. Thatcher, P. H. P. Yeow (eds.), *Ergonomics and Human Factors for a Sustainable Future*, https://doi.org/10.1007/978-981-10-8072-2_3

design of more sustainable socio-technical systems (Lange-Morales, Thatcher, & García-Acosta, 2014). A more delineated topic is the challenges global supply chains pose on sustainability and working conditions. One crucial focus is their possible remediation through the combination of social action, long-term development of business sustainability, and transnational regulation programmes, such as decent work (Hasle & Jensen, 2012). Many of the contributions call for HFE experts to engage professionally in the discussion about sustainability and sustainable development, at the same time accentuating both the potential and the challenges of HFE in the upcoming shift towards a greener economy (e.g. Manuaba, 2007; Martin, Legg, & Brown, 2013; Moore & Barnard, 2012; Thatcher & Yeow, 2016; Zink, 2005; Zink & Fischer, 2013).

The study of work and its consequences is the core subject of human factors and ergonomics (HFE). However, the status of work itself seems somewhat blurred in the discussion about sustainability. Should work be considered a topic in its own right, as is often the case in traditional HFE, or is it more appropriate to consider work and working conditions as tools to achieve sustainable products and production processes?

If work itself is considered the primary objective, what do we then understand by sustainable work? Is sustainable work merely about creating stable jobs that provide people with a livelihood for posterity, without upsetting the balance of the ecosystem, both locally and globally? Is it a question about maintaining work over time, without depleting the individual, the environment, or primary commodities? Or does sustainable work also comprise factors considered vital for the quality of work, as suggested by the International Labour Organization (ILO) in the Decent Work Agenda (ILO, 1999)? An additional question is whether the term sustainability can be carried into discussions about how working conditions affect the sustainability of a workforce as such.

The present chapter seeks to examine such questions, and to provide meaning and content to the term sustainable work, applicable to the context of HFE.

HUMANS AND THE FUNCTION OF WORK

Work in one form or another appears to be pivotal for most humans. Apart from providing subsistence, directly or through income or salaries, work is an important arena for forming social identity, social cohesion, feelings of meaningfulness, and affirmation of personal core values (Jahoda, 1982;

Ros, Schwartz, & Surkiss, 1999; Rosso, Dekas, & Wrzesniewski, 2010). Work also appears to be an important factor in both physical and mental health (Blustein, 2008; Ryff & Singer, 1998; Swanson, 2012). Work is of vital importance, not only for the individual but also for the functioning of organizations and communities. Work is thereby a cornerstone in the development of any society. Work and work-related activities shape much of the foundation of our lives and our perceptions about who we are.

There is little doubt that good work conditions have positive consequences. On the other hand, work and work-related activities also entail potentially negative repercussions that may affect individuals and the society at all levels. At the individual level, the knowledge and study of work-related repercussions has a long history. Suffice to mention Bernardino Ramazzini's (1713/1940) influential treatise "De Morbis Artificatum Diatriba" (Diseases of workers) whereof a first edition was published as early as 1700.

At the societal level, the introduction of "organized work" in the emerging industrial societies by the end of the eighteenth century represents dire examples of the detrimental downside of work. Potential workers moved from poor rural areas to industrial centres, drawn by employment and salaries. The quest for higher profit for owners and investors, and a common view of the population as a never-ending supply of labour, made workers merely a commodity or article of commerce. Poor living and working conditions lead to disasters, both health-wise and socially, in the lives of many industrial workers (Thompson, 2016).

As a response to growing humanitarian concerns, legislation for workers' protection gradually came into being in many countries. The idea of regarding the national workforce as a limited set of input resources gained ground. Governments became more interested in regulating industrial working conditions, especially as compensation for accidents and work-related diseases became national budget issues. Nevertheless, the main part of the contractual situation of labourers was left to the traditional two parties, employers and employees, for many more years (Hepple, 2006; Sengenberger, 2013).

The Decent Work Agenda

The foundation of the International Labour Organization (ILO) in 1919 represents a watershed in the organization of work-life. Labour organizations and unions had admittedly been around for a while, but ILO was the first tripartite collaborative labour organization, with representatives of governments, employers, and workers in its executive bodies. Unlike the former foci on humanitarian and economic aspects, ILO endorsed social justice as the basic premise for the quality of employment and quality of life (Van Daele, 2005). The emphasis on social justice implies that work is more than a safe job with associated income. In addition, work should imply equal rights, dignity, and economic, social, and political empowerment (ILO, 2003). The principle of social justice is, of course, regularly flouted, and exploitation of workers and workers' rights still is a widespread global problem. An ILO report published in 2005, for example, estimated the number of people in forced labour in the world to be at least 12.3 million (Belser, de Cock, & Mehran, 2005).

In 1999, ILO launched the Decent Work Agenda, announcing a modernized amplification of the principle of social justice, with increased emphasis on outcomes (Fields, 2003). Four strategic objectives, coined the four pillars of the Decent Work Agenda, constitute the basis of the initiative:

1. Fundamental principles and rights at work and international labour standards.
2. Employment creation and income opportunities.
3. Social protection.
4. Tripartite dialogue between the social partners: governments, employer organizations, and workers' organizations. (ILO, 2016, p. 2)

This implies that women, on equal terms with men, should have access to work that provides adequate income. Work should be based on egality, providing every worker a prospect of progress and personal development, without any form of discrimination. Child labour, as well as forced or bonded labour, should be abolished. The concept of decent work (DW) also implies the right to safe and satisfactory working conditions, including social protection for workers and their families. Finally, it implies the freedom to form or join unions, the right of unions to bargain collectively, the obligation to reveal unacceptable conditions at the workplace without reprisals, and the encouragement of dialogue between employers, unions, and other stakeholders in the labour market.

The primary goal of ILO is still to combat poverty, inequality, and insecurity globally. The conceptualization of DW provides ILO with the means to operationalize social justice in measurable terms. In the following years, a variety of indices and indicators were developed, and used to monitor the standard of work-life throughout the world (e.g. Bonnet, Figueiredo, & Standing, 2003; ILO, 2013).

DECENT WORK AND SUSTAINABILITY

The notion of DW affords novel and valuable perspectives on the discussion about sustainable work. The professional and political backdrop of the Decent Work Agenda differs somewhat from what is usually found within HFE. The implications and ambitions pursued, however, coincide largely with values regarded as fundamental also within HFE.

Promotion of local or global sustainability is not the primary aim of the DW agenda. As stated by ILO (2016, p. 1), the Decent Work Agenda is: "...a *strategy* to achieve sustainable development that is centred on people". Thus, DW is promoted as a key prerequisite and driving factor for achieving the 17 goals of the 2030 Agenda for Sustainable Development, adopted by the United Nations in 2015. The goals, importance, and implications of DW are emphasized as goal number eight: "Sustainable economic growth, employment and decent work for all" (ILO, 2017). In the Agenda for Sustainable Development, DW is seen as a programme that should be applied to productive work at all levels of a society, both nationally and internationally.

The introduction of DW has engendered transdisciplinary interest in both theoretical and applied research. Reviewing studies addressing DW as a strategy to fight global poverty, Di Fabio and Maree (2016), for example, included five perspectives, rooted in philosophical, juridical, economic, sociological, and psychological frameworks, respectively. Despite some differences, the mutual notion seems to be that if sustainable development is to be achieved, development must be based on DW.

Incorporating DW as a key factor in SD implies that work has a far more important function than only ensuring a livelihood for posterity. According to this line of thinking, the appropriate question is not *if* DW should be included in the discussion about sustainable development. It is rather *how* DW should be interpreted in such a context. Furthermore, it is a question of how a sustainable understanding of work that includes DW fits the framework of HFE, and if this can enhance the transition towards a sustainable future.

Before commencing the discussion of DW and its relation to HFE and sustainable work, a small reminder voiced by Moore and Barnard (2012) seems appropriate. When working with, or discussing, sustainability, one needs to be clear on what we mean by the term sustainability, and what should be sustained.

Sustainability, HFE, and the Understanding of Work

The end goal of HFE is to contribute to good, healthy, and profitable work that benefits both employees and employers (Dul et al., 2012). Macro-ergonomics (or organizational ergonomics) acknowledge that job performance and working conditions are affected by structures and processes at the organizational and societal level, and emphasize that this influence should be included in the analysis of the socio-technical system (Carayon, 2006; Kleiner, 2006). Traditionally, however, most HFE activity seems to be directed towards meso- and macro-level analysis related to the workplace and work itself, such as the prevention of illness and injury, improved well-being at work, and job satisfaction, as well as enhanced productivity and work and systems performance (Thatcher, 2012; Wilson, 2000). Thus, it could be argued that HFE activities to a lesser extent have been directed towards global processes. One reason might be the fact that the strong, applied tradition in the field tends to favour research and practices with immediate utility (Hancock & Diaz, 2010; Meister, 1999). In principle, however, knowledge and methods developed within HFE are applicable to all aspects of human existence, including global issues.

The WCED report (Brundtland et al., 1987) refers to the necessity of addressing not only the environment and the economy, but also social concerns, if one is to achieve the goal of a sustainable future. The importance of these three pillars in the understanding of SD has been elaborated and further specified in later writings and is currently known as the "triple bottom line" (sometimes also referred to as the three-pillar approach) (Elkington, 1997).

At present there appears to be no generally accepted definition of *sustainability* (Johnston, Everard, Santillo, & Robèrt, 2007). Nevertheless, the term seems to implicate that human activity should not compromise the long-term balance between the economic, environmental, and social pillars. The meaning of SD likewise seems to imply that intended utilization of resources, in order to fulfil human demands, should not permanently shift the balance between the same three pillars. This understanding applies to all systems, regardless of size, affecting the triple bottom line, directly or indirectly.

Environmental impact is rarely a main theme in prevailing HFE literature. In cases where this is discussed at all, it is treated mainly in economic terms, like discussions of the potential of energy savings. This may leave the impression that HFE is primarily concerned with relationships existing within or between the economic and social pillars of the triple bottom line (Thatcher, 2012; Zink & Fischer, 2013). Nothing in the content of meaning associated with sustainability and SD seems to prevent the inclusion of environmental concerns into the analysis of work and work systems in HFE.

In addition, most definitions of HFE accord with a systems model based on the triple bottom line (social, economy, and environment) approach (Zink, 2014). Regarding work and sustainability, the dual purpose of HFE should even be considered an advantage compared to other fields, as balancing trade-offs between productivity requirements and human needs are already established as a centrepiece of the HFE profession (Dul et al., 2012).

Incorporation of social and societal thinking in the understanding of SD opens up for the inclusion of work as something that should be organized and developed with sustainability in mind. What this implies is still not obvious, but a sensible starting point would be to capitalize on the platform of shared values that exists between HFE, DW, and the prevailing understanding of sustainability and SD. Thus, it becomes necessary to establish an analytical approach that makes it possible to relate the meaning of sustainability and SD to the function and execution of work. At the same time, the chosen perspective must be able to elucidate possible consequences for the long-term balance of the triple bottom line.

A Resources Perspective on a Nation's Workforce

Both the conceptual apparatus and the terminology used within HFE are well suited for analyses of work and work processes. Unfortunately, other professions and disciplines, with a stake in the triple bottom line, do not necessarily share this understanding. On the contrary, other disciplines frequently emphasize that the views and priorities of HFE differ from their own. Employers and business managers, for example, often criticize HFE for being too attentive to occupational health and safety (OHS) issues while ignoring tasks involving effectiveness, productivity, and financial business goals (Dul & Neumann, 2009; Lee, 2005). Such statements may be considered tendentious simplifications confuting the renowned definition of HFE (International Ergonomics Association, 2017). Nevertheless, this entails prevalent assumptions that may hinder a mutual understanding of sustainable work, and its prerequisites, among the stakeholders of the triple bottom line. Acknowledging Dul and Neumann's (2009) proposition, a possible redress to this problem is to establish a common language frame by utilizing ideas and concepts already familiar to other stakeholders.

A pertinent approach is to capitalize on theory, viewpoints, and practices present in the field of Human Resources (HR) and Human Resource Management (HRM). The terms HR and HRM, as well as writings within this field (e.g. DuBois & Dubois, 2012; Ehnert, 2009; Jackson & Schuler, 1995), clearly point out that a nation's workforce can be seen as a national resource pool and can be treated and managed as such. Nationally, this resource pool can be managed by the use of mandatory health, safety, and environmental (HSE) regulatory measures, of both protective and health promotional natures ruling the activities in the enterprises. The HRM strategy chosen by the enterprise will consequently be of vital importance, not only in shaping the working conditions but also in determining how the enterprise relates to external factors, including the surrounding society and the environment (DuBois & DuBois, 2012; Ehnert, 2009; Jackson & Schuler, 1995). Decisions in HR and HRM will therefore, in many ways, be a determinant for the sustainability of work and the workforce. National legislation relating to sustainability should consequently establish a framework for sustainability of the workforce. Thus, conceptualizing the workforce as a limited resource, to be managed in line with other assets encompassed by the triple bottom line, should make it possible to approach an understanding of the meaning and consequences of sustainable work.

Following this line of thinking, sustainability becomes first and foremost a question of how the resources are managed, regardless of whether in the form of nature, humans, technology, or economy (DuBois & DuBois, 2012). The two main categories of resources, renewable and non-renewable resources, demand different strategies for sustainability. One could describe sustainable use of renewable resources as harvesting no more than what will be replaced (Hilborn, Walters, & Ludwig, 1995; Keohane & Olmstead, 2016; Moldan, Janoušková, & Hák, 2012). A pertinent example is leaving enough fish to enable a large enough remaining pool to spawn and reproduce.

Adopting the perspective of strong sustainability (Neumayer, 2003), sustainable management of non-renewable resources means harvesting resources no faster than alternative resources are developed (Moldan et al., 2012; Reijnders, 2000). The obvious example here is not depleting the remaining store of hydrocarbons before alternative energy sources allow necessary future production by machinery today demanding hydrocarbon-based energy. The common denominator for the treatment of both renewable and non-renewable resources would be sound management. The art of sustainability management is still under development—step-by-step new, common perspectives are being developed and tested on economic, social, and environmental issues of production and business management (DuBois & DuBois, 2012; Ehnert, 2009; Figge, Hahn, Schaltegger, & Wagner, 2002).

In general, production resources are best seen as belonging to one, and only one, of the two classes of resources, renewable or non-renewable. The human workforce, however, defies such easy categorization. All societies known to man so far have been dependent on their workforces. The total workforce of a society consists of "every able and willing man and woman", and constitutes an important part of the resources used in production. Until robots take over completely, man will be a sine qua non ingredient in any production process. Periodically, lack of workforce in one society has been compensated by introduction of workforce from another, be it by import of willing labourers or of slaves, regardless of ethical considerations.

The workforce has obvious traits of being a non-renewable resource. Man is born, lives, and subsequently dies. Every individual element in the total workforce represents a limited amount of work ability, for example, strength and work lifespan. Any individual can become exhausted of work ability. Sudden death by falling from scaffolding creates an immediate lack of work ability. Other, less dramatic, events may also drain the workforce of power. Illness, accidents, and low work morale will reduce the ability to work, eventually reduce output, and thus place the society as a whole in peril. Unless one can find a replacement with at least the same work abilities as those present in the individual who dropped out, the total work ability of the workforce will deteriorate. This constitutes a loss for society (Edwards & Greasley, 2010).

Even when one applies such a resources perspective on individuals, it is obvious that society depends on its inhabitants for more than work. Individuals add to the workforce by reproducing themselves on their way through life. Their offspring take over the tasks of production when the older generation no longer produces. On individual basis, work ability is non-renewable. The workforce as such, however, is renewable, as long as its individual members remain fertile. If a large segment of every generation skips this reproductive pattern, the workforce will cease to renew itself and eventually cease to exist. The society may then come to a productive halt.

The ambiguity of the workforce, seen in the sustainability perspective, thus seems to defy the management parameters of, both renewable and non-renewable resources. This makes it necessary to look further into how this resource should be managed, and by whom.

TRIPARTITE STAKEHOLDERS: POSSIBILITIES, STRATEGIES, AND PITFALLS

The enterprise and the individual worker are the two main stakeholders in the traditional employer-employee model of work (Walliser, 2008). The introduction of the tripartite labour collaboration inducts a third stakeholder—the society, represented by governmental representatives. The tripartite organization is a body established primarily to ensure a frame of equity and cooperation between the parties when negotiating wages and working conditions (Sengenberger, 2013; Van Daele, 2005). It is possible to argue that the tripartite model is valid only for traditional occupations in traditional industrial societies, while other models are more suitable for developing countries and others again for the emerging post-industrial societies of the advanced economies (Gallagher & Sverke, 2005; Hepple, 2006). Nevertheless, the tripartite model reflects general and fundamental differences in the interests of the different stakeholders, which are ubiquitous in almost all forms of paid work. Incorporating the society as the third main stakeholder of regulated work also makes the transition between the dual goal of HFE and the triple bottom line clearer. Changes in productivity and well-being at work always alter the relationship between the pillars of the triple bottom line. With regard to sustainability, the question is whether, and how, these alterations affect the long-term balance of the triple bottom line.

The Society

The first stakeholder is the society. As already shown, society can be seen as the main provider of workforce in the tripartite model. With the emergence of national healthcare systems in more and more countries, the society or the nation becomes an obvious stakeholder in how the working conditions affect its workforce. Bodily damage from work and accidents and medical needs stemming from work related diseases have impact on both the sustainability of the workforce and on national budgets.

Second, society must prepare each and every individual to accommodate the needs of enterprises in different areas of business (Hall & Lansbury, 2006). Annually, a cohort of children sees life, another enters schools and educational institutions, as part of their preparation for adult life. In the HR perspective, they represent raw material for work-life in the future years. Society, at least in the form of some of the modern welfare states, represents a safety net, providing a living for those unable to find employment, or for those who lose employment without finding a new basis for subsistence. The cost of such social interventions and benefits makes society dependent on its members obtaining an education that prepares them for work-life, enabling employment or other economic activity as soon as possible. Society then needs everybody, once employed, to lead a healthy life, with little need for health intervention. Ideally, everybody should provide offspring that will ensure the next-generation raw material for work-life. Again for economic reasons, the society needs everybody to remain employed or in business, also as they grow older. A healthy life will diminish the need for healthcare and social benefits during one's career; employment into old age will ensure a sufficient pension for the no longer employed person to lead a full and healthy life without being a disproportionate burden on the economic shoulders of society. HRM based on the idea of using and discarding employees is a challenge to society, as is the idea of "enjoying life in full, taking a year off".

Whereas enterprises and workers are primarily concerned with their own interests, society will supposedly strive for a common good. A sustainable work-life, understood as providing the individual with the ability to work throughout a normal career span, is in the interests of the collective. At the same time, every enterprise and every individual may profit from strategies that differ and deviate from those leading to the collective well-being.

This propensity is excellently elaborated in the conundrum known as the "tragedy of the commons" (Hardin, 1968). The "tragedy" rests upon the assumption that individual actions rarely are intentionally harmful or make a significant difference. As individuals act rationally, with an intention to increase their own earnings, the total result easily exceeds the environmental tolerance, instigating future impoverishment and loss for all involved. The "tragedy of the commons" was originally laid out with reference to common land and the right to pasture. Hardin (1968) interpreted it figuratively including general pollution, matters of pleasure, as well as breeding, as examples. According to Hardin (1968) the "tragedy of the commons" should be countered by abolishment of commons and by legal regulation of behaviour. Although we are admittedly not there yet, regulatory frameworks (i.e. laws, rules, and regulations, together with regulatory bodies) have become stricter. This does not appear to prevent the fundamental problem from unfolding. Stricter frameworks often seem to prompt stronger resolution to circumvent or overlook the rules. Poaching of endangered species, for example, continues to be a problem in many countries regardless of regulations and harsh punishment, even those including shoot-on-sight policies (Keane, Jones, Edwards-Jones, & Milner-Gulland, 2008; Messer, 2010).

Other measures used to heighten awareness about sustainability are education and targeted public campaigns (Baruch-Mordo, Breck, Wilson, & Broderick, 2011; Sheavly & Register, 2007). The results of such programmes may be discussed, but they do entail ethical problems. Is it right to argue that poor people should reduce their consumption and put their efforts to improve their living conditions aside? In general, people are not content with simply staying alive, on resources that hardly cover basic subsistence. They choose to strive for a better life for themselves and for their family, disregarding how this affects the society. Thus, it may seem that both moral and law enforcement have obvious weaknesses.

The only viable option in our view is facilitating alternative courses of action. Other types of work that do not lead to impoverishment must be available, and seem to be a possible profitable option for those concerned. Increased income, however, will be only a minor factor in a successful transition. Introduction of new work opportunities must create a foundation for developing social cohesion based on new value assumptions while improving well-being and quality of life for other members of the society. In our view, DW entails necessary properties to provide this foundation.

A major obstacle in this regard is the tendency to consider workers like any other commodity (Dick & Hyde, 2006; Lillie & Sippola, 2011). Put bluntly, workers are units of labour and sellers of knowledge and skills, which enterprises buy in a global market. This view signifies the conception of labour markets as a form of commons. In the short term, this entails an individualization of work that benefits groups of people who possess marketable skills. Those who do not possess such skills lose, and have to settle for precarious jobs with unstable employment or no employment at all (Standing, 2011). From a societal point of view, this situation represents wasted or poorly utilized resources. In this perspective, such practices will hardly benefit society. The conceptualization of work as a form of commons thus represents a daunting challenge, not only for the development of sustainable work, but also for the ability to achieve a sustainable future.

The Enterprise

The primary concern of the enterprise is the availability of a sufficient supply of qualified and able workers (Walliser, 2008). These must be available when the enterprise needs to produce. The workers need to have a work ability that allows production to run without unnecessary halts. The main interest of the enterprise is that production runs smoothly at an acceptable cost.

The business of the enterprise will determine what it wishes its staff to provide. Some enterprises seem to thrive on high turnover and a never-ending supply of fresh, "hungry" workers (Flecker, 2009). Others require loyalty and employment permanence, depending heavily on tacit knowledge, and regard employee turnover as a challenge to the future of the enterprise (Winch, 2013). This represents two extremes on a scale where most enterprises can be placed near the centre. Nevertheless, the extremes are good examples. Regarding the first extreme, a "use and discard" mentality in personnel management will suffice as long as there exists a fresh supply of potential employees. In the case of the second extreme, it is important that emphasis be placed on the maintenance and protection of employee health.

Generally, enterprises want to employ individuals who are healthy, willing, and able to contribute amply to production, but wish employment to last only as long as they are both fully able and needed. In difficult times, there is a tendency towards senior employees and those with health problems being laid off (Carden & Boyd, 2014; Iverson & Pullman, 2000; Mastekaasa, 1996; Strully, 2009). This is self-explanatory, as young people are striving to enter work-life, if necessary at lower pay. Junior employees also seem more willing to produce and work limitless hours when demanded. The tendency to lay senior employees off, however, is also linked to a myth regarding the efficiency of junior employees compared with the more laid-back senior employees, a myth repeatedly proven incorrect (e.g. Posthuma & Campion, 2008). Seen both from a workers' perspective as well as a societal perspective, such attitudes and behavioural patterns are problematic and hardly correspond with the main aspects of either sustainable or decent work.

The publication of the WCED report *Our Common Future* (Brundtland et al., 1987) represented in many ways a watershed in the awareness about sustainability. Gradually the global focus on sustainability issues has been growing sharper. Action plans have been developed and national strategies and measures for sustainability have fallen in line with legislative regulations. Through regulatory measures, governments place most of the responsibility for implementation of these measures on enterprises (Randers, 2012). The emphasis on sustainability has led to the development of systems for monitoring and measuring to which degree different solutions contribute to sustainability.

In addition to regulations, other framework conditions affect enterprise attitudes and behaviour. As framework conditions are important for economic success, and liable to change, the main task for enterprises must be to monitor these changes or, at best, to attempt to predict forthcoming changes and align corporate operations and policies accordingly.

Stricter regulations and public awareness regarding sustainability have initiated a discourse about the responsibility of businesses and organizations, which also encompasses the value of work (Budd & Spencer, 2015). A reorientation seems to have occurred whereby sustainability issues and corporate social responsibility (CSR) gradually blend with the agenda of corporate governance. Most likely, this reflects a situation where questions about the organization's purpose, its stakeholders, and how its business activities should be executed are considered to be not quite as simple as previously (Elkington, 2006). Among management and business owners, work-environmental issues have often been regarded to be cost drivers. The incorporation of sustainability and CSR in corporate governance could contribute to change this view. Instead of dwelling upon the negative effects of absence and accidents, commitment to corporate health management programmes, by both management and workers, could create more positive attitudes. This can in turn facilitate improved alliance between employees and employers in matters of sustainability (Zink, 2005, 2014).

An important question in this regard, is how this increased awareness about CSR and sustainability manifests itself in actual workplaces. Studying the way corporations address work-related issues, Brunoro, Bolis, and

Sznelwar (2016) found that work has been included in the social dimension from the earliest documents reporting on sustainable corporate development. This indicates that work and key aspects of DW have been considered an integrated part of what has been understood as corporate sustainability. After studying websites and reports published by large corporations, Bolis, Brunoro, and Sznelwar (2014) confirm that workers increasingly are declared to be decisive stakeholders of the corporations, and that references are made to workers' needs, health, and professional development. Unfortunately, the CSR initiatives are presented exclusively in the top-down perspective. It is therefore possible to question the sincerity of this engagement, and to ponder whether such statements act more as window dressing aimed at public opinion.

Others find that core implications of DW have been challenged in the recent discourse about how to encounter the demands of a globalized economy. A shift has been observed in the writings of influential international non-governmental organizations (INGOs), gradually placing more emphasis on financial considerations at the expense of social responsibility, in the years following the economic crisis in 2008–2009 (Di Ruggiero, Cohen, Cole, & Forman, 2015). Thus, it is difficult to conclude as to whether or not ideas about corporate sustainability and social responsibility have actually led to long-term changes in corporate priorities.

This is supported by the gradual decline in permanent and decent work that has been observed in most countries during the past decade (Blustein, Olle, Connors-Kellgren, & Diamonti, 2016). The reasons may be several, but some researchers point towards an increased acceptance of neoliberal viewpoints. This tends to sway the focus of enterprises towards economic growth and less emphasis on CSR (e.g. Blustein et al., 2016; Hursh & Henderson, 2011; Pouyaud, 2016; Rushton & Williams, 2012; Wisman, 2013).

Regarding the concept of establishing sustainable work for all as a realistic goal for the near future, this is somewhat disconcerting. Reconfiguring work to address the needs of both the employer and the employee in a desired sustainable work-life needs to be based on a common platform of values, norms, and agreements, generally acknowledged in the society. Values and norms undoubtedly vary between nations and cultures. Nevertheless, it should be possible to apply well-known and generally accepted standards as a basis for the formation of a framework for sustainable work. Although the conditions and constraints that constitute the framework may vary somewhat, a minimum standard that is perceived as applicable both for enterprises and societies should be established. DW has from the outset been rooted in ethical principles of equity and justice, and is as such closely connected to the ideas about universal human rights (Pouyaud, 2016). In our view, the ILO agenda of DW represents such a basis, and could be used as framework conditions for a global process towards sustainable work.

The Individual

Individuals have a broad and varied set of possible strategies to deal with the demands of society and of work-life. Thus, individuals can be seen as agents, making most of the possibilities and limitations they perceive in their potential work-life careers[1] (Yerxa, 1998).

There is a general assumption that individuals desire to have, and to maintain, good health and long and meaningful lives. Accordingly, a good job is desirable, providing not only financial and material contingencies, but also a range of other positive gains that extend to the family sphere, and eventually to the community. A rewarding period in work-life should lead to a long and rewarding post-labour life as a pensioner. Individuals using the "homo economicus" view on their work-life will find that the perfect solution is to go through work-life with efforts sufficient to keep employers happy, providing themselves with a fulfilling work situation, without attaining work-related health damage.

To achieve this, the best strategy would appear to be to focus on one's own health. Good health is one of the elements an employer will consider when looking for candidates for continued employment. In a longer per- spective, good health will also enhance one's chances of continued employ- ment until one has accumulated enough work years and a funding to lead a sustainable life as a pensioner.

Nevertheless, there are pitfalls that could make the demands of work difficult to handle. Failing health may become an insurmountable hur- dle. Work-life salutogenesis is complex and depends on sheer luck, individual fulfilment, exposure to damaging substances, physical and psychosocial work environment, and other elements, combined with individual genetic and social dispositions (Antonovsky, 1987). Decline in perceived health is an indicator of a bad match and of possible future undesired destiny, such as loss of employment or early retirement, be it voluntarily or forced.

The emerging post-industrial economy initiates a scope of changes that affect people inequitably, both nationally and internationally. Policy analysts, as well as scientists, seem to agree that future work will be less stable and more difficult to obtain for large parts of the population (e.g. Blustein et al., 2016; Di Ruggiero et al., 2015; Kalleberg, 2009; Ross, 2008). Outsourcing or "offshoring" (i.e. moving business activities abroad) of business functions, often to low-cost countries, is one important factor in this development (Flecker, 2009; Olsen, 2006). Digitalization and atomization of jobs is another, and perhaps even more severe, factor. In Sweden, for example, projections show that up to 52 per cent of those presently employed are facing replacement by digital technology in the coming two decades. This affects approximately 2.5 million jobs (Fölster, 2014). These figures are comparable, although somewhat lower, with prognoses from the USA where 47 per cent of current jobs are seen to be in danger of disappearing (Frey & Osborne, 2013). Adding to this are the apparent problems creating enough new and steady job opportunities (Groshen & Potter, 2003; Shane, 2009). This points cumulatively to a future work-life with less equality, career opportunities for fewer, and the need to create non-work alternatives to make lives fulfilling, as well as a need for finding ways to fund the existence of the unemployed (Standing, 2011).

It may be argued that this is a problem chiefly for industrialized countries. There is, however, little doubt that this also affects countries in the rest of the world. In this perspective, individuals could be considered victims with limited control over their own destiny, compelled to secure a job with no prospects beyond earning a minimum livelihood. Although basic needs could then be considered met, most people will still be dissatisfied with the situation and seek other opportunities to thrive and to fulfil their dreams.

Individual workers need to make some strategic plans regarding their career spans. One usually enters work-life with some kind of stepping stone competence achieved through school, studies, or practical training. Unlike earlier generations of workers, today's workforce can no longer expect their initial competence to last for the whole of their career (Hall & Lansbury, 2006). In situations where additional competence was required, workers could previously expect to receive most of this training at their workplace, organized by the company. Rapid technological and social development seemingly alters this situation, and forces large parts of the workforce to leave work in periods, only to re-enter work later, with new stepping stone competence (Billett, 2014).

New competence consists of not only skills and knowledge taught in the form of a specific curriculum. In employment, one also acquires several useful skills such as learning skills, communication skills, adaptability, group effectiveness, and self-management. Although none of these skills is formally documentable, such skills are still highly appreciated by most employers (Carnevale & Smith, 2013). The result is a potential win-win situation. The employee increases his/her work ability and feelings of competence, mastery, and autonomy. At a general level this is about feeling valued and useful, which is acknowledged as being of crucial importance for perceived meanings of work and life satisfaction (de Lange, Van Yperen, Van der Heijden, & Bal, 2010; Rosso et al., 2010). The employer, on the other hand, gets a more highly motivated and knowledgeable worker, able to handle new challenges more efficiently and at a higher level than previously. Together, this constitutes the foundation for a productive and meaningful life throughout the work career.

Considering the demographic ageing in the industrialized countries, this is highly relevant. The numbers of retirees are increasing alongside a declining working population. Already today, the group of people above 65 years represents more than 15 per cent of the population in many Western countries. As not only recruitment to the group of elderly increases, but also this group's life expectancy, demographic factors entail major economic challenges for the society (Crews & Zavotka, 2006). As protection against an unbearable burden on the social protection system, more people need to stay in the workforce for a prolonged part of their lives. This entails delaying retirement for most people. This requires, however, that accommodations be made to balance the requirements of work and the needs of the individual, both economically and socially. Achieving these objectives depends on the remaining individual ability to master a variety of life challenges, as well as on employers' willingness to participate through workplace facilitation and redistribution of tasks and responsibilities. The employers' willingness again often seems to depend both on assigned government agencies (inspectorates) insisting that work-life rules, sustainability included, are adhered to in all enterprises, and on other government agencies providing incentives for employers to keep marginal employees employed.

Thus, the realization of sustainable decent work, entailing accommodation of living and working conditions enabling workers to engage, feel appreciated, and productive, both at work and in life, must also imply a concerted willingness and effort from the society, the enterprise, and the individual.

Towards an Understanding of Sustainable and Decent Work for HFE

Complexity is a major problem in both sustainability analysis and the implementation of sustainable development. Questions about sustainable work are by no means different in this regard. Work is multifaceted and encompasses a wide range of activities. It is therefore difficult to imagine one single, unified definition of sustainable work that would make it possible to cover the variety of germane situations that exist today, to say nothing of the future. Increasingly conflicting interests among the three main stakeholders of work-life do not improve the situation.

From a societal point of view, the sustainability of work seems to be dependent on strategies that ensure that as large a part as possible of the workforce is able to maintain employment throughout a normal career span. This implies that regulatory measures should include HFE principles in the workplaces and that those ought also to be included in the national programmes for DW. Furthermore, it implies that also future employment produces a value added that benefits the society economically and socially, without overtaxing the environment.

The cardinal interest of the enterprise is, and will assumingly always be, attracting a sufficiently educated and trained workforce to undertake production smoothly and cost-effectively. In addition, the enterprise needs access to fresh capital, commodities, and other means of production in order to evolve and prosper. As already stated, the main provider of both able workers and means of production is the society. To fulfil its obligation to the enterprises, the society must provide its population with a just, safe, healthy, and functional community, both with regard to social and natural environments. Thus, it should be in the interest of any enterprise to facilitate working conditions that benefit the worker and the society, without compromising the environment.

To the individual, work represents not only the means to achieve a satisfactory income but also an arena for social bonding, learning, belongingness, and meaning of life. To accommodate this, the work and the workplace must be perceived as just, safe, and secure, presently and in the future. The enterprise should also be perceived as progressive and in line with society. Finally, neither the society nor the enterprise should be regarded as destructive and non-caring about the natural environment.

In order to give content to the term, *sustainable work should be understood as a totality, where the needs of the individual, the enterprise, and the society are dealt with in a proper and balanced way, at the same time ensuring that the environment remains functional, ecologically and biologically, today as well as in the future.* The implementation of HFE with DW included is in this regard considered a prerequisite to achieve sustainable work.

Pursuing sustainable work as a goal in HFE has implications of both theoretical and practical nature. An obvious consequence is that HFE, and maybe especially the field of macro-ergonomics, must expand its area of interest and to a greater extent reflect global issues in the analyses of work and its consequences. Expanding the socio-technical system to encompass the often conflicting interests of stakeholders involved in the triple bottom line (society, economy, and environment) inevitably means increased complexity, as the number of possible interconnections multiplies.

The systems approach embedded in HFE provides the field with a conceptual framework and a set of analytical tools suitable for investigating and understanding complex interactions between humans and other elements in the system. Zink (2014) argues that this basis makes HFE especially apt to contribute to the development of sustainable work systems.

Current models do not suffice, however, and new developments seem required if one wants to grasp the challenges raised by incorporating the triple bottom line into the analysis of work. A first step, also advocated by Zink (2014), will be to include the dimension of time in the analyses, encompassing a lifecycle perspective on the workforce and the green economics. This corroborates the viewpoints of those who have suggested that time should be included as the fourth pillar of a quadruple-bottom-line of sustainable development (Waite, 2013).

Second, HFE needs to develop a broader understanding of well-being. The focus should no longer be mainly on work-related well-being but also encompass workers' connectedness to family, society, and the environment. In our opinion, HFE can no longer be treated as detached from the societal, economic, and environmental context in which work is embedded, if sustainable work is to be achieved.

A third, but nevertheless demanding, challenge for the field of HFE is the relation to labour unions, employer associations, and the ongoing efforts to establish minimum standards for the work sector (e.g. Hasle & Petersen, 2004; ILO, 2014). A disturbing downside of the current globalization is a decrease in stable and decent employment for many workers around the world (Blustein et al., 2016). Unemployment and exploitation of workers not only affect individuals and families, but also contribute to the impoverishment of local communities, often with pollution and/or depletion of water resources, arable fields, and pasturage as additional miseries (Gray & Moseley, 2005; Liu & Diamond, 2005; Mabogunje, 2002). It is difficult to imagine how sustainable work can be pursued within HFE without advocating international regulations that hinder exploitation of workers and depletion of local communities. Thus, if DW is included as a prerequisite for sustainable work, this will most likely require more time and effort to be invested by people working within HFE, in national and international agendas and agencies aimed at imposing regulations of working life and the labour market. A possibility for HFE could be to take on an evidence-based mediator role in this process.

The increase of precarious work (Standing, 2011) is a problem in this regard. Although DW should apply also for unpaid work, precarious work, and any other form of non-regulated work, this ideal seems difficult to achieve. In the context of HFE, the term sustainable work should consequently carry meaning, mainly in relation to employer-employee relations, that is to say salaried and remunerated work.

Concluding Remarks

Over the past decades, sustainability and sustainable development (SD) have become an issue of utmost importance and have been partially investigated and pursued at the global/planetary level (e.g. Rockström et al., 2009). Other elements within SD are pursued in international politics diplomatically, as in the Paris agreement at the November/December 2015 summit (United Nations Framework Convention on Climate Change, 2015). The idea of *Think globally and act nationally* brings challenges down to the parliamentary levels of the politics of the nations, emerging into national legislation in the areas chosen for national action. Parts of this legislation become the basis for regulation of the enterprises of the same countries. If success is to be achieved, laws and regulations must be adhered to and acted upon. The human tendency to emphasize possible losses higher than possible gains, known as loss aversion (Soman, 2004), may represent a problem in this regard. It is thus important that alternative behaviour, promoting sustainability, appears as attractive or necessary at an individual level.

The idea of sustainable work has been launched both as an integrated part of sustainable development and as a strategic tool to obtain sustainable development. In both cases, the term involves more than only providing enough income to sustain livelihood. In short, the term involves a range of fundamental conditions that enable people to engage and develop in work and life throughout the course of their life. People evolve and change, as they grow older. This means that their working conditions must also change, and that they need to adapt at the same pace if sustainability is to be maintained. Thus, decent work should be considered as an integrated part of sustainable work. In the view of the authors, it is not possible to obtain sustainable work without facilitating decent work.

Decent work is closely connected to human rights, and most people agree principally with the notion that all people have the right to participate in society on equal terms, and that this necessarily requires some adaptations. One should be aware, however, that this is a relatively new way of thinking, closely related to the advanced economies and the development of the modern welfare state (Barnes, 2011). From this viewpoint, active work participation on equal terms is a prerequisite for participation in democratic elections and decision-making processes. Thus, it could be argued that the promotion of sustainable and decent work has political connotations that implicitly endorse some governmental systems over others.

The conditions for short-term implementation of sustainable and decent work in other parts of the world vary. Although one may observe several positive trends towards reduced poverty, less vulnerable employment conditions, and increased social protection, there is still a long way to go before decent work has been achieved, especially in the emerging and developing countries (ILO, 2015). The principles of universal human rights and judicial independence may make the approbation of sustainable and decent work undesirable in some parts of the world. Economic ideologies may also pose a problem in this regard. Nevertheless, the authors doubt that any other solution is possible if a global sustainable future is to be achieved.

Sustainable and decent work obviously contains economic aspects. It is well known that work-life-induced pressures on the welfare system of the advanced economies increase and will continue to do so in years to come. This becomes particularly evident in relation to the large increase in the senior citizen group. One way to meet the demographic and distribution policy challenges facing the welfare states over the next 40–50 years will be to increase the workforce through adaptation for the elderly and disabled, in both workplaces and society. Such a development could also help to reduce sickness absenteeism and to prevent exclusion of the same groups from work-life.

Employee participation in management decision-making processes has been regarded as a key factor in the Nordic model and has been shown to improve productivity and the quality of the working environment (Bhatti & Qureshi, 2007; Busck, Knudsen, & Lind, 2010). Within HFE, employees have been involved in a variety of interventions aimed at identifying and preventing ergonomic problems and planning of new production lines, systems, and schedules (Hignett, Wilson, & Morris, 2005; Vink, Koningsveld, & Molenbroek, 2006). Considering its advantages, increased involvement in the workplace by employee participation might very well be the high road to a more sustainable and decent work.

NOTE

1. It could be wisely noted here that the word career stems from middle Latin "carrāria", meaning cart road, not necessarily made for high-speed and smooth driving.

REFERENCES

Antonovsky, A. (1987). *Unraveling the mystery of health: How people manage stress and stay well.* San Francisco: Jossey-Bass.

Barnes, C. (2011). Understanding disability and the importance of design for all. *Journal of Accessibility and Design for All, 1,* 55–80.

Baruch-Mordo, S., Breck, S. W., Wilson, K. R., & Broderick, J. (2011). The carrot or the stick? Evaluation of education and enforcement as management tools for human-wildlife conflicts. *PLoS One, 6,* e15681.

Belser, P., de Cock, M., & Mehran, F. (2005). *ILO minimum estimate of forced labour in the world,* (7). Retrieved March 2, 2017, from http://digitalcommons.ilr.cornell.edu/nondiscrim/7/

Bhatti, K. K., & Qureshi, T. M. (2007). Impact of employee participation on job satisfaction, employee commitment and employee productivity. *International Review of Business Research Papers, 3*(2), 54–68.

Billett, S. (2014). Conceptualising lifelong learning in contemporary times. In T. Halttunen, M. Koivisto, & S. Billett (Eds.), *Promoting, assessing, recognizing and certifying lifelong learning* (pp. 19–35). New York: Springer.

Blustein, D. L. (2008). The role of work in psychological health and well-being: A conceptual, historical, and public policy perspective. *American Psychologist, 63,* 228–240.

Blustein, D. L., Olle, C., Connors-Kellgren, A., & Diamonti, A. J. (2016). Decent work: A psychological perspective. *Frontiers in Psychology, 7,* 1–10. Retrieved August 30, 2016, from https://doi.org/10.3389/fpsyg.2016.00407

Bolis, I., Brunoro, C. M., & Sznelwar, L. I. (2014). Work in corporate sustainability policies: The contribution of ergonomics. *Work: A Journal of Prevention, Assessment and Rehabilitation, 49*, 417–431.

Bonnet, F., Figueiredo, J. B., & Standing, G. (2003). A family of decent work indexes special issue: Measuring decent work. *International Labour Review, 142*, 213–238.

Brundtland, G., Khalid, M., Agnelli, S., Al-Athel, S., Chidzero, B., Fadika, L., et al. (1987). *Our common future* ('Brundtland report'). Retrieved July 19, 2016, from http://www.citeulike.org/group/13799/article/13602458

Brunoro, C., Bolis, I., & Sznelwar, L. (2016). Exploring work-related issues on corporate sustainability. *Work, 53*, 643–659.

Budd, J. W., & Spencer, D. A. (2015). Worker well-being and the importance of work: Bridging the gap. *European Journal of Industrial Relations, 21*, 181–196.

Busck, O., Knudsen, H., & Lind, J. (2010). The transformation of employee participation: Consequences for the work environment. *Economic and Industrial Democracy.* https://doi.org/10.1177/0143831x09351212

Carayon, P. (2006). Human factors of complex sociotechnical systems. *Applied Ergonomics, 37*, 525–535.

Carden, L. L., & Boyd, R. O. (2014). Age discrimination and the workplace: Examining a model for prevention. *Southern Journal of Business and Ethics, 6*, 58–67.

Carnevale, A. P., & Smith, N. (2013). Workplace basics: The skills employees need and employers want. *Human Resource Development International, 16*, 491–501.

Crews, D. E., & Zavotka, S. (2006). Aging, disability, and Frailty: Implications for universal design. *Journal of Physiological Anthropology, 25*, 113–118.

De Lange, A. H., Van Yperen, N. W., Van der Heijden, B. I. J. M., & Bal, P. M. (2010). Dominant achievement goals of older workers and their relationship with motivation-related outcomes. *Journal of Vocational Behavior, 77*, 118–125.

Di Fabio, A., & Maree, J. G. (2016). Using a transdisciplinary interpretive lens to broaden reflections on alleviating poverty and promoting decent work. *Frontiers in Psychology, 7*, 503. Retrieved December 8, 2016, from https://doi.org/10.3389/fpsyg.2016.00503

Di Ruggiero, E., Cohen, J. E., Cole, D. C., & Forman, L. (2015). Competing conceptualizations of decent work at the intersection of health, social and economic discourses. *Social Science & Medicine, 133*, 120–127.

Dick, P., & Hyde, R. (2006). Consent as resistance, resistance as consent: Re-reading part-time professionals' acceptance of their marginal positions. *Gender, Work & Organization, 13*, 543–564.

DuBois, C. L. Z., & DuBois, D. A. (2012). Strategic HRM as social design for environmental sustainability in organization. *Human Resource Management, 51*, 799–826.

Dul, J., Bruder, R., Buckle, P., Carayon, P., Falzon, P., Marras, W. S., et al. (2012). A strategy for human factors/ergonomics: Developing the discipline and profession. *Ergonomics, 55*, 377–395.

Dul, J., & Neumann, W. P. (2009). Ergonomics contributions to company strategies. *Applied Ergonomics, 40*, 745–752.

Edwards, P., & Greasley, K. (2010). *Absence from work*. The European Foundation for the Improvement of Living and Working Conditions (Eurofound). Retrieved July 22, 2016, from http://www.eurofound.europa.eu/docs/ewco/tn0911039s/tn0911039s.pdf

Ehnert, I. (2009). *Theorising on strategic HRM from a sustainability approach sustainable human resource management: A conceptual and exploratory analysis from a paradox perspective* (pp. 79–121). Heidelberg, Germany: Physica-Verlag HD.

Elkington, J. (1997). *Cannibals with forks: The triple bottom line of 21st-century business*. Oxford, UK: Capstone.

Elkington, J. (2006). Governance for sustainability. *Corporate Governance: An International Review, 14*, 522–529.

Fields, G. S. (2003). Decent work and development policies special issue: Measuring decent work. *International Labour Review, 142*, 239–262.

Figge, F., Hahn, T., Schaltegger, S., & Wagner, M. (2002). The sustainability balanced scorecard – Linking sustainability management to business strategy. *Business Strategy and the Environment, 11*, 269–284.

Flecker, J. (2009). Outsourcing, spatial relocation and the fragmentation of employment. *Competition & Change, 13*, 251–266.

Fölster, S. (2014). *Vartannat jobb automatiseras inom 20 år–utmaningar för Sverige* [Every other job will be automated within 20 years – Challenges for Sweden]. Stockholm, Sweden: Stiftelsen för Strategisk Forskning [Swedish Foundation for Strategic Research]. Retrieved July 28, 2016, from http://stratresearch.se/wp-content/uploads/varannat-jobb-automatiseras.pdf

Frey, C. B., & Osborne, M. A. (2013). The future of employment: How susceptible are jobs to computerisation. The future of employment: How susceptible are jobs to computerisation. *Technological Forecasting and Social Change, 114*, 254–280.

Gallagher, D. G., & Sverke, M. (2005). Contingent employment contracts: Are existing employment theories still relevant? *Economic and Industrial Democracy, 26*, 181–203.

Gray, L. C., & Moseley, W. G. (2005). A geographical perspective on poverty–environment interactions. *The Geographical Journal, 171*, 9–23.

Groshen, E. L., & Potter, S. (2003). Has structural change contributed to a jobless recovery? *Current Issues in Economics and Finance, 9*(8). Retrieved July 1, 2016, from https://ssrn.com/abstract=683258

Hall, R., & Lansbury, R. D. (2006). Skills in Australia: Towards workforce development and sustainable skill ecosystems. *Journal of Industrial Relations, 48*, 575–592.

Hancock, P. A., & Diaz, D. D. (2010). Ergonomics as a foundation for a science of purpose. *Theoretical Issues in Ergonomics Science, 3*, 115–123.

Hardin, G. (1968). The tragedy of the commons. *Science, 162*(3859), 1243–1248.

Hasle, P., & Jensen, P. L. (2012). Ergonomics and sustainability – Challenges from global supply chains. *Work, 41*(Supplement 1), 3906–3913. https://doi.org/10.3233/wor-2012-0060-3906

Hasle, P., & Petersen, J. (2004). The role of agreements between labour unions and employers in the regulation of the work environment. *Policy and Practice in Health and Safety, 2*, 5–23.

Hepple, B. (2006). Rights at work. In D. Ghai (Ed.), *Decent work: Objectives and strategies* (pp. 33–75). Geneva, Switzerland: International Labour Office.

Hignett, S., Wilson, J. R., & Morris, W. (2005). Finding ergonomic solutions—Participatory approaches. *Occupational Medicine, 55*, 200–207.

Hilborn, R., Walters, C. J., & Ludwig, D. (1995). Sustainable exploitation of renewable resources. *Annual Review of Ecology and Systematics, 26*, 45–67.

Hursh, D. W., & Henderson, J. A. (2011). Contesting global neoliberalism and creating alternative futures. *Discourse: Studies in the Cultural Politics of Education, 32*, 171–185.

ILO (International labour Organization). (1999). *Report of the Director-General: Decent work.* Retrieved January 23, 2017, from http://www.ilo.org/public/english/standards/relm/ilc/ilc87/rep-i.htm

ILO (International Labour Organization). (2003). *Working out of poverty.* Report of the ILO Director-General. Retrieved July, 23, 2016, from http://www.ilo.org/public/english/standards/relm/ilc/ilc91/pdf/rep-i-a.pdf

ILO (International Labour Organization). (2013). *Decent work indicators: Concepts and definitions, guidelines for producers and users of statistical and legal framework indicators, manual, ILO.* Retrieved July, 25, 2016 from http://www.ilo.org/wcmsp5/groups/public/---dgreports/---integration/documents/publication/wcms_229374.pdf

ILO (International Labour Organization). (2014). *Rules of the game: A brief introduction to International Labour Standards.* Retrieved January 24, 2017, from http://www.ilo.org/global/standards/information-resources-and-publications/publications/WCMS_318141/lang--en/index.htm

ILO (International Labour Organization). (2015). *World employment and social outlook 2015: The changing nature of jobs 2015.* Retrieved January 24, 2017, from http://www.ilo.org/global/research/global-reports/weso/2015-changing-nature-of-jobs/WCMS_368626/lang--en/index.htm

ILO (International Labour Organization). (2016). *Decent work.* Retrieved December 8, 2016, from http://www.itcilo.org/pt/the-centre/programmes/atividades-dos-trabalhadores/resources/decent-work-booklet_pt/view

ILO (International Labour Organization). (2017). *Decent work and the 2030 agenda for sustainable development #8- Decent work and Economic growth.* Retrieved February 2, 2017, from http://ilo.org/global/topics/sdg-2030/resources/WCMS_436923/lang--en/index.htm

International Ergonomics Association (IEA). (2017). *Definition and domains of ergonomics.* Retrieved January 12, 2017, from http://www.iea.cc/whats/index.html

Iverson, R. D., & Pullman, J. A. (2000). Determinants of voluntary turnover and layoffs in an environment of repeated downsizing following a merger: An event history analysis. *Journal of Management, 26,* 977–1003. https://doi.org/10.1177/014920630002600510

Jackson, S. E., & Schuler, R. S. (1995). Understanding human resource management in the context of organizations and their environments. *Annual Review of Psychology, 46,* 237–264.

Jahoda, M. (1982). *Employment and unemployment: A social-psychological analysis* (Vol. 1). New York: Cambridge University Press Archive.

Johnston, P., Everard, M., Santillo, D., & Robèrt, K.-H. (2007). Reclaiming the definition of sustainability. *Environmental Science and Pollution Research – International, 14,* 60–66.

Kalleberg, A. L. (2009). Precarious work, insecure workers: Employment relations in transition. *American Sociological Review, 74,* 1–22.

Keane, A., Jones, J. P., Edwards-Jones, G., & Milner-Gulland, E. J. (2008). The sleeping policeman: Understanding issues of enforcement and compliance in conservation. *Animal Conservation, 11,* 75–82.

Keohane, N. O., & Olmstead, S. M. (2016). *Markets and the environment* (2nd ed.). Washington, DC: Island Press/Center for Resource Economics.

Kleiner, B. M. (2006). Macroergonomics: Analysis and design of work systems. *Applied Ergonomics, 37,* 81–89.

Lange-Morales, K., Thatcher, A., & García-Acosta, G. (2014). Towards a sustainable world through human factors and ergonomics: It is all about values. *Ergonomics, 57,* 1603–1615.

Lee, K. S. (2005). Ergonomics in total quality management: How can we sell ergonomics to management? *Ergonomics, 48,* 547–558.

Lillie, N., & Sippola, M. (2011). National unions and transnational workers: The case of Olkiluoto 3, Finland. *Work, Employment & Society, 25,* 292–308.

Liu, J., & Diamond, J. (2005). China's environment in a globalizing world. *Nature, 435*(7046), 1179–1186. Retrieved July 18, 2016, from http://www.nature.com/nature/journal/v435/n7046/suppinfo/4351179a_S1.html

Mabogunje, A. L. (2002). Poverty and environmental degradation: Challenges within the global economy. *Environment: Science and Policy for Sustainable Development, 44,* 8–19.

Manuaba, A. (2007). A total approach in ergonomics is a must to attain humane, competitive and sustainable work systems and products. *Journal of Human Ergology, 36,* 23–30.

Martin, K., Legg, S., & Brown, C. (2013). Designing for sustainability: Ergonomics – Carpe diem. *Ergonomics, 56,* 365–388.

Mastekaasa, A. (1996). Unemployment and health: Selection effects. *Journal of Community & Applied Social Psychology, 6,* 189–205.

Meister, D. (1999). *The history of human factors and ergonomics.* Mahwah, NJ: Lawrence Erlbaum Associates, Publishers.

Messer, K. D. (2010). Protecting endangered species: When are shoot-on-sight policies the only viable option to stop poaching? *Ecological Economics, 69,* 2334–2340.

Moldan, B., Janoušková, S., & Hák, T. (2012). How to understand and measure environmental sustainability: Indicators and targets. *Ecological Indicators, 17,* 4–13.

Moore, D., & Barnard, T. (2012). With eloquence and humanity? Human factors/ergonomics in sustainable human development. *Human Factors: The Journal of the Human Factors and Ergonomics Society, 54,* 940–951.

Neumayer, E. (2003). *Weak versus strong sustainability: Exploring the limits of two opposing paradigms.* Cheltenham, UK: Edward Elgar Publishing.

Olsen, K. B. (2006). *Productivity impacts of offshoring and outsourcing a review. A review.* OECD Science, Technology and Industry Working Papers, 2006/01. Paris: OECD Publishing. Retrieved September 1, 2016, from https://doi.org/10.1787/685237388034

Posthuma, R. A., & Campion, M. A. (2008). Age stereotypes in the workplace: Common stereotypes, moderators, and future research directions. *Journal of Management.* https://doi.org/10.1177/0149206308318617

Pouyaud, J. (2016). For a psychosocial approach to decent work. *Frontiers in Psychology, 7,* 1–14. Retrieved December 8, 2016, from https://doi.org/10.3389/fpsyg.2016.00422.

Ramazzini, B. (1713/1940). *Diseases of workers.* [De Morbis Artificatum Diatriba] (W. C. Wright, Trans.). Chicago: The University of Chicago Press.

Randers, J. (2012). *2052: A global forecast for the next forty years.* Chelsea, UK: Green Publishing.

Reijnders, L. (2000). A normative strategy for sustainable resource choice and recycling. *Resources, Conservation and Recycling, 28,* 121–133.

Rockström, J., Steffen, W., Noone, K., Persson, Å., Chapin III, F. S., Lambin, E., et al. (2009). Planetary boundaries: Exploring the safe operating space for humanity. *Ecology and Society, 14*(2), 32. Retrieved January 27, 2017, from http://www.ecologyandsociety.org/vol14/iss2/art32/

Ros, M., Schwartz, S. H., & Surkiss, S. (1999). Basic individual values, work values, and the meaning of work. *Applied Psychology, 48,* 49–71.

Ross, A. (2008). The new geography of work: Power to the precarious? *Theory, Culture & Society, 25,* 31–49.

Rosso, B. D., Dekas, K. H., & Wrzesniewski, A. (2010). On the meaning of work: A theoretical integration and review. *Research in Organizational Behavior, 30,* 91–127.

Rushton, S., & Williams, O. D. (2012). Frames, paradigms and power: Global health policy-making under neoliberalism. *Global Society, 26,* 147–167.

Ryff, C. D., & Singer, B. (1998). The contours of positive human health. *Psychological Inquiry, 9,* 1–28.

Sengenberger, W. (2013). *The International Labour Organization: Goals, functions and political impact.* Retrieved February 2, 2017, from http://library.fes.de/pdf-files/iez/10279.pdf

Shane, S. (2009). Why encouraging more people to become entrepreneurs is bad public policy. *Small Business Economics, 33,* 141–149.

Sheavly, S. B., & Register, K. M. (2007). Marine Debris & Plastics: Environmental concerns, sources, impacts and solutions. *Journal of Polymers and the Environment, 15,* 301–305.

Soman, D. (2004). Framing, loss aversion, and mental accounting. In D. J. Koehler & N. Harvey (Eds.), *Blackwell handbook of judgment and decision making* (pp. 379–398). Oxford, UK: Blackwell Publishing.

Standing, G. (2011). *The precariat. The new dangerous class.* New York/London: Bloomsbury Academic.

Strully, K. W. (2009). Job loss and health in the U.S. labor market. *Demography, 46,* 221–246.

Swanson, J. L. (2012). Work and psychological health. In N. A. Fouad, J. A. Carter, & L. M. Subich (Eds.), *APA handbook of counseling psychology, Practice, interventions, and applications* (Vol. 2, pp. 3–27). Washington, DC: American Psychological Association.

Thatcher, A. (2012). Early variability in the conceptualisation of "sustainable development and human factors". *Work, 41*(Supplement 1), 3892–3899.

Thatcher, A., & Yeow, P. H. P. (2016). Human factors for a sustainable future. *Applied Ergonomics, 57,* 1–7.

Thompson, E. P. (2016). *The making of the English working class.* New York: Open Road Integrated Media.

United Nations Framework Convention on Climate Change. (2015). *Adoption of the Paris Agreement.* Retrieved February 20, 2017, from https://unfccc.int/resource/docs/2015/cop21/eng/l09r01.pdf

Van Daele, J. (2005). Engineering social peace: Networks, ideas, and the founding of the International Labour Organization. *International Review of Social History, 50,* 435–466.

Vink, P., Koningsveld, E. A. P., & Molenbroek, J. F. (2006). Positive outcomes of participatory ergonomics in terms of greater comfort and higher productivity. *Applied Ergonomics, 37,* 537–546.

Waite, M. (2013). SURF framework for a sustainable economy. Journal of Management and Sustainability, 3(4), 25–40.

Walliser, B. (2008). Cognitive economics. Leipzig, Germany: Springer.

Wilson, J. R. (2000). Fundamentals of ergonomics in theory and practice. Applied Ergonomics, 31, 557–567.

Winch, C. (2013). Learning at work and in the workplace: Reflections on Paul Hager's advocacy of work-based learning. Educational Philosophy and Theory, 45, 1205–1218.

Wisman, J. D. (2013). Wage stagnation, rising inequality and the financial crisis of 2008. Cambridge Journal of Economics, 37, 921–945.

Yerxa, E. J. (1998). Health and the human Spirit for occupation. American Journal of Occupational Therapy, 52, 412–418.

Zink, K. J. (2005). From industrial safety to corporate health management. Ergonomics, 48, 534–546.

Zink, K. J. (2014). Designing sustainable work systems: The need for a systems approach. Applied Ergonomics, 45, 126–132.

Zink, K. J., & Fischer, K. (2013). Do we need sustainability as a new approach in human factors and ergonomics? Ergonomics, 56, 348–356.

CHAPTER 4

Human Factors Issues in Responsible Computer Consumption

Paul H. P. Yeow, Wee Hong Loo, and Uchenna Cyril Eze

INTRODUCTION

Computers are very much a part of our modern life as there are more than one billion computers worldwide (PC Energy Report, 2009). This research focuses on the individual user's responsible computer consumption including the use of power-management settings such as shut down, sleep and hibernation modes, and turning off computers when not in use. This is crucial because a computer consumes considerable energy in idle mode, that is, 65 watts (without a power-management setting and excluding the power used by the monitor), compared with 3–5 watts in the off mode. The energy utilised can be huge based on the hours at idle, and the number of devices a consumer owns. For instance, a computer consumes 189,800 watt-hours energy per year (65 watts × 8 hours per day × 365 days) if it stands idle throughout the night (Barnatt, n.d.). The PC Energy Report (2009) highlighted that if the world's one billion PCs were shut

P. H. P. Yeow (✉) • W. H. Loo
School of Business, Monash University Malaysia, Bandar Sunway, Malaysia

U. C. Eze
Division of Business and Management, United International College,
Beijing Normal University-Hong Kong Baptist University, Zhuhai, China

© The Author(s) 2018   77
A. Thatcher, P. H. P. Yeow (eds.), *Ergonomics and Human Factors for a Sustainable Future*, https://doi.org/10.1007/978-981-10-8072-2_4

down for just one night when not in use, it would save energy required to light up New York City's Empire State Building—inside and outside—for more than 30 years. Additionally, Gartner's findings (as presented in PC Energy Report, 2009, p. 14) estimated that a computer's energy consumption will reduce by more than 90% (from 988,026 kWh to 91,203 kWh, based on a 2500-user survey) if users choose to use power-management settings. Therefore, responsible computer consumption behaviour (RCCB) offers a huge opportunity for cost savings and a greener environment from energy conservation.

There are several definitions of responsible consumption behaviour (RCB) in previous literature. Webster (1975) called it "socially responsible consumption" and defined it as consumer products that bring a positive impact on society such as buying second-hand clothes from charity shops, which would benefit mankind. Henion (1976) and Antil (1984) called it "environmentally responsible consumption", for example, consuming products with less plastic packaging, which will have less negative effects on the environment. Thatcher's (2013) green ergonomics concept merged the two concepts; both social and environmental aspects should be addressed on the basis that there is a bi-directional relationship (i.e. responsible human behaviour such as conservation, preservation and restoration of nature will benefit nature, which in turn will benefit mankind from sustained ecosystem services such as providing food, pharmaceutical products and energy, water purification, carbon sequestration, habitat services, recreational opportunities, aesthetic beauty, etc.). Several authors included the economic aspect (thus forming the Triple Bottom Line (TBL), i.e. social, environmental and economic aspects) and the consumption life cycle, that is, the purchase, use and disposal of products (Prothero et al., 2011; Stern, 2000; Thatcher, 2013; Zink, 2014). In the present research, we define responsible consumption behaviour as RCCB, that is. using computers without wastage such as switching them off when not in use or using power-management settings, with consideration of the TBL.

An individual system such as the use of computers at work and leisure is not an island system as it affects higher-level systems (Costanza & Patten, 1995; Thatcher & Yeow, 2016a, 2016b; Wilson, 2014). Energy wastage from idling computers in individual system contributes to increased CO_2 from the energy production, which depends on the electricity generation matrix. This affects the environment system as excessive CO_2 disrupts the natural system, causing global warming, and the rise of sea levels, flooding and acidification of oceans (Cox, Betts, Jones, Spall, & Totterdell, 2000; Munasinghe, 2012). RCCB is related to sustainability because it affects the TBL goals (Elkington, 1997; Thatcher & Yeow, 2016a, 2016b; Zink, 2014). Firstly, it impacts the environment through CO_2 emissions. Electricity consumed through computer use contributes to 2% of the world's total CO_2 emissions, that is, 35 million tonnes (due to the energy production)—as much as the aviation industry's contribution (Hopkinson & James, 2011; The Economist, 2009). Secondly, it affects economic and social aspects from the huge electricity costs and prevalent behaviour of wastage. The National Energy Foundation, for instance, reported that in 2006, 18% of 12.6 million working adults in the United Kingdom did not switch off their computers at night or weekends (social aspect), resulting in 1,500,000,000 (1.5 bn) kilowatt hours (kWh) of electricity wastage per annum (economic aspect); this would contribute 700,000 metric tonnes of CO_2 emissions (environmental aspect) based on fossil fuel electricity generation (National Energy Report, 2007). Another study highlighted that approximately 60% of 104 million workers in the United States contributed 19.82 bn kWh of electricity wastage (which translates to 14.4 million metric tonnes of CO_2 emissions) due to not always/never shutting down their computers (National Energy Foundation, 2007). Not practising RCCB also has secondary effects. It will increase a computer's cumulative operational time thus shortening its lifespan. This results in more discarded computers, which will drive up the demand for computer production and increase the amount of electronic waste (environmental aspect). This is a serious environmental problem as the production of a computer generates a huge amount of CO_2 (e.g. production of a 1.4 kg computer generates 410 kg of CO_2 [Andrae & Andersen, 2010]) and electronic waste contains toxic materials like lead, chromium, cadmium and mercury.

The International Ergonomics Association (2015) defines HFE as:

Ergonomics (or human factors) is the scientific discipline concerned with the understanding of interactions among humans and other elements of a system, and the profession that applies theory, principles, data and methods to design in order to optimise human well-being and overall system performance.

The study of individual RCCB comes under the HFE domain as it is about users' interaction with elements of a system, that is, computers, with the aim of optimising human well-being and overall system performance in the three aspects of sustainability, that is, social, economic and environmental (Zink, 2014). The HFE issues represent human well-being, that is, they relate to social aspects such as inhibitions and motivations to practise RCCB when interacting with computers. These aspects are to be "optimised" so that the overall system performance, which includes the economic and environmental aspects, can be optimised. In other words, the social aspects of energy wastage have to be addressed in individual system so that the environmental threats and economic loss can be addressed. In this chapter, we seek to understand the social aspects, that is, HFE issues in RCCB. This understanding will provide input to the design for aspirational ideology (Thatcher, 2012), that is, the design of the individual system to persuade users to change their behaviour to consume computers responsibly (which is the aspirational ideology). This research used the Theory of Planned Behaviour (TPB) and customised the theory accordingly in the context of responsible computer consumption.

Research Gap and Objectives

Moray (1994, 1995) highlighted a research gap, i.e. HFE concerns should extend from safety, productivity and usability to behavioural changes that support sustainable development, which is in the interest of HFE since we are interested to optimise the overall system performance. He stressed that by understanding the inhibitions and motivations in sustainable behaviour (such as the HFE issues of RCCB), HFE can provide inputs/ recommendations to the design of behavioural-change interventions (such as policies, education and government social messages) to facilitate environmental sustainability (Hanson, 2013, p. 405). Several researchers (including Flemming, Hilliard, & Jamieson, 2008; Haslam & Waterson, 2013; Martin, Legg, & Brown, 2013; Radjiyev, Qiu, Xiong, & Nam, 2015) supported Moray's argument. For example, Flemming et al. (2008) provided a literature review on how reductions in energy consumption can be achieved through behavioural-change programmes/interventions. Additionally, many researchers (e.g. Durugbo, 2013; Harvey, Thorpe, & Fairchild, 2013; Kobus, Mugge, & Schoormans, 2013; Lockton, Harrison, & Stanton, 2010) have attempted to identify such HFE issues in order to provide input to designing appropriate behavioural-change interventions. For example, Kobus et al. (2013) conducted interviews with households who had used the Energy Management System (EMS) "Smart Wash". The findings indicated that user behaviour was influenced by HFE issues such as users' motivation, contextual factors and the design of the EMS. Durugbo's (2013) study explored user interactions with recycling facilities in public buildings. The HFE issues identified were related to the effective use of the facilities, including the (1) appropriate location to encourage recycling, (2) attitudes of intended users to recycling, and (3) presentation of information to users to guide their use. Based on the identification of the HFE issues, interventions were recommended to encourage behavioural change. This study investigates the HFE issues in RCCB, which are the inhibitions and motivations to practise energy-saving behaviour such as turning off computers when not in use or using power-management settings. The research questions are (1) why do/don't consumers practise RCCB and (2) how to encourage adoption of RCCB? The research objectives are (1) to identify the HFE issues that influence RCCB and (2) to make recommendations to the HFE practitioners on how to encourage RCCB.

Context of Study

eMarketer (2012) reported that China had the highest number of Internet users in Asia Pacific at 453.8 million (33.9% of China's population) in 2011, followed by India with 91.4 million (7.6% of Indian population), Indonesia with 36.5 million (15.5% of Indonesian population) and Malaysia with 17.5 million (61.2%). Malaysia had the highest Internet penetration (61.2%) among these countries. The penetration of PCs such as desktop computers, laptop computers, tablet computers and smartphones in Asian developing countries is on the rise in conjunction with citizen access computers in order to engage in online activities such as checking e-mail, surfing and accessing social networks (Asia Pacific, n.d.). This is reflected in the Department of Statistic Malaysia's (2015) survey, as per which, the percentage of individuals using computers increased from 56.0% in 2013 to 68.7% in 2015. Meanwhile, TrendMicro (2012) did find that over half of the Malaysians own three or more computers. As such we chose Malaysia as the host country as the penetration of computer is very high similar to other Asian developing countries where RCCB is important to mitigate environmental problems.

METHOD

The research adopts a quantitative approach through the development of a research framework and a questionnaire instrument followed by the data collection and analysis to validate the model.

Research Framework and Hypotheses

Figure 4.1 shows the research framework of the study. It consists of HFE issues in the form of the TPB and contextual antecedents, which are posited to predict responsible computer consumption intention (RCCI) and RCCB.

Theory of Planned Behaviour (TPB)

Since this study investigates the HFE issues related to RCCB, the use of behavioural theory is appropriate. Human behavioural theories such as the Technology Acceptance Model, Attitude-Behaviour Model, Theory of Planned Behaviour, Unified Theory of Acceptance and Use Technology have been used in many ergonomics studies (Burkotler, Weyers, Kluge, & Luther, 2014; Chen & Huang, 2016; Szalma, 2014; Yeow, Yuen, & Loo, 2013) to understand the inhibitors or motivators of human behaviour so as to provide input to designing a system acceptable by users (Dillon, 2001). Among the various behavioural theories, we chose the TPB

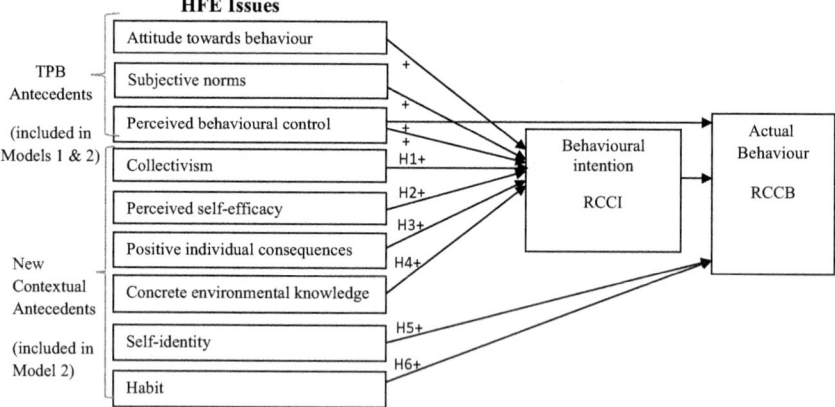

Fig. 4.1 HFE issues influencing responsible computer consumption behaviour (RCCB) (Note: Model 1 consists of only the Theory of Planned Behaviour [TPB] antecedents; Model 2 consists of the TPB and new contextual antecedents; + refers to a positive relationship; H refers to the hypotheses tested in this research; *RCCI* responsible computer consumption intention, *HFE* human factors and ergonomics)

(Ajzen, 1991) as the foundation model to identify the HFE issues because it has received empirical support in predicting pro-environmental behaviour such as electricity/water conservation, recycling, travel-mode choice, food choice and ethical investment (Egmond & Bruel, 2007; Tonglet, Philips, & Read, 2004). In ergonomics research, Kowalewski et al. (2014), for instance, adopted the TPB to understand the underlying mechanisms that influence acceptance of renewable energies, that is, geothermic. In the TPB, three factors, that is, attitude, subjective norms and perceived behavioural control, are the predictors of behavioural intention, and behavioural intention and perceived behavioural control will affect actual behaviour. (Note that these independent and dependent factors in the TPB are presented in the following sections.) However, the TPB has been found to only predict a variance of 27% and 39% in behaviour and intention, respectively (Armitage & Conner, 2001; Rise, Sheeran, & Hukkelberg, 2010, p. 1086). Several researchers (e.g. Ajzen, 1991; Conner & Armitage, 1998; Davies, Foxall, & Pallister, 2002) have suggested that the predictive powers of the TPB can be improved with the inclusion of contextual variables (as cited in Tonglet et al., 2004, p. 198). In view of this, this study incorporates HFE issues such as perceived self-efficacy, collectivism, concrete environmental knowledge, positive individual consequences, self-identity and habit into the TPB. These social factors have been found in prior literature to influence environmentally friendly behaviour, but their applicability in predicting RCCB has not been tested empirically.

Dependent Variables

This research has two main factors of interest, that is, intention (RCCI) and behaviour (RCCB).

Responsible Computer Consumption Behaviour (RCCB)

Responsible consumption behaviour is defined as "the degree to which the usage and consumption of a product could minimise adverse environmental effects, and thus benefit society and people in the end" (Webb, Mohr, & Harris, 2008, p. 92). Based on this, the study operationalises RCCB as consumer behaviour in using computers in a responsible manner to reduce the harmful effects on the environment. RCCB is measured by four items adapted from Murugesan (2008) as shown in Table 4.4, question numbers 1–4. Specifically, this study conceptualises RCCB with actions such as turning off idle computers, using power-management settings such as shutdown, sleep and hibernation modes, and using a blank screen to save energy, reduce contributions towards CO_2 emission and extending the lifespan of computers.

Responsible Computer Consumption Intention (RCCI)

Ajzen (1991, p. 181) defined intention as "the likelihood of an individual performing the behaviour in the future". Intention indicates a person's efforts and willingness in trying to perform a particular behaviour. This study conceptualises RCCI as a consumer's plan to engage in using computers responsibly in the future. The attributes of behavioural intention in using computers responsibly are presented in Table 4.3: question numbers 1–4, which were adapted from Murugesan (2008). Ajzen and Fishbein (1980) found that if an individual has a strong intention to engage in certain behaviour (e.g. RCCI), he/she is likely to perform the behaviour (e.g. RCCB).

Independent HFE Issues

These are factors believed to be directly or indirectly related to RCCB based on the TPB, HFE and other literatures.

Attitudes Towards Behaviour (ATB)

Attitude towards behaviour refers to "a person's evaluation of a specific behaviour" (Ajzen, 1991, p. 188), which affects his/her feelings towards that behaviour. Similarly, Kaufmann, Panni and Orphanidou (2012, p. 53) defined attitude as "the enduring positive and negative feeling about some person, object or issue". This study conceptualises ATB as consumers' positive and negative feelings about practising responsible computer consumption. The attributes of ATB are shown in Table 4.2, question numbers 1–4, which were adapted from Ajzen (1991). ATB was found to be the most consistent explanatory factor in positively predicting the behavioural intention to perform a particular behaviour (Ajzen & Fishbein 1980) as found in several ergonomics studies (Durugbo, 2013; Harvey et al., 2013; Kowalewski et al., 2014); therefore, this relationship is classified under the control model (or Model 1), that is, the model with the traditional TPB variables.

Subjective Norms (SN)
Ajzen (1991, p. 188) defined subjective norms as "the individual's perception of social pressure to perform the particular behaviour". SN is a function of a person's perception of important referents' evaluations of a behaviour and a person's motivation to conform to those evaluations (Fishbein & Ajzen, 1975). This study conceptualises SN as a person's perception of the social pressure to practise responsible computer consumption. The items used to measure SN are presented in Table 4.2, question numbers 24–26, which were adapted from Ajzen (1991). The SN factor was found to have positive effects on behavioural intentions in many studies (e.g. Ajzen & Fishbein, 1980; Mida, 2009), including the HFE study conducted by Holden (2012); therefore, this relationship is classified under the control model.

Perceived Behavioural Control (PBC)
PBC refers to "people's perceptions of the ease or difficulty of performing the behaviour of interest" (Ajzen, 1991, p. 188). Armitage and Conner (2001) and Valle, Rebelo, Reis and Menezes (2005) reported that PBC reflects past experience of impediments while practising a behaviour. In this study, PBC is operationalised as a consumer's perceptions of the difficulties/obstacles in practising responsible computer consumption, for example, inconvenience and insufficient information. The attributes of PBC are shown in Table 4.2, question numbers 15–17, which were adapted from Ajzen (1991). PBC was found to positively predict both intention and actual behaviour in a TPB model (Armitage & Conner, 2001). Ajzen (1991, p. 188) stated that the relative importance of PBC to predict intention is expected to vary across behaviour and situations, "The addition of perceived behavioural control should become increasingly useful as volitional control over behaviour decreases" (Ajzen, 1991, p. 185). This is reflected in Durugbo's (2013, p. 414) findings, that is, the inadequate locations of recycle bin (called "chimney") makes recycling behaviour uncontrollable (volitional control decreases) and thus affects consumers' intention of practising the behaviour. In another study, Klockner's (2013) study yields similar findings as the TPB model, that is, PBC affects behavioural intention and actual behaviour, in the context of environmentally relevant behaviours such as energy use, car purchase, water use and so on. Since these relationships (i.e. PBC predicts intention and behaviour) have already been validated, they are classified under the control models.

Collectivism (CO)

CO is defined as a value that reflects concern for the welfare of others (Kim & Choi, 2005). Rivera, Ganey, Dalton and Hancock (2004) describe CO as the prioritisation of a reference group's agenda or benefits such as willingness to help others, being cooperative and emphasis on group goals. In this study, items measuring CO are shown in Table 4.2, question numbers 5–8, which were adapted from Kim and Choi's (2005) study. In HFE literature, CO was found to have strong effects on multiple behaviours ranging from teamwork performance (Rivera et al., 2004) to workplace behaviour (Jackson & Johnson, 2002; Skarżyńska, 2002), and technology acceptance (Kothaneth, 2010). In other literatures in the environment and psychology fields, several studies have found that CO has an effect on ecological behaviour (Poortinga, Steg & Vlek, 2004; Schultz & Zelezny, 1998; Urien & Kilbourne, 2011). For example, Urien and Kilborne (2011) found that American respondents who scored high in collectivism were more likely to have eco-friendly intentions. Their concern for society's welfare drove them to have good intentions towards environmental sustainability. Since Malaysia is a collectivist society (Hofstede, 2001), Malaysians' intention to use computer responsibly (a pro-environment intention) may be driven by their concern for the welfare of the society. This is a new relationship that extends the TPB model; it is tested with the following hypothesis:

H1: CO positively influences the intention of using computers responsibly (RCCI)

Perceived Self-Efficacy (PSE)

PSE refers to "the extent to which individuals believe that their actions make a difference in solving a problem" (Kim & Choi, 2005, p. 593). PSE emphasises people's perceptions of their ability to make a difference, which is distinct from PBC, which focuses on constraints that inhibit the behaviour (Conner & Armitage, 1998). Hence, PSE is operationalised as one's belief in one's own ability to make a difference in environmental sustainability through responsible computer consumption. This contextual variable was selected because individuals in developing countries may not have perceived self-efficacy due to a lack of the technical knowledge and skills to practise responsible computer consumption. The items measuring PSE are presented in Table 4.2, question numbers 12–14, which were adapted from Kim and Choi's (2005) study. Prior studies reveal that an individual is likely to engage in a certain behaviour when he/she feels capable (Conner & Armitage, 1998). For example, Kowalewski et al.'s (2014) study highlighted that self-efficacy will motivate an individual to adopt renewable energy. According to Kim and Choi (2005), those with strong beliefs that their environmentally friendly behaviour (i.e. RCCB in this context) will make a difference in mitigating problems such as climate change will likely engage in such behaviour. Since this a new relationship, it is tested with the following hypothesis:

H2: PSE positively influences the intention of using computers responsibly (RCCI).

Positive Individual Consequences (PIC)
Follows and Jobber's (2000) study defined individual consequences as the impact of a product on individual consumers who are affected by using it. This study operationalises PIC as the positive effect of a consumer's engagement in the responsible computer consumption, for example, reduced electricity bills. The dimensions of PIC are shown in Table 4.2, question numbers 18–20, which were adapted from Follows and Jobber (2000). This contextual variable is chosen because computer use requires high power consumption, which may drive individuals to practise responsible behaviour to reduce their electricity bills. Several ergonomics studies have identified this contextual factor as important in determining behavioural change (Harvey et al., 2013; Szalma, 2014). For example, Harvey et al.'s (2013) study found that drivers will opt for eco-driving to save fuel. This new relationship is tested with the following hypothesis:

H3: PIC positively influences the intention of using computers responsibly (RCCI).

Concrete Environmental Knowledge (CEK)
CEK is "the specific solution-oriented behavioural knowledge, such as knowing the ways of recycling, which can allow an individual to act and take the right action to protect the environment" (Schahn & Holzer, 1990) (as cited in Lee, 2011, p. 24). In this study, it is operationalised

as consumers' concrete knowledge of the impact of their responsible computer consumption on the environment, for example, turning off idle computers indirectly reduces CO_2 emissions. The attributes of measuring CEK are presented in Table 4.2, question numbers 27–29, which were adapted from Lee's (2011) study. Many HFE studies (e.g. Haslam, 2002; Robertson et al., 2009; Sauer, Wiese, & Ruttinger, 2002; Voorbij & Steenbekkers, 2002) discovered that an increase in knowledge among users will translate to behavioural intention. The present study tests if this relationship applies to computer usage with the following hypothesis:

H4: CEK positively influences the intention of using computers responsibly (RCCI).

Self-Identity (SI)
Lee (2008) defined SI as one's conception of one's self. Self-identity reflects the extent to which an individual sees himself/herself as fulfilling the criteria set by society, for example, someone who is concerned about green issues. This study operationalises SI as an individual's view of himself/herself as a responsible computer user to gain social status. The items measuring self-identity were adapted from Lee's (2008) study, which are shown in Table 4.2, question numbers 9–11. Several studies revealed that SI has a direct influence over individual behaviour (Baker & Ozaki, 2008; Lee, 2008; McDonagh, Bruseberg & Haslam, 2002; Ries, Hein, Pihu & Armenta, 2012; Zeitlin, 1994). Oyserman, Elmore and Smith (2012) and Hagger, Anderson, Kyriakaki and Darkings (2007) provided a reason for this relationship, that is, once a certain behaviour reflects an individual's self-identity, an individual will act/behave in accordance with his/her identity spontaneously, which does not require rational decision making (via behavioural intention). As Oyserman et al. (2012, p. 93) put it, "If it (a certain behaviour) feels identity-syntonic, it feels right (to perform the behaviour) and does not require further reflection". This explains why SI has an influence on behaviour but not on intention. In the acquisition cycle, Lee (2008) found that SI was a predictor of green purchasing behaviour. The present study tests if this relationship applies to the use cycle with the following hypothesis:

H5: SI positively influences responsible use of computer behaviour (RCCB).

Habit (HA)
Limayem, Hirt and Cheung (2007) defined HA as "the extent to which people tend to perform the behaviours automatically because of learning", while Kim et al. (2005) equated habits with automaticity (as cited in Venkatesh, Thong, & Xu, 2012, p. 161). Steg and Vlek (2009) mentioned that in many cases, behaviour is governed by automated cognitive processes instead of being preceded by reasoning. This study operationalises HA as the extent to which people automatically use computers responsibly. The items that measure HA are presented in Table 4.2, question numbers 21–23, which were adapted from Venkatesh et al.'s (2012) study. Prior studies have shown that HA influences green consumer behaviour (Dahlstrand & Biel, 1997; Eriksson, Garvill & Nordlund, 2008; Moller & Thogersen, 2008). Ouellette and Wood (1998), Ajzen (2011) and Bissell, Duda and Young (1998) highlighted that habits will have direct influence on behaviour and not via intention, in such conditions that the behaviours are performed on a daily or weekly basis in a stable and predictable supporting context, for example, computer use, alcohol and coffee consumption, most types of exercise, seat belt use and so on. The rationale is as the behaviour is performed regularly, the conscious reasoning (intention) process is not required. In HFE literature, Hanson (2013) has highlighted many studies that examined how habits can influence individuals in reducing energy usage at home. The present study tests if this relationship applies to computer use with the following hypothesis:

H6: HA positively influences responsible use of computer behaviour (RCCB).

Research Instrument
The questionnaire was designed based on the research framework and has four sections: (1) respondents' demographic information (see Table 4.1), (2) 29 questions (Table 4.2: Nos. 1–29) related to factors (HFE issues comprising the research framework) affecting respondents' intention (RCCI) and behaviour (RCCB), (3) four questions (Table 4.3: Nos.1–4) measuring the extent of respondents' RCCI and (4) four questions (Table 4.4: Nos. 1–4) relating to respondents' RCCB. The items of each factor were adapted from previous studies as presented in the research

Table 4.1 Demographic information and types of computer ownership

Gender	Age	Total	Types of computer ownership				
			Laptop	Smartphone	Desktop computer	Tablet PC	Netbook
Male	17–32 years	47	44	32	26	13	4
	33–47 years	9	6	7	8	2	4
	48 years and above	2	1	0	1	1	2
	Total (male)	58	51	39	35	16	10
	17–32 years	32	27	20	13	7	7
Female	33–47 years	7	7	5	2	4	1
	48 years and above	3	3	3	2	0	0
	Total (female)	42	37	28	17	11	8
	Total	100	88	67	52	27	18

Sample size = 100

framework section (Section 2.1). Varied scales were used. The questions in Section 1 were measured by nominal and ordinal scales. The questions in Sections 2–4 were measured by a five-point Likert scale ranging from strongly disagree (one point) to strongly agree (five points). The questionnaire was initially tested using a small sample size of 11 respondents. The purpose was to get the respondents' feedback on the questionnaire on any biases and errors such as double-barreled questions, socially desirable questions, ambiguous items, typo errors and so on. The questionnaire was then corrected and finalised.

Data Collection and Analysis
Convenience sampling was used with 100 personally administered questionnaires distributed and collected in public places such as shopping complexes and bus stations in Kuala Lumpur (KL). The respondents were asked if they owned a computer before the questionnaire was distributed. This study was confined to KL as over 50% of the 16.9 million Malaysians who access the Internet via computers are from the city (Ng, 2011).

The data was analysed through the use of descriptive and inferential statistics such as factor analysis and hierarchical regression analysis to test the hypotheses. A varimax rotation method was used in the factor analysis. This method is widely adopted across research literature because it produces a simpler, easier to interpret factor solution (Costello and Osborne, 2005).

Table 4.2 Factor analysis for the attributes measuring the independent variables

	ATB	CO	SI	PSE	PBC	PIC	HA	SN	CEK
1. I like the idea of using computers responsibly	**.875**	-.070	.220	.056	-.002	.178	.033	.046	.147
2. Using computers responsibly is a good idea	**.871**	.005	.198	.101	.040	.229	-.013	.030	.071
3. Using computers responsibly is pleasant	**.848**	.058	.125	.196	.054	.109	.106	-.005	.158
4. Using computers responsibly makes me feel satisfied	**.821**	-.064	.265	.097	.152	.125	.110	.080	.064
5. I support my group, whether they are right or wrong	-.070	**.808**	-.102	.034	-.127	.024	-.053	.056	.066
6. I respect the decisions made by my social group	-.017	**.786**	.209	.123	.033	.074	.058	-.002	.119
7. I respect the majority's wish	.096	**.725**	.262	.165	.137	.070	.247	-.090	-.004
8. I maintain harmony in my social group	-.044	**.661**	.198	.254	.033	.004	.287	.198	-.029
9. RUC will enhance my self-image	.334	.116	**.851**	.138	.046	.167	.086	-.047	.114
10. My involvement in RUC is a status symbol	.271	.175	**.803**	.218	.153	.150	.002	.001	.087
11. I feel I'm better than others if I practise RUC	.383	.317	**.730**	.056	.005	.148	.105	.020	.118
12. I can protect the environment by using computers responsibly	.135	.180	.075	**.835**	.085	.087	.201	-.042	.109
13. I feel I can help solve environmental problems by using computers responsibly	.074	.081	.159	**.830**	.200	.095	.124	.021	.106
14. I feel I can help solve natural resource problems (e.g. saving electricity from overusing computers) by using computers responsibly	.293	.305	.155	**.680**	.040	.106	.093	-.157	.068
15. I use computers responsibly although it is incompatible with my lifestyle	-.001	.070	-.010	.162	**.882**	.125	.023	-.062	.001
16. I use computers responsibly although it is inconvenient for me	.049	-.003	.034	.257	**.880**	-.012	-.028	-.098	.016
17. I use computers responsibly although insufficient information is provided by the media (such as radio and TV) on how to use it responsibly	.164	-.062	.156	-.132	**.711**	.100	.201	.103	.104
18. It is important to me that practising RUC helps reduce any direct/indirect hazardous effect on me	.218	-.046	.167	.152	.107	**.804**	-.031	.048	.059

(continued)

Table 4.2 (continued)

	ATB	CO	SI	PSE	PBC	PIC	HA	SN	CEK
19. It is important to me that practising RUC helps reduce my expenses	.134	.056	.110	.024	.037	.779	.207	.005	.126
20. It is important to me that practising RUC doesn't hurt my relationship with others such as friends or family members	.190	.161	.077	.087	.092	.729	.184	-.168	.098
21. Practising RUC has become natural to me	.097	.118	-.018	.123	.171	.132	.790	.071	.295
22. RUC has become a habit for me	-.057	.169	.072	.303	.085	.045	.753	-.024	.202
23. I must practise RUC	.255	.133	.121	.031	-.063	.310	.663	.007	-.180
24. People who influence my behaviour think I should use computers responsibly	.184	.010	-.050	-.226	.069	-.54	.044	.838	-.031
25. People who are important to me think I should use computers responsibly	.097	.061	-.004	.057	-.054	-.121	.110	.830	.061
26. The Malaysian government encourages citizens to use computers responsibly	-.382	.048	.052	.059	-.135	.186	-.232	.638	.142
27. I often read to obtain more information about how to use computers to save the environment	.109	.222	-.032	.288	.121	.146	-.042	.044	.721
28. I am knowledgeable about how to use computers to protect the environment	.030	-.171	.313	.112	-.019	.053	.325	-.028	.704
29. I can list at least three ways of using computers responsibly to protect the environment in our daily lives	.329	.127	.093	-.054	.021	.117	.122	.121	.703
Eigen value	7.93	3.09	2.37	1.97	1.71	1.48	1.37	1.22	1.05
Variance (percentage)	27.4	38.0	46.2	52.9	58.8	64.0	68.7	72.9	76.5
Cronbach alpha coefficient	.939	.805	.851	.922	.805	.787	.745	.819	.687
Mean rating (/5)	3.92	3.62	3.46	3.70	2.89	3.40	3.34	3.48	3.32
SD	.90	.46	.76	.84	.54	.57	.57	.48	.50

Notes: *ATB* attitude towards behaviour, *CO* collectivism, *SI* self-identity, *PSE* perceived self-efficacy, *PBC* perceived behavioural control, *PIC* positive individual consequences, *HA* habit, *SN* social norms, *CEK* concrete environmental knowledge, *RUC* responsible use of computers/responsible computer consumption

Table 4.3 Factor analysis for attributes measuring responsible computer consumption intention (RCCI)

	RCCI
1. I intend to use a lower-power consumption mode, such as shutdown, hibernation, sleep or standby mode, when the computers are not in use	.913
2. I intend to use computers responsibly in the near future	.891
3. I intend to use a blank screensaver instead of a screensaver that displays moving images	.843
4. I intend to turn off my computer when not in use	.839
Eigen value	3.04
Variance (percentage)	76.04
Cronbach alpha coefficient	.893
Mean rating(/5)	3.78
Standard deviation	0.83

Table 4.4 Factor analysis for attributes measuring responsible computer consumption behaviour (RCCB)

	RCCB
1. I have used a lower-power consumption mode, such as shutdown, hibernation, sleep or standby mode, when the computers are not in use	0.869
2. I have used computers responsibly	0.860
3. I have turned off my computer when not in use	0.842
4. I have used a blank screensaver instead of a screensaver that displays moving images	0.819
Eigen value	2.87
Variance (percentage)	71.85
Cronbach alpha coefficient	0.869
Mean rating (MR) (/5)	3.76
Standard deviation (SD)	0.83

RESULTS

This section presents the descriptive statistics, exploratory factor analyses, discriminant validity analysis and hierarchical regression analyses. The key results of the hypotheses test are presented.

Table 4.1 presents the percentage of respondents who own computer devices. They consist of 58 males and 42 females of whom 88% and 52% own a laptop and desktop computer, respectively, while 67% own a smartphone. It is noted that 79% of the respondents were aged between 17 and 32, which is consistent with the population pyramid presented in Index Mundi's survey where the majority of the country's population is in that age group (Index Mundi, 2014).

Table 4.2 presents the results of the exploratory factor analysis. With eigen values greater than 1.00 and 76.08% of variance explained, the table shows the 29 items with factor loadings greater than 0.40. The items can be grouped under nine HFE issues (constructs) as per the research framework in Fig. 4.1 with Cronbach alpha values ≥ 0.7.

Tables 4.3 and 4.4 show the results of the factor analysis for the items measuring RCCI and RCCB. The items in each table are grouped into RCCI and RCCB, respectively, with Cronbach alpha values ≥ 0.7.

Table 4.5 presents the square root of the average variance extracted (AVE) (diagonal elements in parenthesis) and the correlations between constructs (off-diagonal elements). The results demonstrate that the AVE square root is of a higher value than the correlations in the same row; therefore, discriminant validity at the construct level was achieved. Additionally, the AVE value of each construct meets the acceptable requirement, that is, 0.5 (Hair, Ringle & Sarstedt, 2011); therefore, the convergent analysis for these constructs was achieved.

Hierarchical regression, with RCCI as the dependent variable, was used to determine whether the additional contextual factors (i.e. collectivism, perceived self-efficacy, positive individual consequences and concrete environmental knowledge) will enhance the predictive power of behavioural intentions beyond that provided by the TPB constructs. The prediction of the dependent variables was tested in two blocks. The first block (Model 1) comprised the control variables (TPB variables, i.e. ATB, SN and PBC). In addition to the control variables, the second block (Model 2) included the contextual variables. Table 4.6 shows that Model 1 explained *30.7%* of the variance in RCCI with ATB as the only significant factor, while Model 2 explained 59.7% with ATB, CO, PSE and PIC as significant; therefore, H1, H2 and H3 were supported.

Hierarchical regression, with RCCB as the dependent variable, was used to determine whether additional factors, that is, self-identity and habits, will increase the prediction of RCCB, beyond that engendered by the TPB constructs. The prediction of the dependent variable was tested in two blocks. Table 4.7 shows that Model 1 explained *51.5%* of the variance in RCCB with RCCI as the only significant predictor, while Model 2 explained 64.0% with RCCI, SI and HA being significant; therefore, H5 and H6 were supported.

Table 4.5 Discriminant validity analysis

	ATB	SN	PBC	PIC	PSE	CEK	CO	HA	SI	RCCI	RCCB
ATB	(0.890)										
SN	0.144	(0.707)									
PBC	0.137	0.005	(0.919)								
PIC	0.486	0.041	0.160	(0.852)							
PSE	0.606	0.022	0.150	0.471	(0.896)						
CEK	0.465	0.001	0.135	0.447	0.469	(0.707)					
CO	0.121	0.042	0.137	0.263	0.454	0.308	(0.721)				
HA	0.249	0.015	0.197	0.417	0.281	0.550	0.444	(0.774)			
SI	0.351	0.198	0.354	0.374	0.436	0.449	0.514	0.463	(0.819)		
RCCI	0.613	0.066	0.064	0.580	0.757	0.433	0.539	0.299	0.492	(0.828)	
RCCB	0.419	0.161	0.228	0.610	0.553	0.397	0.489	0.552	0.684	0.790	(0.792)

Notes: The square root of the average variance extracted (AVE) is in parenthesis diagonally; below is the estimated correlation between the constructs; *ATB* attitude towards behaviour, *SN* social norms, *PBC* perceived behavioural control, *PIC* positive individual consequences, *PSE* perceived self-efficacy, *CEK* concrete environmental knowledge, *CO* collectivism, *HA* habit, *SI* self-identity, *RCCI* responsible computer consumption intention, *RCCB* responsible computer consumption behaviour

Table 4.6 Hierarchical regression analysis for responsible computer consumption intention (RCCI)

Dependent variables	Mean	Std. deviation	Responsible computer consumption intention (RCCI)		
			Model 1 Beta weight	Model 2 Beta weight	Hypothesis
R^2			0.307	0.597	
ΔR^2 between models				0.290	
Attitude towards behaviour	3.92	0.90	0.534*	0.229*	
Perceived behavioural control	2.89	0.54	0.075	0.004	
Subjective norms	3.48	0.48	0.032	−.026	
Collectivism (H1)	3.62	0.46		0.243*	Supported
Perceived self-efficacy (H2)	3.70	0.84		0.393*	Supported
Positive individual consequences (H3)	3.40	0.57		0.015*	Supported
Concrete environmental knowledge (H4)	3.32	0.50		0.009	Rejected

Notes: *$p < .05$

R^2 refers to the proportion of the variance in the dependent variable accounted for by independent variables
Beta weight refers to the relative contribution of each independent variable in explaining the variance in the dependent variable
Model 1: TPB antecedents
Model 2: TPB and contextual antecedents

Table 4.7 Hierarchical regression analysis for responsible computer consumption behaviour (RCCB)

Dependent variables	Mean	Std. deviation	Responsible computer consumption behaviour (RCCB)		
			Model 1 Beta weight	Model 2 Beta weight	Hypothesis
R^2			0.515	0.640	
ΔR^2 between models				0.125	
Perceived behavioural control	2.89	0.54	0.154	0.064	
Responsible computer consumption intention (RCCI)	3.78	0.83	0.675*	0.474*	
Self-identity (H5)	3.46	0.76		0.291*	Supported
Habit (H6)	3.34	0.57		0.214*	Supported

Notes: $*p < .05$

R^2 refers to the proportion of the variance in the dependent variable accounted for by independent variables

Beta weight refers to the relative contribution of each independent variable in explaining the variance in the dependent variable

Model 1: TPB variables

Model 2: TPB variables and contextual antecedents

DISCUSSIONS

The implications to HFE theory will be discussed in the first section. The new factors introduced in TPB were significant because they increased the variance of the original TPB model; therefore, this is our new contribution to HFE theory. The recommendations to HFE practitioners will be given in the second section.

Predictors of RCCI and RCCB

Overall, the respondents' ratings for RCCI (mean rating $[MR]$ = 3.78; SD = 0.83; see Table 4.3) and RCCB (MR = 3.76; SD = 0.83; see Table 4.4) do not indicate a high tendency towards using computers responsibly.

Model 1: original TPB—consistent with the original TPB (Ajzen, 1991), this study found that ATB was the predictor of RCCI (see Table 4.6). Respondents agree that using computers responsibly is a good idea and pleasant and makes them feel satisfied (MR = 3.92; SD = 0.9; see Table 4.2). This is consistent with Harun, Lim and Othman's (2011) study, which found that the majority of Malaysians have a positive environ- mental attitude due to exposure to environmental campaigns such as the Go Green Campaign, Color Me Green Campaign and education in school. Conversely, according to the TPB, SN should be one of the predictors of behavioural intention. This study's finding is consistent with Trafimow and Finlay's (2001), which found that SN did not always influence behav- ioural intention if individuals were confident in their own opinion (or "attitudinally controlled"). This means the respondents were not influ- enced by friends, family or governments; instead, they were influenced by their own attitudes, that is, favourable and unfavourable feelings. Besides, the TPB states that PBC exerts an influence on behavioural intention or actual behaviour, which was not found to be true in this study (see Tables 4.6 and 4.7). One explanation, as suggested in Armitage and Conner's study (2001, p. 472), is that in a situation where attitude is a significant factor, PBC may be less predictive of intention. The respondents' prefer- ences (attitudes) formed their RCCI regardless of the impediments, i.e. Malaysian consumers opt to shut down/activate the sleep mode because they think responsible consumption is a good idea despite the hassle of the time taken to restart their computers. The low $MR2.89$ (SD = 0.54) of PBC further supports this finding. Additionally, the study concurs with the original TPB, which found that RCCI is a predictor of RCCB (see Table 4.7).

Model 2: Model 2 is the key theoretical contribution in this research as it identified the new variables in the TPB in the context of responsible computer consumption. (1) For behavioural intention (see Table 4.6), PSE was a predictor and it had the greatest influence on RCCI (with the largest beta weight value). Consistent with previous studies (i.e. Conner & Armitage, 1998; Rice, Wongtada & Leelakulthanit, 1996), respondents showed a lack of PSE (*MR* = 3.40; *SD* = 0.57; see Table 4.2), that is, they feel their actions will not make a difference. Additionally, users in a developing country like Malaysia may lack the skills and knowledge to practise responsible computer consumption, for example, understanding how power-management settings work and technical terms such as the Electronic Product Environmental Assessment Tool (EPEAT) standard. Collectivism was also a predictor of RCCI, consistent with prior literature (Hansla, 2011; Nordlund & Garvill, 2002). Since Malaysia is a collectivist society (Hofstede, 2001), Malaysians' intention to shut down idling computers and employ power-management settings was considered for the benefit of society as a whole. PIC is also a predictor; however, the respondents gave a neutral rating for PIC, which indicates that they did not perceive the benefits of practising responsible computer consumption. This may be due to the relatively low electricity tariff in Malaysia (RM0.218 or USD0.061/kWh for the first 200kWh); thus, using computers responsibly will not reduce electricity bills significantly. The findings also reveal that CEK is not a predictor of RCCI. One explanation is that Malaysians could not see the personal benefits of practising responsible consumption as electricity savings would be small, thus their CEK did not translate into behavioural intention. This is similar to studies by Ottman (2000) and the Massachusetts Department of Environmental Protection (2002), which found that consumers were knowledgeable about how to protect the environment but personal interest made them unwilling to act. (2) For actual behaviour (see Table 4.7), SI and HA were significant predictors; however, the respondents rated SI low (MR of 3.46; SD = 0.76). This is not surprising as turning off computers and using power-management settings are actions not very visible to others; therefore, SI failed to influence RCCB. As for HA, even though it is a predictor of behaviour, the low MR of 3.34 (SD = 0.57) indicates that Malaysians have yet to develop the habit of responsible use. This is because they do not have a high tendency to practise RCCB (see the MR of RCCB in Table 4.4), which does not support the forming of a habit.

To summarise, the HFE issues influencing RCCB are intention (RCCI), self-identity and habit. Intention is influenced by perceived self-efficacy, collectivism, attitude towards the behaviour and positive individual consequences.

Recommendations to Encourage Responsible Computer Consumption

Based on the findings of this research, we make the following recommendations for the HFE practitioners. Since this research provides input to the design for sustainable development (Thatcher, 2012, p. 10) which involves a community and collective approach and balances the interests of the economic, social and natural capital, the recommendations are relevant to HFE practitioners in government and non-governmental organisations, for example, policy makers, non-governmental organisations social workers, media producers, educators and so on.

1. This study found that consumers' intention to perform responsible computer consumption was for the gain of society (collectivist reason) but they did not have PSE. Moreover, SN was not a predictor. It is therefore recommended that PSE should be the key focus of education (at talks, exhibitions and seminars, and in schools and education centres) and in the mass media. Government social messages should convince consumers that they can make a difference to the environment through their individual practice of using computers responsibly for the collective benefit and not rely on SN (Gupta & Ogden, 2009, p. 387). For instance, WWF-Malaysia in collaboration with Microsoft Malaysia has jointly developed a mobile application "Earth Hour Malaysia" to encourage consumers to pledge to switch power off for one hour during Earth Hour. This message has an impact on both PSE and collectivism. Consumers can enhance their PSE as they can perceive how their actions can be accounted for through online statistics of pledges made. They cannot rely on their parents or friends to influence them as such behaviour (e.g. switching off computers and using power-management options) is not obvious. Guidelines on using power-management options can be provided in the application to equip them with the necessary knowledge to act responsibly. Additionally, the messages can provide information on how individual actions collectively benefit society, for example, in energy conservation, the mitigation of global warming and reduction of hazardous electronic waste.

2. This study found that attitudes towards behaviour were important (i.e. consumers formed their intention through positive attitudes [ATB] regardless of the impediments [PBC], like turning off computers although it takes time to restart computers). As such, it is recommended that emotional and cognitive appeals be emphasised in social advertisements/campaigns, for example, by portraying the setbacks associated with energy wastage and by protecting and loving nature and the environment through responsible consumption despite the inconveniences.

3. The emphasis in government social messages should also be on enhancing the user's self-image since SI is an important factor; however, this should be done with consideration to the collectivist value, which is an important factor. To achieve this, the messages should convince consumers to practise RCCB for the benefit of society, and there should be a form of social recognition for those who practise such "good behaviour" to enhance their self-image (Steg & Vlek, 2009). Campaigns such as "green computer user" badges can be a status symbol to encourage RCCB.

4. Concrete environmental knowledge did not predict behavioural intention; thus, education in Malaysian schools to inculcate concrete environmental knowledge is insufficient (Thang & Kumarasamy, 2006). Since PIC is an important factor, the public should also be informed and convinced of the benefits of responsible computer consumption, for example, longer computer lifespan and energy cost savings. Campaigns can be introduced such as an activity/quiz to calculate electricity usage from using computers responsibly versus not doing so. The stepwise increase in the electricity tariff should be highlighted, especially the fact that those in the highest consumption bracket (RM0.571 or USD0.160/kWh) end up paying more than double the amount paid by the lowest bracket (RM0.218 or USD0.061/kWh).

5. Habit is an important factor but the findings indicate that users do not have a habit of using computers responsibly. Egmond and Bruel's (2007) study pointed out that getting rid of old habits (irresponsible use) can be achieved by making the new desired habit (responsible use) more rewarding. Therefore, the advantages of RCCB should be emphasised through education and government social messages to inculcate such a habit. The introduction of policies for a steeper stepwise increase in electricity tariff can also make such a habit more rewarding.

Conclusion

This study has identified the inhibitors and motivators in responsible computer consumption, that is, attitudes, perceived self-efficacy, collectivism, self-identity, positive individual consequences and habit. These encompass the social aspects, that is, the key HFE issues faced by individuals in practising RCCB, which indirectly affect the environmental aspects, that is, climate change in natural systems (through the emission of CO_2 during energy production in computer consumption) and the economic aspect as energy is wasted through idling computers. As in prior HFE studies, through the understanding of the inhibitors and motivators in sustainable behaviour, this study has provided inputs on how to encourage behavioural change among individuals to adopt RCCB. These include introducing a steeper stepwise electricity tariff; government social messages, for example, saving money by practising RCCB, prolonging computer life, making a difference and attaining collective benefits in mitigation of global warming and reduction of hazardous waste, loving nature and the environment, making pledges to be responsible users through mobile applications and awarding social badges; and education in the use of power-management options and calculation of monetary savings from energy saved. Lastly, this research has added to the HFE theory by enhancing the TPB with the addition of the HFE issues leading to increased explanatory powers in intention and behaviour.

Limitations and Future Studies

Even though the study and data samples are from Malaysia, we believe the results may be replicated in developing countries such as India, China and Thailand where growth in computer use is exponential due to expanding economies and a collectivist culture. However, future studies should be conducted to confirm this. Although the sample size for this study was fairly modest, the total responses are considered appropriate for the statistical analysis conducted (Field, 2009), and the results are statistically significant. Future researchers may consider conducting a larger sample size study with a broader scope, taking into consideration computer design and usability factors such as ease of setting power-management options and switching off and clarity of switch-off reminders. With a larger sample, it will allow more robust analyses (such as structural equation modelling) and the test of the entire model with all the variables concurrently. We adopted a non-random sampling approach (collecting data in public places) because there was no population frame of potential participants to enable the use of the probability sampling method. We did not collect samples of minors (below 17 years old) as that would require consent from parents (which would have been complicated) and minors may not have the maturity to be able to understand the questionnaire. The self-reported results on actual usage in the questionnaire survey might differ from the actual situations but this is an inherent limitation of any questionnaire survey. We have taken steps to reduce this limitation by conducting reliability and validity tests on the instrument. The validation of the effectiveness of recommendations (of interventions) in addressing wastage behaviour and analysis of demographic differences such as age, gender and occupation are beyond the scope of the research funded and can be potential future studies. We have used cross-sectional research design to collect the data at a single time point since such design enables us to achieve our objectives, that is, examine the correlation between the variables, that is, ATB, SN, PBC, CO, PSE, PIC and CEK vs. RCCI and PBC, RCCI, SI and HA vs. RCCB. Although cross-sectional studies allow the authors to examine the correlation between variables at a single point of time, future studies can adopt longitudinal studies to examine the extent change of the causal relationships between variables over different time periods.

This study's behavioural intention construct (RCCI) does not act as a proxy for actual behaviour (RCCB). In fact, the study aims to know if the intentions are translated into actual behaviour, that is, RCCI → RCCB. Many researchers (e.g. Wee et al., 2014) proposed that we should measure both intention and actual behaviour because there may be a gap between intention and behaviour constructs. We measured intentions and actual behaviour constructs at the same time (cross-sectional) by asking respondents if their intention (cognition) translates into what they do (actual behaviour). Owing to cross-sectional study was used, the researcher did explain to the respondents "actual behaviour" refers to what they do (i.e. I have used computers responsibly), while "intention" refers to what they plan to do previously (i.e. I intended to use computers responsibly in the near future) in order to know if their actual behaviour is consistent with their intentions. Ideally, we should have measured it at a different point of time, but it was not possible to collect data from the same respondents who may not be the same public places a few months later.

Acknowledgement The authors thank the Ministry of Higher Education, Malaysia, for funding this research under the Fundamental Research Grant Scheme (FRGS)
(Ref: FRGS/1/2012/SS05/MUSM/02/2). The authors also thank the respondents who participated in this research.

REFERENCES

Ajzen, I. (1991). The theory of planned behavior. *Organizational Behavior and Human Decision Processes, 50*, 179–211.

Ajzen, I. (2011). The theory of planned behaviour: Reactions and reflections. *Psychology & Health, 26*, 1113–1127.

Ajzen, I., & Fishbein, M. (1980). *Understanding attitudes and predicting social behavior.* Englewood Cliffs, NJ: Prentice Hall.

Andrae, A. S. G., & Andersen, O. (2010). Life cycle assessments of consumer electronics – Are they consistent. *International Journal of Life Cycle Assessment, 15*, 827–836.

Antil, J. H. (1984). Socially responsible consumers: Profile and implication for public policy. *Journal of Macromarketing, 4*(Fall), 18–30.

Armitage, C. J., & Conner, M. (2001). Efficacy of the theory of planned behavior: A meta-analytic review. *British Journal of Social Psychology, 40*, 471–499.

Asia Pacific. (n.d.). *Internet world stats – Usage and population statistics.* Retrieved from: http://www.internetworldstats.com/stats3.htm

Baker, J. P., & Ozaki, R. (2008). Pro-environmental products: Marketing influence on consumer purchase decision. *Journal Consumer Market, 25*, 281–293.

Barnatt, C. (n.d.). *Green computing.* http://www.explainingcomputers.com/green.html. Accessed 5 Sept 2014.

Bissell, S. J., Duda, M. D., & Young, K. C. (1998). Recent studies on hunting and fishing participation in the United States. *Human Dimensions of Wildlife, 3*, 75–80.

Burkotler, D., Weyers, B., Kluge, A., & Luther, W. (2014). Customization of user interfaces to reduce errors and enhance user acceptance. *Applied Ergonomics, 45*, 346–353.

Chen, N. H., & Huang, S. C. T. (2016). Domestic technology adoption: Comparison of innovation adoption models and moderators. *Human Factors and Ergonomics in Manufacturing & Service Industries, 26*, 177–190.

Conner, M., & Armitage, C. J. (1998). Extending the theory of planned behavior: A review and avenues for further research. *Journal of Applied Social Psychology, 28*, 1429–1464.

Costanza, R., & Patten, B. C. (1995). Defining and predicting sustainability. *Ecological Economics, 15*, 193–196.

Costello, A. B., & Osborne, J. W. (2005). Best practices in exploratory factor analysis: Four recommendations for getting the most from your analysis. *Practical Assessment, Research & Evaluation, 10*(7), 1–9.

Cox, P. M., Betts, R. A., Jones, C. D., Spall, S. A., & Totterdell, I. J. (2000). Acceleration of global warming due to carbon-cycle feedbacks in a coupled climate model. *Nature, 408*, 184–187.

Dahlstrand, U., & Biel, A. (1997). Pro-environmental habits: Propensity levels in behavioral change. *Journal of Applied Social Psychology, 27*, 588–601.

Davies, J., Foxall, G. R., & Pallister, J. (2002). Beyond the intention–behavior mythology: An integrated model of recycling. *Market Theory, 2*, 29–113.

Department of Statistics Malaysia. (2015). *ICT use and access by individuals and households survey report, Malaysia*. https://www.statistics.gov.my/index.php?r=column/cthemeByCat&cat=395&bul_id=Q3l3WXJFbG1PNjRwcHZQTVlSR1UrQT09&menu_id=amVoWU54UTl0a21NWmdhMjFMMWcyZz09

Dillon, A. (2001). User acceptance of information technology. In W. Karwowski (Ed.), *Encyclopedia of human factors and ergonomics*. London: Taylor and Francis.

Durugbo, C. (2013). Improving information recognition and performance of recycling chimneys. *Ergonomics, 56*, 409–421.

Egmond, C., & Bruel, R. (2007). *Nothing is as practical as a good theory, analysis of theories and a tool for developing interventions to influence energy-related behavior*. SenterNovem, 16 pp. http://www.cres.gr/behave/pdf/paper_final_draft_CE1309.pdf. Accessed 13 July 2015.

Elkington, J. (1997). *Cannibals with forks: The triple bottom line of 21st century business*. Oxford: Capstone Publishing.

eMarketer. (2012). *China's key internet stats rival the US and Japan*. http://www.emarketer.com/Article.aspx?R=1008899&ecid=a6506033675d47f88165194 3c21c5ed4

Eriksson, L., Garvill, J., & Nordlund, A. M. (2008). Interrupting habitual car use: The importance of car habit strength and moral motivation for personal car use education. *Transportation Research Part F: Traffic Psychology and Behaviour, 11*, 10–23.

Field, A. P. (2009). *Discovering statistics using SPSS: And sex and drugs and rock 'n' roll*. London: Sage.

Fishbein, M., & Ajzen, I. (1975). *Belief, attitude, intention and behavior: An introduction to theory and research*. Reading, MA: Addison-Wesley.

Flemming, S. A. C., Hilliard, A., & Jamieson, G. A. (2008). *The need for human factors in the sustainability domain*. Proceedings of the Human Factors and Ergonomics Society 52nd Annual Meeting, 748–752.

Follows, S. B., & Jobber, D. (2000). Environmentally responsible purchase behavior: A test of a consumer. *European Journal of Marketing, 34*(5/6), 723–746.

Gupta, S., & Ogden, D. T. (2009). To buy or not to buy? A social dilemma perspective on green buying. *Journal of Consumer Marketing, 26*(6), 376–391.

Hagger, M. S., Anderson, M., Kyriakaki, M., & Darkings, S. (2007). Aspects of identity and their influence on intentional behavior: Comparing effects for three health behaviors. *Personality and Individual Differences, 42*, 355–367.

Hair, J. F., Ringle, C. M., & Sarstedt, M. (2011). PLS-SEM: Indeed a silver bullet. *Journal of Marketing Theory and Practice, 19*, 139–151.

Hansla, A. (2011). Value orientation, awareness of consequences, and environmental concern. *University of Gothenburg, Sweden, 2011*, 1–30.

Hanson, M. A. (2013). Green ergonomics: Challenges and opportunities. *Ergonomics, 56*, 399–408.

Harun, R., Lim, K. H., & Othman, F. (2011). Environmental knowledge and attitude among students in Sabah. *World Applied Sciences Journal, 14*, 83–87.

Harvey, J., Thorpe, N., & Fairchild, R. (2013). Attitudes towards and perceptions of eco driving and the role of feedback systems. *Ergonomics, 56*, 507–521.

Haslam, R., & Waterson, P. (2013). Editorial: Ergonomics and sustainability. *Ergonomics, 56*, 343–347.

Haslam, R. A. (2002). Targeting ergonomics interventions-learning from health promotion. *Applied Ergonomics, 33*, 241–249.

Henion, K. E. (1976). *Ecological marketing*. Columbus, OH: Grid.

Hofstede, G. (2001). *Culture's consequences: Comparing values, behaviors, institutions, and organizations across nations*. Thousand Oaks, CA: SAGE.

Holden, R. J. (2012). Social and personal normative influences on healthcare professionals to use information technology: Towards a more robust social ergonomics. *Theoretical Issues in Ergonomics Science, 13*, 546–569.

Hopkinson, L., & James, P. (2011). Lifecycle energy and environmental impacts of computing equipment: A June 2011 update to a 2009 SusteIT Report. http://www.goodcampus.org/files/files/57LCA_of_computing_equipment_v7_final_June_2011.pdf

Index Mundi. (2014). *Malaysia age structure – Population pyramid*. http://www.indexmundi.com/malaysia/age_structure.html. Accessed 8 July 2015.

International Ergonomics Association. (2015). *Definitions and domain ergonomics*. http://www.iea.cc/whats/

Jackson, T. L. S., & Johnson, A. E. (2002). Cultural ergonomics in Ghana, West Africa: A descriptive survey of industry and trade workers' interpretations of safety symbols. *International Journal of Occupational Safety and Ergonomics, 8*, 37–50.

Kaufmann, H. R., Panni, M. F., & Orphanidou, Y. (2012). Factors affecting consumers' green purchasing behavior: An integrated conceptual framework. *Amfiteatru Economic, 14*, 50–69.

Kim, S. S., Malhotra, N. K., & Narasimhan, S. (2005). Two competing perspectives on automatic use: A theoretical and empirical comparison. *Information Systems Research, 16*, 418–432.

Kim, Y., & Choi, S. R. (2005). Antecedents of green purchase behavior: An examination of collectivism, environmental concern and PCE. *Advances in Consumer Research, 32,* 592–599.

Klockner, C. A. (2013). A comprehensive model of the psychology of environmental behaviour—A meta-analysis. *Global Environmental Change, 23,* 1028–1038.

Kobus, C. B., Mugge, R., & Schoormans, J. P. (2013). Washing when the sun is shining! How users interact with a household energy management system. *Ergonomics, 56,* 451–462.

Kothaneth, S. (2010). *A pilot study on the cross-cultural acceptance of technology.* Human Factors and Ergonomics Society Annual Meeting Proceedings 09/2010 54, 1951–1955.

Kowalewski, S., Borg, A., Kluge, J., Himmel, S., Trevisan, B., Erabme, D., Ziefle, M., & Jakobs, M. (2014, July 19–23). *Modeling the influence of human factors on the perception of renewable energies.* Taking geothermics as example. Proceedings of the 5th International Conference on Applied Human Factors and Ergonomics AHFE 2014, Krakow, Poland, 1884–1891.

Lee, K. (2008). Opportunities for green marketing: Young consumers. *Marketing Intelligence & Planning, 26,* 573–586.

Lee, K. (2011). The green purchase behavior of Hong Kong young consumers: The role of peer influence, local environmental involvement, and concrete environmental knowledge. *Journal of International Consumer Marketing, 23,* 21–44.

Limayem, M., Hirt, S. G., & Cheung, C. M. K. (2007). How habit limits the predictive power of intentions: The case of IS continuance. *MIS Quarterly, 31,* 705–737.

Lockton, D., Harrison, D., & Stanton, N. A. (2010). The design with intent method: A design tool for influencing user behavior. *Applied Ergonomics, 41,* 382392.

Martin, K., Legg, S., & Brown, C. (2013). Designing for sustainability: Ergonomics – Carpe diem. *Ergonomics, 56,* 365–388.

Massachusetts Department of Environmental Protection. (2002). *Barrier/motivation inventory.* http://www.state.ma.us. Accessed 10 Aug 2013.

McDonagh, D., Bruseberg, A., & Haslam, C. (2002). Visual product evaluation: Exploring users' emotional relationships with products. *Applied Ergonomics, 33,* 231–240.

Mida, S. (2009). *Factors contributing in the formation of consumer's environmental consciousness and shaping green purchasing decision.* Symposium on Computers & Industrial Engineering, Moncton, 957–962.

Moller, B., & Thogersen, J. (2008). Car-use habits: An obstacle to the use of public transportation? In C. Jensen-Butler, B. Madsen, O. A. Nielsen, & B. Sloth (Eds.), *Road pricing, the economy, and the environment.* Amsterdam: Elsevier.

Moray, N. (1994, August). *Ergonomics and the global problems of the 21st century*. Keynote Address Presented at the 12th Triennial Congress of the International Ergonomics Association, Toronto.

Moray, N. (1995). Ergonomics and the global problems of the twenty first century. *Ergonomics, 38,* 1691–1707.

Munasinghe, M. (2012). Millennium consumption goals (MCGs) for Rio+20 and beyond: A practical step towards global sustainability. *Natural Resources Forum: A United Nations Sustainable Development Journal, 36,* 202–212.

Murugesan, S. (2008). Harnessing green IT: Principles and practices. *IT Professional Journal, 10,* 24–33.

National Energy Foundation, (2007). *Assessment of potential for energy savings from PC software management.* http://www.1e.com/energycampaign/downloads/1ENEFReport.pdf. Accessed 10 Aug 2013.

Ng, Y. C. (2011). *Comscore Malaysia segmentation: Which areas have the most internet users?* http://www.ibizclicks.com/blog/comscore-malaysia-segmentation/. Accessed 10 Aug 2013.

Nordlund, A. M., & Garvill, J. (2002). Value structure behind pro-environmental behavior. *Environment and Behavior, 34,* 740–756.

Ottman, J. A. (2000). *It's not just the environment, stupid.* http://www.greenmarketing.com/articles/IB_Sept00.html. Accessed 2 Aug 2013.

Ouellette, J. A., & Wood, W. (1998). Habit and intention in everyday life: The multiple processes by which past behavior predicts future behavior. *Psychological Bulletin, 124,* 54–74.

Oyserman, D., Elmore, K., & Smith, G. (2012). Self, self-concept, and identity. In M. Leary & J. P. Tangney (Eds.), *Handbook of self and identity* (2nd ed., pp. 69–104). New York: The Guilford Press.

PC Energy Report. (2009). *The power to save money.* http://www.1e.com/EnergyCampaign/downloads/PC_EnergyReport2009-US.pdf. Accessed 29 July 2015.

Poortinga, W., Steg, L., & Vlek, C. (2004). Values, environmental concern, and environmental behavior: A study into household energy use. *Environment and Behavior, 36,* 70–93.

Prothero, A., Dobscha, S., Freund, J., Kilbourne, W. E., Luchs, M. G., Ozanne, L. K., et al. (2011). Sustainable consumption: Opportunities for consumer research and public policy. *Journal of Public Policy & Marketing, 30,* 31–38.

Radjiyev, A., Qiu, H., Xiong, S., & Nam, K. H. (2015). Ergonomics and sustainable development in the past two decades (1992–2011): Research trends and how ergonomics can contribute to sustainable development. *Applied Ergonomics, 46A,* 67–75.

Rice, G., Wongtada, N., & Leelakulthanit, O. (1996). An investigation of self-efficacy and environmentally concerned behavior of Thai consumers. *Journal of International Consumer Marketing, 9,* 1–19.

Ries, F., Hein, V., Pihu, M., & Armenta, J. M. S. (2012). Self-identity as a component of the theory of planned behaviour predicting physical activity. *European Physical Education Review, 18*(3), 322–334. http://kodu.ut.ee/~vello/Ries. Hein.EPER2012TPB.pdf

Rise, J., Sheeran, P., & Hukkelberg, S. (2010). The role of self-identity in the Theory of Planned Behavior: A meta-analysis. *Journal of Applied Social Psychology, 40,* 1085–1105.

Rivera, M. E. K., Ganey, H. C. N., Dalton, J., & Hancock, P. A. (2004). *Worldview and acculturation as predictors of performance: Addressing these variables in human factors/ergonomics research.* Proceedings of the Human Factors and Ergonomics Society 48th Annual Meeting, 1223–1227.

Robertson, M., Amick, B. C., III, DeRango, K., Rooney, T., Bazzani, L., Harrist, R., et al. (2009). The effects of an office ergonomics training and chair intervention on worker knowledge, behavior and musculoskeletal risk. *Applied Ergonomics, 40,* 124–135.

Sauer, J., Wiese, B. S., & Ruttinger, B. (2002). Improving ecological performance of electrical consumer products: The role of design-based measures and user variables. *Applied Ergonomics, 33,* 297–307.

Schahn, J., & Holzer, E. (1990). Studies of individual environmental concern. The role of knowledge, gender, and background variables. *Environment and Behavior, 22,* 767–786.

Schultz, P. W., & Zelezny, L. C. (1998). Values and pro-environmental behavior: A five country survey. *Journal of Cross-Cultural Psychology, 29,* 540–558.

Skarżyńska, K. (2002). Work as a cultural and personal value: Attitudes towards work in Polish society. *International Journal of Occupational Safety and Ergonomics, 8,* 195–208.

Steg, L., & Vlek, C. (2009). Encouraging pro-environmental behavior: An integrative review and research agenda. *Journal of Environmental Psychology, 29,* 309–317.

Stern, P. C. (2000). Toward a coherent theory of environmentally significant behavior. *Journal of Social Issues, 56,* 407–424.

Szalma, J. L. (2014). On the application of motivation theory to human factors/ergonomics: Motivational design principles for human-technology interaction. *Human Factors, 56*(8), 1453–1471.

Thang, S. M., & Kumarasamy, P. (2006). Malaysian students' perceptions of the environment contents in their English language classes. *Electronic Journal of Foreign Language Teaching, 3,* 190–208.

Thatcher, A. (2012). Affect in designing for sustainability in human factors and ergonomics. *International Journal of Human Factors and Ergonomics, 1,* 127–147.

Thatcher, A. (2013). Green ergonomics: Definition and scope. *Ergonomics, 56*(3), 389–398.

Thatcher, A., & Yeow, P. H. P. (2016a). Human factors for a sustainable future. *Applied Ergonomics, 57,* 1–7.

Thatcher, A., & Yeow, P. H. P. (2016b). A sustainable system of systems approach: A new HFE paradigm. *Ergonomics, 59,* 167–178.

The Economist. (2009). *Computing climate change.* http://www.economist.com/node/14297036. Accessed 10 Sept 2013.

Tonglet, M., Philips, P. S., & Read, A. D. (2004). Using the Theory of Planned Behavior to investigate the determinants of recycling behavior: A case study from Brixworth, UK. *Resources, Conservation and Recycling, 41,* 191–214.

Trafimow, D., & Finlay, K. A. (2001). The importance of traits and group memberships. *European Journal of Social Psychology, 31,* 37–43.

TrendMicro. (2012). *Trend Micro Study shows increasingly usage behaviour across multiple devices in APAC.* http://apac.trendmicro.com/apac/about-us/newsroom/releases/articles/20130314012909.html

Urien, B., & Kilbourne, W. (2011). Generativity and self-enhancement values in eco-friendly behavioral intentions and environmentally responsible consumption behavior. *Psychology & Marketing, 28,* 69–90.

Valle, P. O. D., Rebelo, E., Reis, E., & Menezes, J. (2005). Combining behavioral theories to predict recycling involvement. *Environment and Behavior, 37,* 364–396.

Venkatesh, V., Thong, J. Y. L., & Xu, X. (2012). Consumer acceptance and use of information technology: Extending the unified theory of acceptance and use of technology. *MIS Quarterly, 36,* 157–178.

Voorbij, A. I. M., & Steenbekkers, L. P. A. (2002). The twisting force of aged customers when opening a jar. *Applied Ergonomics, 33,* 105–109.

Webb, D. J., Mohr, L. A., & Harris, K. E. (2008). A re-examination of socially responsible consumption and its measurement. *Journal of Business Research, 61*(2), 91–98.

Webster, F. E. (1975). Determining the characteristics of socially responsible consumer. *Journal of Consumer Research, 2,* 188–196.

Wee, S. C., Mohd Soki, B. M. A., Norhayati, Z., Tajudin, M. N. M., Khalid, I., & Ishak, N. (2014). Consumers' perception, purchase intention and actual purchase behavior of organic food products. *Review of Integrative Business & Economics Research, 3,* 378–397.

Wilson, J. R. (2014). Fundamentals of systems ergonomics/human factors. *Applied Ergonomics, 45,* 5–13.

Yeow, P. H. P., Yuen, Y. Y., & Loo, W. H. (2013). Ergonomics issues in national identity card for homeland security. *Applied Ergonomics, 44,* 719–729.

Zeitlin, L. R. (1994). Failure to follow safety instructions: Faulty communication or risky decisions? *The Human Factors, 36,* 172–181.

Zink, K. J. (2014). Designing sustainable work systems: The need for a systems approach. *Applied Ergonomics, 45,* 126–132.

CHAPTER 5

Human Factors and Ergonomics in Interactions with Sustainable Appliances and Devices

Kirsten M. A. Revell and Neville A. Stanton

INTRODUCTION

The majority of appliances and devices found in the home were originally designed with little thought for conserving the energy required for operation. Their primary purpose was focussed on reducing the time and effort needed to perform domestic tasks (e.g. washing machines, tumble dryers, dishwashers, vacuum cleaners, ovens), prolong food life (e.g. fridge, freezer), provide a comfortable environment (e.g. controls for heating, cooling, lighting, and hot water), or provide entertainment (e.g. radio, television set top boxes). Domestic energy consumption contributes 25% of total energy emissions in the UK, and with government targets to reduce CO_2 emissions by 80% by 2050 (Climate Change Act, 2008), the push towards the sustainable design of appliances and devices is highly pertinent.

Energy efficiency is an aspect of sustainable design that presents particular challenges for appliances and devices that rely on appropriate user

K. M. A. Revell (✉) • N. A. Stanton
Human Factors Engineering, Transportation Research Group, University of Southampton, Southampton, UK

A. Thatcher, P. H. P. Yeow (eds.), *Ergonomics and Human Factors for a Sustainable Future*, https://doi.org/10.1007/978-981-10-8072-2_5

interaction to be realised. For example, the Energy Savings Trust in the UK (Energy Saving Trust, 2017) advises tumble driers require full loads and filters to be regularly cleared; washing machines and dishwashers must be fully loaded and specific settings to be selected; traditional electric kettles demand users to fill only required amount of water and to use this immediately on boiling; and entertainment systems need to be deliberately turned off rather than left on standby when not in use. Heating and cooling systems require timing and set points to be optimally selected, demanding users to expend time and effort in programming devices, as well as finding agreement amongst multiple occupants on a programmed temperature (Peffer, Pritoni, Meier, Aragon, & Perry, 2011). These principles are backed up by consumption figures in MacKay (2009).

Because of this symbiotic relationship between the design of an appliance or device and user behaviour for achieving efficiency, the onus on reducing consumption is frequently put at the feet of users. Lutzenhiser and Bender (2008) found significant variations in domestic energy consumption attributed solely to behavioural differences in householders. Lutzenhiser (1993) argues that technology-based efficiency improvements are amplified or dampened by human behaviour, yet Glad (2012) found when technology is introduced into homes, it does not always meet the user requirements, both in terms of the type of technology and the design of the interface, hindering energy-efficient behaviour. Pierce, Schiano, and Paulos (2010) argue that energy-consuming behaviour is unconscious, habitual, and irrational, and users ignore visible options. The perception of 'irrationality', however, relies on how human judgements and decision making are viewed. In Tversky and Kahneman's seminal 1974 paper, heuristics, such as representativeness, availability, and adjustment from an anchor, were identified as three highly economical 'rules of thumb', heuristics that are usually effective but lead to systematic and predictable errors. What is more, Crossman and Cooke (1974) showed slow response systems (such as home heating) are difficult to learn and control effectively, and Sauer, Wastell, and Schmeink (2009) argued that some users do not have adequate strategies available to manage a system more effectively, even if aware of undue consumption.

Government initiatives in the UK have focussed on addressing variations in consumption through smart meter rollouts (to 25 million homes by 2020) under the premise, it seems, that being 'informed' of consumption is sufficient for behaviour change. However, this type of feedback provided by smart meters (not to be confused with energy monitors) does not tell people how they should interact, only the results of the interactions that have been made. This chapter hopes to illustrate that this type of linear 'cause and effect' approach to reduced consumption belies the true complexity of managing domestic energy. We approach this topic by first understanding the home heating domain, then make the case for a 'mental models' approach to understanding user behaviour that impacts its sustainable use. Core issues in this domain are summarised followed by recommendations for practitioners, designers, and policy makers. An illustration of a mental models approach at different levels in a system is provided for a control device. Whilst the evidence provided in this chapter is often UK centric, the sustainable use of appliances and devices is a global issue. The core issues and recommendations are equally applicable to other countries.

Home Heating

The largest contribution to energy consumption in the UK is not due to the use of self-contained appliances but from space heating, which accounts for 66% of all domestic consumption (Department for Business, Energy, and Industrial Strategy, 2016) and is considered the most complex system in the domestic domain (Sauer et al., 2009). The UK government has initially looked to drive down heating demand through thermal efficiency, smart meters, and heating controls. Energy companies are also held to 'obligations' focussed primarily, but not restricted to insulation measures in the home (The Electricity and Gas Order 2014) where targets must be met to avoid government-imposed financial penalties. Similar initiatives are found in other countries (e.g. Federal Energy Management Program's (FEMP) Energy Incentive Program in the USA), although this approach clearly represents a 'conflict of interest' for energy companies, whose profits increase with greater consumption and have competing obligations to their shareholders. Often targets can be met in a variety of ways such as offering subsidised insulation, installation of smart meters, and provision of energy-saving advice, rather than quotas for consumption reduction. With such clearly competing goals, it is unrealistic, therefore, to expect energy companies to lead the way in ensuring optimisation of energy use and continued reduction of overall domestic consumption.

The Renewable Heat Incentive (RHI) has been introduced in the UK to encourage householders to move to more sustainable low carbon heating alternatives, facilitating heat networks in urban areas and heat pumps in rural areas where a change in fuel source would present a clear economic benefit to residents. Fifty-nine per cent of housing stock in the UK falls under 'suburban residential', however (English Housing Survey, 2008). For residents in this group, substantial investment would be required to move to a more sustainable heating source, despite minimal economic benefit compared to on-grid fossil fuel with a high efficiency condensing boiler. In addition, lack of awareness and lack of trust of sustainable heating systems means this group will be expected to transition to more sustainable technology far later, with gas heating being an important heat energy source well into the 2030s (Department of Energy and Climate Change, 2013, p. 9). In terms of sustainable devices, it is more likely that this user group will be exposed to and invest in smarter heating control devices (e.g. programmers and thermostats), or networked home management systems than change to a non-carbon energy source.

A large body of research has been undertaken on the challenges householders have found with heating controls. Glad (2012) found thermostats were not used as intended, negatively affecting performance and user satisfaction, making an astute observation that smart technology is not that smart unless the user can effectively use it. Brown and Cole (2009) support the assertions of Glad (2012) and cite responsiveness and lack of immediate feedback (or lack of relevant feedback) were responsible for poor comfort levels in green buildings. Peffer et al. (2011) found nearly half of the users do not use programming features in home heating devices, and when they do, only half are programmed to make adjustments to correspond with night time or unoccupied times, limiting the energy-saving benefits. They place the cause of misunderstandings of terminology (e.g. set point) and programming difficulties firmly as a result of poor design by designers and engineers. Vastamäki, Sinkkonen, and Leinonen (2005) state that most existing temperature controllers do not provide initial feedback in a way that is understandable to the user, and temperature change feedback is delayed, resulting in trial and error behaviour, reducing comfort and wasting energy, and reducing motivation to use the control again. They describe the temperature control as a seemingly simple everyday device that is difficult to use because 'everyday thinking' leads to the wrong conclusions about its way of working. This lends support to the thinking laid out by Cooper, Reimann, Cronin, and Noessel (2014) whereby intuitive interfaces, particularly metaphorical ones, are limited in utility when applied to complex scenarios. It also echoes the views of Kempton (1986) who found householders frequently developed an inappropriate 'mental model' of the thermostat, resulting in less appropriate behaviour with the device.

Operating energy-consuming technology in the home presents substantial difficulties to householders, providing barriers to reaping energy efficiencies intended in the design of appliances and devices.

Mental Models

Progressing the work of Kempton (1986), the authors have undertaken extensive research exploring how the mental models held by householders of heating devices impact energy consumption and fulfilment of their primary goal to provide comfort (see Revell & Stanton, 2012, 2014, 2017; K. M. A. Revell & N. A. Stanton, 2016; Kirsten et al., 2015). Mental models can be thought of as internal constructs that explain human behaviour (Kempton 1986; Wickens, Hollands, Banbury, & Parasuraman, 2015). Kahneman and Tversky (1982, p. 201) class the deliberate manipulation of mental models as an important and distinct 'simulation heuristic' used particularly in: (1) prediction, (2) assessing the probability of a specified event, (3) assenting conditioned probabilities, (4) counterfactual assessments, and (5) assessments of causality.

However, the term 'mental model' is used in different domains to mean different things (Wilson & Rutherford, 1989) and even within a domain can be used to describe internal constructs that differ significantly in terms of content, function, or perspective (Revell & Stanton, 2012; Richardson & Ball, 2009). The form of mental model descriptions may have similarities to the way other types of models (e.g. process models or logic models) that do not depict internal constructs are represented, resulting in confusion when interpreting outputs. This chapter refers to mental models in terms of its (1) function, (2) source, and (3) individuality. When the term 'mental model' is used, it refers to the following specification. In terms of function, the definition most fitting for sustainable appliances and devices is a 'device model'. Kieras and Bovair (1984) adopted this terminology to describe a mental model held by a user of how a device works. It includes a set of conceptual entities and their interrelationships (Payne, 1991). In terms of its source, this author adopts Norman's (1983) definition of a 'user mental model' (UMM). He describes this as 'the actual mental model (e.g. of a home heating system) a user might have' (p. 11) that can only be gauged by undertaking observations or experimentation with the user. In terms of the individuality of mental models, we embrace Kempton's (1986) 'shared theory', which is derived by an analyst through the identification of similarities in separate UMMs of individuals.

The notion of mental models has formed the basis of strategies to improve interface design (Baxter, Besnard, & Riley, 2007; Carroll & Olson 1987; Jenkins, Salmon, Stanton, & Walker, 2010; Norman, 2002; Williges, 1987), to promote usability (Branaghan, Covas-Smith, Jackson, & Eidman, 2011; Jenkins, Salmon, Stanton, Walker, & Rafferty, 2011; Larsson, 2012; Mack & Sharples, 2009; Norman, 2002), and to encourage sustainable behaviour (Kempton, 1986; Lockton, Harrison, & Stanton, 2010; Sauer et al., 2009).

In 1986, Kempton described two distinct 'forms' of mental models of thermostat function that were prevalent in the population of that time, which he termed 'Valve' and 'Feedback' models. The former considers the thermostat analogous to a 'gas valve' and the latter describes a simplified version of the actual functioning of the device. Since Kempton (1986), additional shared models have been proposed in the literature such as 'Timer' (Norman, 2002) and 'Switch' (Peffer et al., 2011). Revell and Stanton (2014) provide further insights on present-day models of home heating devices. Kempton (1986) proposed that the form of model held could result in significant variations in the amount of energy consumed due to home heating, by promoting different patterns of manual thermostat adjustment. In line with Kempton, Norman argues for appropriate interaction with an interface that the user mental model is compatible with the design model of the underlying system. Norman (1986) proposed that the designer can promote compatible user mental models through the choices they make when constructing the system image, which in turn influences user mental models.

The authors combined these concepts to highlight the link between design of devices and sustainability (see Fig. 5.1). From Norman (1986), the design of energy-consuming domestic devices influences the mental model held by users of those devices (top of Fig. 5.1). From Kempton (1986), the mental model held influences patterns of device use (left side of Fig. 5.1), and also from Kempton (1986), patterns of device use over time influence how optimally energy is used (bottom of Fig. 5.1). Here, the authors recognise that the primary goal for most people using (rather than avoiding use of) space heating is unlikely to be 'saving energy' but 'comfort'. Interviews with householders by Revell and Stanton (2014) identified goals related also to the health of children, drying laundry, and house maintenance.

Researchers in the field endorse a mental models approach to home heating design. Sauer et al. (2009) when considering instructional displays

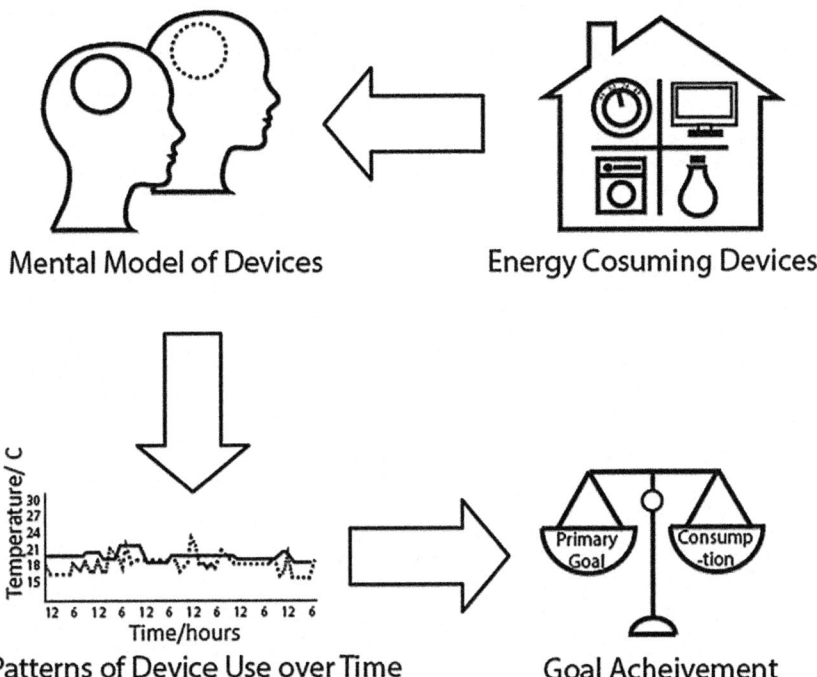

Mental Model of Devices **Energy Cosuming Devices**

Patterns of Device Use over Time **Goal Acheivement**

Fig. 5.1 A simplified diagram exploring the link between device design, mental models, interaction patterns, and goal achievement. Arrows denote hypothesised influence that has directed the authors' body of research

emphasised that a poor mental model of system functioning would prevent an operator from knowing how to interpret the information available. Shipworth et al. (2009) echo this sentiment by stating that operating controls without understanding how to use them is counterproductive, though this is often a reality. They propose that new controls should be developed that appeal to householders, are more intuitive to use, and make it easy to reduce consumption. Peffer et al. (2011) propose that user misconceptions that encourage incorrect usage cannot be easily overcome by better interfaces. Improved usability is essential but insufficient to encourage correct use, and the authors advocate that a mental models approach to interface design that encourages a compatible user mental model at the system and device level could go some way to promoting appropriate understanding.

For the remainder of this chapter, the authors would like to share the core issues and recommendations emerging from their body of work relating to mental model of home heating. These insights have relevance to future sustainable technology relating to home heating, and other sustainable appliances in a domestic setting.

CORE ISSUES

Existing Technology Does Not Support a 'Systems User Mental Model' of Home Heating

The home heating systems typically found in UK homes were designed to provide 'space heating', rather than an 'optimal balance between comfort and consumption'. However, through rising fuel prices and environmental concerns, householders are being tasked with operating a system designed for one purpose, to achieve another purpose. To achieve optimal control, householders need to hold mental models of technology at a system level, not device level. However, the modular nature of home heating devices hinders this goal. The controls, radiators, and boilers found in typical households are manufactured by different companies. The companies themselves do not know the range or type of controls that will integrate with their device, so provide specific user manuals relating to the device in question, but can only provide generic guidance outside the feature of their product. Boilers vary in their efficiency, whilst controllers vary in their interface, features, and location within the home. In the UK, heating systems are generally inherited, rather than chosen. Unless aware of specific controls through prior experience, it is unsurprising some householders are wholly unaware of their existence. User manuals are also often lost or misplaced. Wiese, Sauer, and Rüttinger (2004) have found that awareness of energy-saving information in user manuals, the complexity of manuals, and the proximity of information to the interaction device were all factors that influenced users' actual behaviour with devices but are not given due consideration. In a small-scale case study of a set of six houses matched in terms of heating controls, room layout, and insulation levels, Revell and Stanton (2014) found half of the householders omitted a key heating control from their mental model of their heating system. One was even unaware the house had a thermostat, and controlled heating using the programmer and boiler on/off switch. They were confused as to why the heating kept 'turning off by itself' within programmed 'on' times (as house temperature had reached the thermostat 'set' level) and resorted to additional electrical portable heaters to achieve comfort levels. The absence of key devices in a user mental model prevents the formation of appropriate strategies to ensure energy efficiency or achieve comfort goals. Another barrier to users forming an appropriate systems mental model of home heating is the limited form of feedback that controls and components provide. Often constrained to the set point, status, and temperature measurements, feedback is insufficient to allow an understanding of cause and effect of user behaviour, on the functioning of the heating system as a whole, and the impact on individual heating goals. Programmable thermostats have been adopted to help save energy at night or when the house is unoccupied, but do not emphasise the conditional nature of set point choice for boiler function. For the boiler to emit heat, both the thermostat has to have a higher set point than current room temperature (so it will send a message to the boiler that more heat is needed), and the programmer needs to be set to an 'on' period (allowing the boiler to receive the message from the thermostat, and respond by generating heat). Programmers also do not facilitate users choosing appropriate start and end times that fit their lifestyle. They require the householder to do this without prompts or advice, nor understanding of the heat lags associated with the specific layout and insulation levels of individual homes. The smart meters and energy monitors that have been introduced to make householders aware of their consumption aggregate consumption. In this way they essentially rely on the user being able to interpret data at a system level in order to specify an appropriate behaviour strategy. This is difficult without a sufficient user mental model and a highly routine lifestyle. Even learning thermostats, such as Nest or Ecobee, which learn user schedules, based on adjustment behaviour over time (rather than one-off intentional set up) still demand on the householders' decisions for target set points. Choice of set points driven by a potentially faulty mental model means these 'intelligent' systems may still support wasted consumption. However, the benefit of these devices is seen in their ability to take into consideration the thermodynamic properties of the house, as well as sensing when a home is empty, to better optimise (rather than necessarily reduce) consumption. By failing to help householders interpret these strategies at the 'system level', householders will necessarily refer to and amend their own mental models of their heating system (Johnson-Laird, 1983) to inform what ultimately will lead to a less than energy-efficient behaviour strategy. Strategies to reduce consumption positioned at the 'device' level are at risk of compromising, rather than optimising the balance between comfort and consumption. It is not surprising that the Energy Star rating for programmable thermostats was suspended by the Environmental Protection Agency (EPA) in 2009 and the benefit of a smart meter rollout in the UK is being questioned (BBC, 2014). Householders currently have the wrong tools for the job of optimal consumption, and the right tools require a shift in approach (see K. M. A. Revell & N. A. Stanton, 2016 for a design specification of a more optimal home heating system).

Optimal Home Heating Control Is a Complex Task

Sauer et al. (2009) described central heating as the most complex system in the domestic domain. When users try to interpret complex systems, they access previously developed mental models, rather than apply simple rules (Moray, 1990). The central heating system is a slow responding system, and, inherent in systems of this type, cause and effect is difficult to gauge by observation alone as the link between action and reaction is hidden (Crossman & Cooke, 1974). In addition, the user is faced with multiple distributed controls that vary between households in their location, interface, and functionality. The heating system under observation in a research study by K. M. A. Revell and N. A. Stanton (2014, 2015, 2016) had control devices distributed across the house, with different levels of prominence. The thermostatic radiator valves (TRVs) were positioned at the ankle level adjoined to each radiator in each room. The central thermostat was positioned at the eye level in the hall, exposed to high levels of traffic by occupants in the house, yet this prominent position is at odds with manufacturers' recommendations to be left at a static set point rather than frequently adjusted. The programmer and master power switch were just below the eye level and positioned in the kitchen near the boiler, and the Boost button was a sub feature of the programmer. This discrete location of the Boost button is at odds with its intended function, to allow easy override of the programmer schedule when lifestyle variations require heat at additional times. Optimal comfort and consumption levels are dependent on the compatible adjustment of integrated controls; however the layout does not provide appropriate prominence for correct operation.

Optimal comfort and consumption levels are also affected by variables relating to the environmental setting. These include static variables such as house structure and level of insulation, as well as changing variables within the control of the user (infiltration of air due to door and window positions) and outside the control of the user (external temperatures varying throughout the day and over changing seasons). For optimal heating control, users have a number of different levels of understanding to navigate: (1) awareness of controls and the correct mental model at the device level, (2) awareness of how controls are integrated at the system level, (3) awareness of house structure characteristics on comfort and consumption, and (4) awareness of climate characteristics on consumption (e.g. where heat loss increases as the difference between internal and external temperatures increases). But this is only one side of the story. For optimal heating control, householders also need to match this understanding to an understanding about their own lifestyle. This means that they need to have an appreciation of their occupancy levels, and those of others within the household, as well as the different needs of different members of the household (e.g. greater comfort levels for vulnerable occupants; lower room temperatures when cooking, exercising, or doing housework; night time comfort for those studying late). This is a challenging task, and Sauer et al. (2009) found it was far more difficult to conserve energy with variable daily routines. Given the barriers to forming an appropriate user mental model, householders are faced with matching their unique lifestyle goals with the demands of home heating control, by referring to what is often an incomplete mental model of the system. It is therefore unsurprising that the behaviour specifications that result are far from optimal.

We Cannot Control All the Variables That Affect Optimal Home Heating Control

There are a number of broader system variables that influence comfort and consumption levels including people's lifestyles, the local climate, and the structure and insulation levels of the house. Whilst householders have some control over their lifestyles, they may not be able to dictate the regularity of their occupancy if this is linked to work and school obligations and other commitments outside their home. Although they can make choices relating to internal and external infiltration of air around the home, and the installation of insulation, it is more difficult for householders to make adjustments to the structure of the house to better support thermodynamics and evenly distributed heat. Without moving some distance, householders have no control over the external world climate to which they are subjected. Householders will always be subject to variations that make it difficult to ensure comfort or avoid wasting energy with existing heating systems, although the reality of this does not assure its inevitability. Similarly there is a limit to what government initiatives can do to control these broader variables, although subsidised insulation and the building of more energy-efficient buildings are positive steps. Whilst this chapter has focussed on the problem of influencing householders' *interaction* with home heating controls, it is important to appreciate that appropriate householder behaviour with controls is subject to many other variables. A good choice of set point at one time of day is a poor choice at another time of day. A programmed set of times optimally fits a lifestyle one week but wastes energy the next week. What is considered a comfortable temperature for occupants sitting and watching TV in the evening is different when doing the housework. If there is a regular routine in a home, learning thermostats such as NEST and Ecobee can be trained to adjust the schedule to reflect when users typically want a higher or lower set point, as well as reduce consumption of unoccupied house, but they are only as good as the inputs they receive (from proximity sensors and deliberate adjustment by a user). The provision of 'one-size-fits-all', prescriptive advice on how to manage home heating systems, whether through energy monitors, government campaigns, or interface design, is therefore unrealistic.

RECOMMENDATIONS FOR PRACTITIONERS, DESIGNERS, AND POLICY MAKERS

Recognise the Complexity of the Task for Householders, When Embarking on Strategies to Reduce Consumption

The complexity of the task that householders face to optimise consumption needs to be recognised before effective guidance can be provided. Whether technology-driven guidance or government campaigns, simplistic, generic advice is inappropriate given the variations affecting householders. Tailored guidance that takes into account differences in householders' lifestyles and the influence of broader variables is more likely to result in appropriate home heating management. Certification (e.g. Energy Star) of devices that rely on the adoption of specific behaviour habits should be tested in the 'context of use' to gain robust indications of energy savings. The specified behaviour may be unachievable or unrealistic in the context of use for many users, highlighting where claims of certification may be misleading.

In the human factors literature, a functional approach to control, such as Hollnagel's Contextual Control Model (COCOM) (Hollnagel, 1993), provides an insight into the 'demand characteristics of a situation' that contribute to the complexity of the task faced by householders when forming a strategy to reduce consumption. Context of use can be viewed through the perspective of: (1) competence (the possible actions a system (device and user) can apply to a situation according to needs/demands), (2) control (performance and manifestation of competence, based on four control modes), and (3) constructs (e.g. mental models, including what is known or assumed about the situation where action takes place). Variations not only in the context but also the time available to form a strategy are proposed by Hollnagel (1993) to have a large impact not only on how well a strategy can be planned but also the type of action that is carried out, ultimately affecting consumption optimising performance.

As time available increases and the context improves, the user is likely to progress up the four control modes identified by Hollnagel below:

'Scrambled' control mode, which denotes irrational or random action with minimal reflection. The behaviour in this mode amounts blind 'trial and error' strategies.

'Opportunistic' control model reflects when the user takes the salient features of the current context to determine the next action. Rather than planning, heuristics are employed. This occurs due to inadequate constructs resulting in inefficient actions with many useless attempts undertaken.

'Tactical' control mode, which occurs in situation that allows sufficient time for a limited level of planning beyond immediate need.

Strategic control results in the best performance for achieving higher-level goals but requires a broader time period for both planning and action. For this type of control, the dominant features of the interface are less significant as a predictor of performance.

Hollnagel (1993) believes that if users are provided not only feedback on their actions but feedforward information to help predict the consequences of their actions, this will support their capacity to rise above 'scrambled' mode and operate in opportunist and tactical control modes. To reach the performance 'strategical control', which in this context represents fully optimised energy consumption, a recognition of cognitive effort and extensive implementation time is necessary. Framing energy consumption in this way could help direct the form of government and designer strategies, as well as direct more realistic expectations about the level of control householders can be expected to achieve. What is considered by policy makers to be an acceptable level of waste by users, by rights, should be dictated by these factors.

Use System-Level Strategies for Encouraging Appropriate Home Energy Consumption

The focus of past literature on mental models of home heating and associated behaviour focussed on the thermostat device. Revell and Stanton (2014) revealed other devices were often used as the main control (e.g. programmer, on/off switch, or TRVs), resulting in static or very infrequent thermostat set point adjustments. To understand the home heating behaviour strategy adopted by a householder, set point adjustments need to be viewed for the whole heating system, not just a single device. Home heating controls are integrated in function, so set points on one device affect operation of other devices. However, other appliances in the home (e.g. washing machines and tumble driers, or ovens and dishwashers) are also linked in usage.

Before embarking on a strategy targeted at the single device or appliance level, an understanding of the interdependency of this device with other devices in the system is necessary. Ensuring the 'user' understands these dependencies is crucial for success. This applies not only for the redesign of existing home heating controls (e.g. improving the usability of programmable thermostats) but also the introduction of new technologies designed to aid energy reductions. For example, the introduction of networked energy monitors that effectively communicate to householders how consumption feedback relates to the chosen settings of key controls in different contexts.

Promote the Impact of Broader System Variables on Optimal Consumption in Sustainable Technology

To balance comfort and consumption with home heating controls, expert recommendations revealed that users' mental models needed to include concepts relating to the context of use (K. M. A. Revell & N. A. Stanton, 2016). To select an appropriate thermostat set point, householders need to be aware of variations in comfort levels throughout their home. To select appropriate programmer start and end times, they need to be aware of, and able to quantify, lag times in heating and cooling. To appreciate that over a particular time period, that energy consumption is greater at night, or when internal doors are open, the user needs an understanding of how household thermodynamics are affected by infiltration of air throughout the home, and temperature differentials (Revell & Stanton, 2015). These types of concepts extend beyond the heating system as they relate to variables associated with house structure and weather conditions. Kempton (1986) highlighted how the feedback 'folk model' of the thermostat could result in wasted energy because it did not incorporate these broader variables. Typical home heating technology in the UK does not make this visible nor communicate the influence of these variables on comfort or consumption (K. M. A. Revell & N. A. Stanton, 2016). System-level strategies that go beyond the central heating system controls to include broader variables are likely to be even more effective to householders, as they would promote a mental model that enables appropriate consumption. In addition, providing householders with a system-level view that considers heat loss rates could have the 'knock-on' effect of making explicit the benefits of investing in low-tech improvements such as insulation and draft excluders. It may even positively influence behaviour by making explicit the effect of leaving doors and windows open for longer than necessary. This approach has relevance for the design of other domestic energy-consuming appliances (e.g. tumble driers).

Use a Mental Models Approach When Seeking to Encourage Appropriate Interaction in Complex Systems

A mental models approach to system design has the benefit of aiding learning and facilitating troubleshooting (Norman, 1983). Where a user holds an appropriate mental model of a system, variations in their goals can be accommodated (Moray, 1990). This in turn can facilitate appropriate behaviour when undertaking tasks. In the case of home heating, this could lead to systematic improvements in goal achievement. Householders whose goals include reducing waste (e.g. energy or money) could systematically reduce consumption. Helping users to hold an appropriate 'picture in the mind' of cause and effect at the point of set point adjustment is possible through design driven by mental models research.

Design Sustainable Appliances and Technology with Optimal Consumption as a Primary Goal

Ultimately, the way in which legacy home heating technology is presented to householders is no longer 'fit for purpose'. Designers for devices, such as home heating controls, which currently rely heavily on human behaviour for energy efficiency, have a responsibility for enabling energy-efficient behaviour in the context of use. The same applies for other domestic appliances and devices around the home. Thatcher (2013) argues that when devices are not designed for sustainability, it is difficult to 'design in' this feature. Devices and systems need to be designed, so optimal consumption is an 'equal-weighted' goal to its primary goal (e.g. perform domestic tasks, provide comfort, entertainment, etc.).For example, work by Oi, Yanagi, Tabata, and Tochihara (2011) exploring how to optimise consumption and driver comfort in electric cars identified how heated seats provided a better design solution than air heaters (the most prevalent approach to car heating).

Make Use of the Existing Body of Knowledge

There is a body of knowledge in the literature to direct a metals model approach to design. Manktelow and Jones (1987) provide 24 principles, including evoking existing schemata through analogy, providing a clear task structure, and emphasising appropriate dominance of controls. Norman (2002) provides seven key principles that can be summarised as: (1) use knowledge in the head and the world; (2) simplify the structure of a task; (3) change the nature of a task; (4) use technology to make visible the invisible; (5) make the outcomes of an action obvious; (6) make actions match intentions; and (7) make the system state easily interpretable. K. M. A. Revell and N. A. Stanton (2016) have also produced a set of 53 design requirements based on these principles to 'bridge the gap' between the 'designer' and the 'user' of home heating systems. This was achieved through a systematic comparison of expert recommended home heating behaviour and system model, to that of a novice user. The gaps identified through behaviour and model held directed the focus for design features to encourage a mental model that promotes sustainable interaction.

Findings from qualitative and quantitative research approaches provide insights and direction to target design decisions at different levels of an overall system. Sauer, Wiese, and Rüttinger (2002, 2003, 2004) and Sauer, Schmeink, and Wastell (2007) has undertaken substantial amount of empirical research in this area investigating the environmental benefit of different types of feedback in central heating systems (Sauer et al., 2007), the benefit of different types of design features (Sauer et al., 2004) in electronic consumer products, as well as a comparison between automation and information-based measures on ecological performance (Sauer et al., 2004).

Shove (2007) provides a social practice view to consumption and design analysis of energy-consuming domestic products. Katzeff, Nyblom, Tunheden, and Torstensson (2012) and Kobus, Mugge, and Schoormans (2013) have highlighted the barriers to the usability of energy management systems including the broader perspective of compatibility with 'everyday life'. Lockton et al. (2010) have produced the 'design with intent' toolkit offering a range of different strategies for sustainable behaviour change.

Human factors practitioners, designers, and policy makers will benefit considerably by accessing the tremendous source of knowledge and a variety of approaches to drive sustainable interaction design with devices and appliances.

EXAMPLE APPROACH

To give the reader a practical example of a mental models approach to design at different levels of a system, Figs. 5.2 and 5.3 show how a 'Boost' button can be redesigned at a 'device' level and 'system of controls' level, respectively. This design was used by the authors in a control panel-style interface for home heating driven by a mental model multi-system

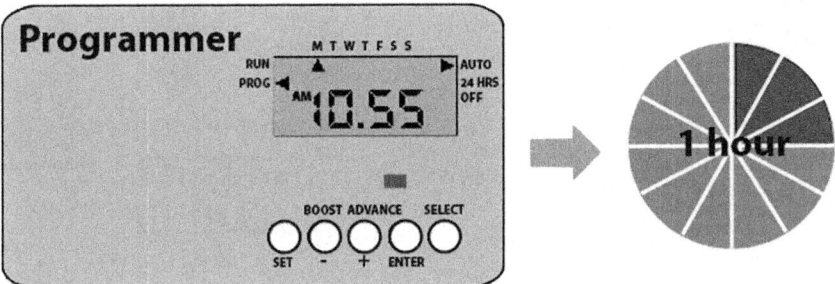

Fig. 5.2 Boost redesign at the device level, to promote an appropriate mental model of device function that communicates a pre-defined time period of operation, using the analogy of a clock

perspective (see K. Revell and N. Stanton (2016) for more details, including further integration at higher system levels). The 'Boost' button is a common feature on traditional heating devices, often integral to a programmer device. On models similar to that found on the CentaurPlus 17, 'BOOST' text is discretely shown on the LED display when a standard button is pressed (see Fig. 5.2, left). This remains for the duration of its one hour operation. The function of the 'Boost' button is typically poorly communicated, nor is its conditional link to the thermostat for boiler operation explicit. Research by K. M. A. Revell and N. A. Stanton (2015, 2016) found novice householders using this type of device control had the 'Boost' button absent from their UMMs or this device was avoided in behaviour strategies, resulting in strategies at greater risk of energy waste (such as boiler override without 'automatic off' features). A redesign of this control in Fig. 5.2 (right) was undertaken to raise the prominence of the Boost button in the hierarchy of controls, as well as to promote an appropriate 'timer' mental model of device function, using the analogy of a clock. In Fig. 5.2, both the system state (progress around the clock face) and the time period of operation (one hour) are highlighted.

At the system level, a lack of clear visible connections between devices prevents interdependency of devices being emphasised and cause and effect rules being developed. Figure 5.3 shows how the conditional rule is emphasised to users, by grouping key controls by their relationship to

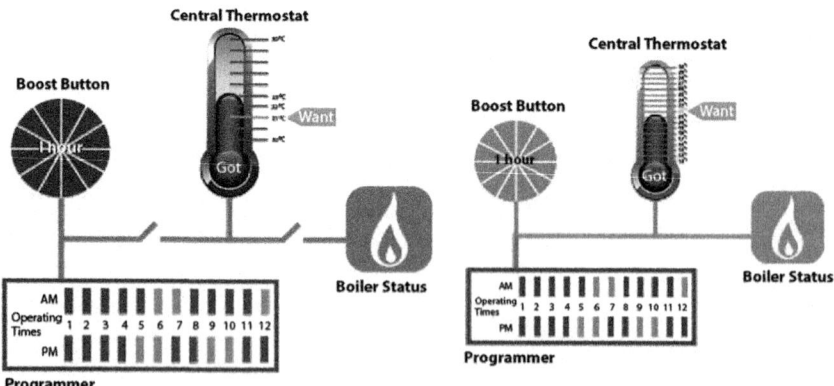

Fig. 5.3 Boost redesign at the control system level, to promote an appropriate mental model of design function that communicates the 'conditional rule' for boiler operation using an electrical circuit analogy

boiler activation. An analogy of a 'switches in a circuit' is used to trigger a schema that guides the development of a functional user mental model. Boiler activation is made visible (highlighted green) when the 'Boost' button, or programmer 'on' period is active (closing the left switch), and the thermostat is actively calling for heat (closing right switch). If one or both of these conditions are not met, one or both switches will be open and the boiler icon will remain inactive (grey). Empirical studies undertaken by the authors have shown how changes in design led to improved mental models of the Boost button and conditional rule and will form the subject of future papers.

Final Remarks

A system-level approach is prevalent in human factors research in a wide range of domains. Thatcher and Yeow (2016) emphasise the need to consider a 'systems of system' approach that the authors endorse. The core issues and recommendations described also allude to factors relating to economic, environmental, and social systems that are interwoven with the challenges of understanding and operating devices. As has been shown within the specific example of home heating, there are multiple layers of interaction that a user has to navigate to interact sustainably (a control on a device, within a system of devices, within a domestic context of use involving, all within a broader external environment). Rather than baulk and turn the other way, the authors urge designers and engineers to embrace the inherent complexity in the design of sustainable devices and appliances, to be truly effective.

References

Baxter, G., Besnard, D., & Riley, D. (2007). Cognitive mismatches in the cockpit: Will they ever be a thing of the past? *Applied Ergonomics, 38*, 417–423.

BBC. (2014). Smart meters will save only 2% on energy bills, say MPs [Online]. *BBC.* http://www.bbc.co.uk/news/business-29125809. Accessed 25 Nov 2014.

Branaghan, R. J., Covas-Smith, C. M., Jackson, K. D., & Eidman, C. (2011). Using knowledge structures to redesign an instructor-operator station. *Applied Ergonomics, 42*, 934–940.

Brown, Z., & Cole, R. J. (2009). Influence of occupants' knowledge on comfort expectations and behaviour. *Building Research and Information, 37*, 227–245.

Carroll, J. M., & Olson, J. R. (Eds.). (1987). *Mental models in human-computer interaction: Research issues about what the user of software knows.* Washington, DC: National Academy Press.

Climate Change Act. (2008). http://www.legislation.gov.uk/ukpga/2008/27/pdfs/ukpga_20080027_en.pdf. Accessed 20 Feb 2017.

Cooper, A., Reimann, R., Cronin, D., & Noessel, C. (2014). *About face: The essentials of interaction design.* Indianapolis, IN: John Wiley & Sons.

Crossman, E. R. F. W., & Cooke, J. E. (1974). Manual control of slow-response systems. In E. Edwards & F. Lees (Eds.), *The human operator in process control* (pp. 51–66). London: Taylor & Francis.

Department for Business, Energy, and Industrial Strategy. (2016). *Energy consumption in the UK.* https://www.gov.uk/government/uploads/system/uploads/attachment_data/file/541163/ECUK_2016.pdf. Accessed 20 Feb 2017.

Department of Energy and Climate Change. (2013). *The future of heating: Meeting the challenge,* 9. https://www.gov.uk/government/uploads/system/uploads/attachment_data/file/190151/16_04_DECC-The_Future_of_Heating-Evidence_Annex_ACCESSIBLE.pdf. Accessed 20 Feb 2017.

Energy Saving Trust. (2017). *Energy saving quick wins.* http://www.energysavingtrust.org.uk/home-energy-efficiency/energy-saving-quick-wins. Accessed 20 Feb 2017.

English Housing Survey. (2008). *Housing stock report 2008.* https://www.gov.uk/government/uploads/system/uploads/attachment_data/file/6703/1750754.pdf. Accessed 20 Feb 2017.

Glad, W. (2012). Housing renovation and energy systems: The need for social learning. *Building Research and Information, 40,* 274–289.

Hollnagel, E. (1993). *Human reliability analysis: Context and control.* London: Academic Press.

Jenkins, D. P., Salmon, P. M., Stanton, N. A., & Walker, G. H. (2010). A new approach for designing cognitive artefacts to support disaster management. *Ergonomics, 53,* 617–635.

Jenkins, D. P., Salmon, P. M., Stanton, N. A., Walker, G. H., & Rafferty, L. (2011). What could they have been thinking? How sociotechnical system design influences cognition: A case study of the Stockwell shooting. *Ergonomics, 54,* 103–119.

Johnson-Laird, P. N. (1983). *Mental models: Towards a cognitive science of language, inference and consciousness.* Cambridge, UK: Cambridge University Press.

Kahneman, D., & Tversky, A. (1982). Variants of uncertainty. *Cognition, 11*(2), 143–157.

Katzeff, C., Nyblom, A., Tunheden, S., & Torstensson, C. (2012). User-centred design and evaluation of EnergyCoach—An interactive energy service for households. *Behaviour & Information Technology, 31,* 305–324.

Kempton, W. (1986). Two theories of home heat control. *Cognitive Science, 10,* 75–90.

Kieras, D. E., & Bovair, S. (1984). The role of a mental model in learning to operate a device. *Cognitive Science, 8,* 255–273.

Kirsten M.A. Revell & Neville A. Stanton (2015) When energy saving advice leads to more, rather than less, consumption, International Journal of Sustainable Energy, 36:1, 1–19, https://doi.org/10.1080/14786451.2014.999071

Kobus, C. B. A., Mugge, R., & Schoormans, J. P. L. (2013). Washing when the sun is shining! How users interact with a household energy management system. *Ergonomics, 56,* 451–462.

Larsson, A. F. (2012). Driver usage and understanding of adaptive cruise control. *Applied Ergonomics, 43,* 501–506.

Lockton, D., Harrison, D., & Stanton, N. A. (2010). The design with intent method: A design tool for influencing user behaviour. *Applied Ergonomics, 41*(3), 382–392.

Lutzenhiser, L. (1993). Social and behavioral aspects of energy use. *Annual Review of Energy and the Environment, 18,* 247–289.

Lutzenhiser, L., & Bender, S. (2008). *The average American unmasked: Social structure and difference in household energy use and carbon emissions.* ACEEE Summer Study on Energy Efficiency in Buildings.

Mack, Z., & Sharples, S. (2009). The importance of usability in product choice: A mobile phone case study. *Ergonomics, 52,* 1514–1528.

MacKay, D. J. (2009). *Sustainable energy – Without the hot air.* Cambridge, UK: UIT Cambridge.

Manktelow, K., & Jones, J. (1987). Principles from the psychology of thinking and mental models. In M. M. Gardiner & B. Christie (Eds.), *Applying cognitive psychology to user-interface design* (pp. 83–117). Chichester, UK: Wiley.

Moray, N. (1990). Designing for transportation safety in the light of perception, attention, and mental models. *Ergonomics, 33,* 1201–1213.

Norman, D. A. (1983). Some observations on mental models. In D. Gentner & A. L. Stevens (Eds.), *Mental models* (pp. 7–14). Hillsdale, NJ: Lawrence Erlbaum Associates.

Norman, D. A. (1986). Cognitive engineering. In D. A. Norman & S. W. Draper (Eds.), *User centered system design: New perspectives on human-computer interaction* (pp. 31–61). Hillsdale, NJ: Lawrence Erlbaum Associates.

Norman, D. A. (2002). *The design of everyday things.* New York: Basic Books.

Oi, H., Yanagi, K., Tabata, K., & Tochihara, Y. (2011). Effects of heated seat and foot heater on thermal comfort and heater energy consumption in vehicle. *Ergonomics, 54,* 690–699.

Payne, S. J. (1991). A descriptive study of mental models. *Behaviour & Information Technology, 10,* 3–21.

Peffer, T., Pritoni, M., Meier, A., Aragon, C., & Perry, D. (2011). How people use thermostats in homes: A review. *Building and Environment, 46,* 2529–2541.

Pierce, J., Schiano, D. J., & Paulos, E. (2010, April). *Home, habits, and energy: Examining domestic interactions and energy consumption.* Proceedings of the SIGCHI Conference on Human Factors in Computing Systems. ACM, 1985–1994

Revell, K., & Stanton, N. (2016). Change the mental model, change the behavior: Using interface design to promote appropriate energy consuming behavior in the home. In *Advances in ergonomics in design* (pp. 769–778). Cham, Switzerland: Springer International Publishing.

Revell, K. M. A., & Stanton, N. (2017). When energy saving advice leads to more, rather than less, consumption. International Journal of Sustainable Energy, 36, 1–19.

Revell, K. M. A., & Stanton, N. A. (2012). Models of models: Filtering and bias rings in depiction of knowledge structures and their implications for design. *Ergonomics, 55,* 1073–1092.

Revell, K. M. A., & Stanton, N. A. (2014). Case studies of mental models in home heat control: Searching for feedback, valve, timer and switch theories. *Applied Ergonomics, 43,* 363–378.

Revell, K. M. A., & Stanton, N. A. (2016). Mind the gap: Deriving a compatible user mental model of the home heating system to encourage sustainable behaviour. *Applied Ergonomics, 57,* 48–61.

Richardson, M., & Ball, L. J. (2009). Internal representations, external representations and ergonomics: Toward a theoretical integration. *Theoretical Issues in Ergonomics Science, 10,* 335–376.

Sauer, J., Schmeink, C., & Wastell, D. G. (2007). Feedback quality and environmentally friendly use of domestic central heating systems. *Ergonomics, 50,* 795–813.

Sauer, J., Wastell, D. G., & Schmeink, C. (2009). Designing for the home: A comparative study of support aids for central heating systems. *Applied Ergonomics, 40,* 165–174.

Sauer, J., Wiese, B. S., & Rüttinger, B. (2002). Improving ecological performance of electrical consumer products: The role of design-based measures and user variables. *Applied Ergonomics, 33,* 297–307.

Sauer, J., Wiese, B. S., & Rüttinger, B. (2003). Designing low-complexity electrical consumer products for ecological use. *Applied Ergonomics, 34,* 521–531.

Sauer, J., Wiese, B. S., & Rüttinger, B. (2004). Ecological performance of electrical consumer products: The influence of automation and information-based measures. *Applied Ergonomics, 35,* 37–47.

Shipworth, M., Firth, S. K., Gentry, M. I., Wright, A. J., Shipworth, D. T., & Lomas, K. J. (2009). Central heating thermostat settings and timing: Building demographics. *Building Research and Information, 38,* 50–69.

Shove, E. (2007). *The design of everyday life.* Oxford/New York: Berg.

Thatcher, A. (2013). Green ergonomics: Definition and scope. *Ergonomics, 56,* 389–398.

Thatcher, A., & Yeow, P. H. (2016). A sustainable system of systems approach: A new HFE paradigm. *Ergonomics, 59,* 167–178.

The Electricity and Gas (Energy Company Obligation) Order. (2014). http://www.legislation.gov.uk/uksi/2014/3219/pdfs/uksi_20143219_en.pdf. Accessed 20 May 2017.

Tversky, A., & Kahneman, D. (1974). Judgment under uncertainty: Heuristics and biases. *Science, 185*(4157), 1124–1131.

Vastamäki, R., Sinkkonen, I., & Leinonen, C. (2005). A behavioural model of temperature controller usage and energy saving. *Personal and Ubiquitous Computing, 9,* 250–259.

Wickens, C. D., Hollands, J. G., Banbury, S., & Parasuraman, R. (2015). *Engineering psychology & human performance.* Abingdon, Oxford: Psychology Press.

Wiese, B. S., Sauer, J., & Rüttinger, B. (2004). Consumers' use of written product information. *Ergonomics, 47,* 1180–1194.

Williges, R. C. (1987). The society's lecture 1987 the use of models in human-computer interface design. *Ergonomics, 30,* 491–502.

Wilson, J. R., & Rutherford, A. (1989). Mental models: Theory and application in human factors. *Human Factors, 31,* 617–634.

CHAPTER 6

Human Factors and Ergonomics in the Individual Adoption and Use of Electric Vehicles

Thomas Franke, Franziska Schmalfuß, and Nadine Rauh

THE USER FACTOR IN SUSTAINABLE ELECTRIFICATION OF TRANSPORT

The electrification of individual road transport is one of the greatest transformations in the field of sustainable development (Capros, Tasios, De Vita, Mantzos, & Paroussos, 2012; McCollum, Krey, Kolp, Nagai, & Riahi, 2014), with the potential to address many sustainability challenges, such as decarbonisation, global warming, and air pollution (e.g. Hawkins, Gausen, & Strømman, 2012; Pietzcker et al., 2014). As in many systems, however, the ultimate sustainability effect is largely dependent on the user factor, namely, (a) whether potential customers accept sustainable vehicle layouts and (b) whether users interact with the system

T. Franke (✉)
Engineering Psychology and Cognitive Ergonomics, Institute for Multimedia and Interactive Systems, University of Lübeck, Lübeck, Germany

F. Schmalfuß • N. Rauh
Cognitive and Engineering Psychology, Technische Universität Chemnitz, Chemnitz, Germany

© The Author(s) 2018
A. Thatcher, P. H. P. Yeow (eds.), *Ergonomics and Human Factors for a Sustainable Future*, https://doi.org/10.1007/978-981-10-8072-2_6

135

in an optimal way—that is, support the sustainability potential of the system with appropriate usage behaviour. In the present chapter, we focus on one core issue of this sustainability challenge: the individual adoption and use of electric vehicles for individual transport.

In order to realise the full potential of sustainable individual road transport, two objectives must be achieved: (1) the proportion of renewable energy used for propulsion as well as for production needs to be maximised and (2) energy efficiency as well as the protection of natural resources during production and usage need to be maximised. For the case of individual mobility, this translates into four user-related topics that are relevant to maximising the sustainability of electrification:

1. Customers have to accept (i.e. prefer) sustainable vehicle layouts.
2. Users have to use the valuable battery resources efficiently (i.e. use the full available battery range) with an optimal user experience.
3. Users have to charge the vehicle in an optimal way.
4. Users have to optimise general energy efficiency in usage (i.e. apply ecodriving strategies).

These four user-related challenges provide the structure for the following chapter. The focus lies on battery electric vehicles (BEVs) as the prototype class of electric vehicles (EVs) that can also be assumed to be the most prominent electric vehicle type in the future electric vehicle market. However, in some sections exemplary human factors-related findings regarding hybrid electric vehicles (HEVs), plugin hybrid electric vehicles (PHEVs), and extended range electric vehicles (EREVs) are added.

Acceptance of BEVs

Although the BEV market has recently expanded, the market share of BEVs is still relatively low in many countries (Sierzchula, Bakker, Maat, & van Wee, 2014). Some years ago, researchers in several countries started to intensify their investigation of the potential market share of BEVs, including Australia (Higgins, Paevere, Gardner, & Quezada, 2012), Japan (Kuwano, Tsukai, & Matsubara, 2012), Canada (Ewing & Sarigöllü, 1998), the USA (Hidrue, Parsons, Kempton, & Gardner, 2011), and Germany (Lieven, Mühlmeier, Henkel, & Waller, 2011). Various authors concluded that the demand was weak (e.g. Achtnicht, Bühler, & Hermeling, 2012) and BEVs were 'not fully competitive'

(Dagsvik, Wennemo, Wetterwald, & Aaberge, 2002). To date, however, the global BEV stock has increased steadily (International Energy Agency, 2016). The availability of charging infrastructure and financial incentives positively correlated with the increase of BEV market shares but are not the only factors influencing BEV adoption (e.g. Sierzchula et al., 2014).

Drivers and Barriers for BEV Acceptance

In BEV acceptance research, BEV adoption *intention* is often in the focus of research due to the relatively low distribution of BEVs and the resulting low possibility of investigating actual behaviour. Thereby, intention is often operationalised as the willingness to pay for (e.g. Hidrue et al., 2011), purchase (e.g. Barth, Jugert, & Fritsche, 2016; Bühler, Cocron, Neumann, Franke, & Krems, 2014; Bühler, Franke, et al., 2014; Peters & Dütschke, 2014; Schmalfuß, Mühl, & Krems, 2017), and/or use a BEV (e.g. Barth et al., 2016; Peters & Dütschke, 2014).

Several concerns of potential consumers including limited range, high costs, limited charging infrastructure, and lengthy charging times (e.g. Egbue & Long, 2012; Hidrue et al., 2011; Ziegler, 2012) present potential reasons for the relatively low acceptance intention of BEVs in many countries. However, Schuitema, Anable, Skippon, and Kinnear (2013) showed that not only do instrumental attributes, such as limited range, determine adoption intention but that also symbolic (i.e. characteristics that reflect driver's identity) and hedonic attributes (i.e. characteristics that evoke emotions) have an influence on adoption intention of BEVs and PHEVs.

Other authors have focused more on psychological variables such as norms (personal and social), perceived behavioural control or environmental motives for explaining BEV acceptance, and confirmed the influence of these variables (e.g. Barth et al., 2016; Klöckner, 2014; Moons & De Pelsmacker, 2012; Peters & Dütschke, 2014). Thereby, the Theory of Planned Behaviour (TPB, Ajzen, 1991) was often chosen, extended (e.g. Moons & De Pelsmacker, 2012), and/or combined with other theories such as diffusion of innovation (DOI—Rogers, 2010; see Peters & Dütschke, 2014) or norm-activation theory (NAT—Schwartz & Howard, 1981; see Klöckner, 2014). Apart from that, the Unified Theory of Acceptance and Use of Technology (UTAUT, Venkatesh, Morris, Davis, & Davis, 2003), that combines several theories including TPB, has been applied for HEVs (e.g. Riga, 2015). In a comprehensive review, Rezvani, Jansson, and Bodin (2015) identified diverse categories of factors associated with BEV adoption: technical (e.g. instrumental, functional BEV attributes), contextual (e.g. policy, charging infrastructure), cost (e.g. purchase price, fuel costs), as well as individual and social factors (e.g. knowledge, perceived behavioural control, emotions, symbolic meaning of BEVs). None of the reviewed studies had implemented all factors. Nevertheless, it seems that many people who adopt a BEV highly value BEV-specific attributes such as environmental benefits and low noise emission, appreciate the symbolic meaning, derive enjoyment from the evoked emotions, and evaluate the given range, performance, and charging requirements as compatible with their daily routine and needs. Furthermore, BEV adopters possess the required resources (e.g. cost-related, infrastructure-related) and are most likely supported by their social environment. Right now, industry and stakeholders are reducing barriers such as limited range, limited charging infrastructure, and the uncertainty connected with an innovative technology. Examples such as Tesla and the Opel Ampera-e (i.e. Chevrolet Bolt) give a first impression of the future of BEVs. In the near future, when the barriers have been overcome, BEV acceptance might rely more on perceived social factors.

Strategies for enhancing BEV acceptance should focus, of course, on reducing actual barriers due to limited resources via prolonging available range or providing additional and faster charging opportunities. Yet, stakeholders should also consider potential negative effects from a sustainability perspective, such as the ecological footprint of large batteries (McManus, 2012; Yuan, Li, Gou, & Dong, 2015), when deciding which strategies to implement. That is, strategies should place considerable emphasis on reducing *psychological* barriers based on false conceptions regarding BEVs.

Practical Experience and BEV Acceptance

Practical experience with a BEV has been repeatedly argued to be important in overcoming prejudices and convincing people that BEVs are fun and convenient vehicles (Bakker & Trip, 2013; Burgess, King, Harris, & Lewis, 2013; Ozaki & Sevastyanova, 2011; Rezvani et al., 2015). As many people are still unfamiliar with BEVs, and knowledge concerning performance, technology and specific aspects such as charging is limited, providing experience seems to be a potential strategy for increasing BEV acceptance. Empirical evidence of the effect of BEV experience exists

(Bühler, Cocron, et al., 2014; Nayum, Klöckner, & Mehmetoglu, 2016; Peters & Dütschke, 2014; Schmalfuß, Mühl, & Krems, 2017; Vilimek, Keinath, & Schwalm, 2012).

As some examples, BEV experienced people were willing to pay more (e.g. Larson, Viáfara, Parsons, & Elias, 2014; Schmalfuß, Mühl, et al. 2017), and had higher purchase intentions for BEVs (e.g. Peters & Dütschke, 2014) and fuel-efficient cars in general (e.g. Nayum et al., 2016) compared to BEV inexperienced respondents. Furthermore, they perceived their ability to buy an environment-friendly car (e.g. Nayum et al., 2016) and evaluated their social environment as more supportive than BEV inexperienced drivers (e.g. Peters & Dütschke, 2014). In *pre-post studies* with early adopters, changes in the evaluation of car attributes such as low noise (Cocron & Krems, 2013), perceived advantages and disadvantages (Bühler, Cocron, et al., 2014), attitudes towards BEVs (e.g. Carroll, 2010; Wikström, Hansson, & Alvfors, 2014), and purchase intentions (Turrentine, Garas, Lentz, & Woodjack, 2011) were reported after drivers had tested a BEV for three months or longer. Schmalfuß, Mühl, and Krems (2017) investigated the relevance of practical experience in a BEV adoption model including traditional factors of the Theory of Planned Behaviour (Ajzen, 1991) and the evaluation of BEV attributes. Experience proved to have positive effects on the attitudes towards BEVs and the willingness to purchase. Providing relevant experience, therefore, should reside at the focus of any strategy (i.e. green ergonomics intervention; Thatcher, 2013) aimed at increasing BEV acceptance.

The Role of Range in BEV Adoption

In some BEV adoption models (e.g. Noppers, Keizer, Bolderdijk, & Steg, 2014; Schuitema et al., 2013), range is explicitly considered as one instrumental attribute. The evaluation of instrumental BEV attributes proved to be a relevant factor for the prediction of BEV purchase intention within these studies. Range as a single attribute, however, has not been a separately influencing factor in models that aim to explain BEV adoption.

Stated preference studies (e.g. Daziano, 2013; Dimitropoulos, Rietveld, & van Ommeren, 2013; Jensen, Cherchi, & Mabit, 2013) have examined the role of range in vehicle choice in greater depth. A typical finding in this field of research is that range preferences of car drivers are typically markedly above their actual everyday range needs (e.g. Bunzeck, Feenstra, & Paukovic, 2011; Giffi, Vitale Jr., Drew, Kuboshima, & Sase, 2011).

Additionally, Franke and Krems (2013c) identified experience as an important variable that can lead to reduced range preferences (i.e. smaller preferred range values). Franke, Schneidereit, Günther, and Krems (2015) reported similar findings referring to EREVs; potential customers with greater prior BEV experience tend to prefer smaller battery ranges of an EREV indicating a decreasing disparity between range needs and range preferences with increasing practical experience with limited range.

INTERACTION WITH BEV RANGE

The available driving range (in combination with recharging duration) is a key difference between BEVs and conventional combustion vehicles. Surprisingly, there is still limited research with a psychological or human factors and ergonomics background reporting data on actual user-range interaction in daily BEV usage. When considering the broader literature on BEVs, range anxiety (i.e. users' fear to strand with an empty battery; Tate, Harpster, & Savagian, 2008) is a highly prominent topic and one could conclude that interaction with BEV range leads to frequent stressful encounters.

Research shows, however, that experiencing range stress is relatively seldom in everyday usage of BEVs, with around one stressful range situation per month occurring for users driving approximately 38 km per day in a metropolitan usage setting with home charging and some public charging opportunities (Franke & Krems, 2013a; Franke, Neumann, Bühler, Cocron, & Krems, 2012). Yet, there is considerable evidence that BEV drivers adopt certain range safety buffers in their daily BEV usage and therefore *avoid* stressful situations. Hence, the daily interaction with range is rather characterised by the avoidance and not the experience of range stress. This pattern constitutes the basis for the following model.

Adaptive Control of Range Resources

The adaptive control of range resources (ACOR) model (see, e.g. Franke & Krems, 2013a) integrates conceptions from several control theoretic models on stress (Hancock & Warm, 1989; Lazarus & Folkman, 1984) and driver behaviour (Fuller, 2005; Summala, 2007). The basic proposition is that whenever users interact with limited resources they perceive the currently available range resource buffer (i.e. available versus required mobility resources) and compare it with their preferred range safety buffer (their individual comfortable range; similar to the notion of safety margins and comfort zones in the field of safe driving; Summala, 2007).

The model further postulates, resembling the stress model of Lazarus and Folkman (1984), that this individual comfortable range is dependent on factors like personality (e.g. control beliefs; Franke & Krems, 2013a) and coping skills (e.g. range competence; Franke & Krems, 2013a). The more critical drivers appraise the discrepancy between available and preferred range safety buffer, the more likely they will experience range stress (see also Rauh, Franke, & Krems, 2015a) as well as adopt coping strategies to resolve the situation (e.g. charge the vehicle or intensify ecodriving efforts). Consequently, similar to general stress, the experience of range stress is highly subjective; that is, different drivers can have a highly diverse experience of objectively identical range situations. Comfortable range (i.e. the user's individual range comfort zone) constitutes a key variable in this respect.

The Individual Range Comfort Zone

Each BEV is characterised by its technical range—the range assessed with a standardised driving cycle (Franke, Neumann, et al., 2012). However, similar to the discrepancy between technical and real-world fuel efficiencies of conventional combustion vehicles (Duarte, Gonçalves, & Farias, 2016; Pathak, Sood, Singh, & Channiwala, 2016), it is the psychological range (the actual usable range) that constitutes the truly meaningful range figure for users. The ACOR model proposes three psychological range levels: competent range (the range a driver can achieve under optimal conditions), performant range (the range that is available/displayed with a full battery under everyday conditions), and comfortable range (the range that users are actually comfortable to use).

While competent range should be particularly dependent on drivers' ecodriving skills, performant range should be particularly dependent on individual driving style (Franke & Krems, 2013a) and drivers' ecodriving motivation (i.e. apart from environmental conditions), while comfortable range relates to drivers' individual stress resistance. From a green ergonomics perspective (Thatcher, 2013), a high competent range can be supported by, for instance, a tutor system that helps drivers to develop skills to control the energy consumption. Performant range can be supported by gamified systems that increase ecodriving motivation in addition to changes in interface design that help reduce the workload for drivers' continuous energy management. Finally, highly trustworthy range information user interfaces could be one possibility to enhance comfortable range (Franke, Trantow, et al., 2015). Research has also shown that comfortable range increases with experience (Franke, Cocron, Bühler, Neumann, & Krems, 2012; Franke, Günther, Trantow, Rauh, & Krems, 2015), and there is some indication that active exploration of range can amplify this effect (Franke, Cocron, et al., 2012). Hence, there is a considerable potential to help users develop a high actually usable range with systems that support active exploration.

Range Stress in BEV Usage

Although research indicates that range stress is not experienced frequently in daily BEV usage, the general phenomenon of range anxiety has received considerable attention in the literature. Since it was first featured in the literature (Tate et al., 2008), the term appeared often both in media coverage (Almasy, 2010; BBC, 2013; Seeking Alpha, 2013) and in scientific literature (e.g. Egbue & Long, 2012; Neubauer & Wood, 2014; Zhang, Wang, Kobayashi, & Shirai, 2012). It has become a widely used term due to its intuitiveness (Nilsson, 2011a), yet a good psychological foundation relating to the concept and a comprehensive, empirically based understanding of range anxiety is rare to find (Nilsson, 2011b). Most of the scientific literature refers to range anxiety as the fear of becoming stranded with a BEV (Nilsson, 2011b; see also Eberle & von Helmolt, 2010; Tate et al., 2008). However, several studies examining the phenomenon of range anxiety in greater depth suggest that there are more facets that should be considered, like experienced range discomfort, range workload, range stress, avoidance behaviours, or even anticipation of such states (e.g. Franke, Rauh, Günther, Trantow, & Krems, 2016; Franke, Rauh & Krems, 2015; Rauh et al., 2015a).

One highly prototypic and crucial facet of range anxiety is the users' experience when driving a BEV in critical range situations. To refer to this facet of range anxiety, the concept range stress has been developed (Rauh et al., 2015a). Range stress is defined as a stressful experience of a range situation where the range resources, in addition to the personal resources available to effectively manage the situation, are subjectively estimated to be insufficient (primary and secondary appraisal, according to Lazarus' Transactional model of stress and coping; Lazarus, 1995). Based on similar classifications in the fields of general anxiety/stress symptoms (e.g. Clark & Beck, 2011) and range anxiety (Nilsson, 2011b), it has been argued that the stress experienced in a critical range situation can typically be described on four different levels: (1) a *cognitive* level (i.e. negative cognitions associated with range like concerns about running out of energy and being unable to reach the destination), (2) an *emotional* level (i.e. changes in affect associated with a range situation, like feelings of nervousness or even fear), (3) a *behavioural* level (i.e. behaviours such as changing driving style to save energy or frequent checking of relevant displays, such as the range and navigation displays), and (4) a *physiological* level (i.e. increased arousal such as an increased heart rate).

As stated above, actual studies demonstrated that BEV drivers do not frequently experience range stress (Franke, Neumann, et al., 2012). Nevertheless, it was shown to be negatively related to the likelihood of purchasing a limited-range BEV (Franke & Krems, 2013c), range satisfaction (Franke, Rauh, et al., 2016), and users' confidence in using a BEV for longer trips (Carroll, 2010). Parallel to this, it has also been suggested that range stress should be considered as one potential barrier to the widespread adoption of BEVs (e.g. Luettringhaus & Nilsson, 2012; Nilsson, 2011a), similar to the discussions regarding BEVs' limited range as an important purchase barrier (e.g. Bühler, Cocron, et al., 2014; Egbue & Long, 2012).

Considerable resources have been invested in finding ways to reduce range stress in BEV usage (e.g. Lundström & Bogdan, 2012) and to examine factors, which may be responsible for variance in range stress (e.g. Franke, Rauh, et al., 2015; Rauh et al., 2015a). Various factors that could encourage a better adaptation to BEVs limited range were examined and found to have the potential to reduce range stress. On the one side, individual differences like personality traits (e.g. internal control beliefs in dealing with technology, emotional stability), domain-specific knowledge (e.g. system knowledge, higher route familiarity), trust in the range estimation system, as well as drivers' subjective range competence and practical driving experience with BEVs (e.g. experience over several years or even the experience of one trip with a critical ratio of available driving range and remaining trip length) could all alleviate range stress (e.g. Burgess et al., 2013; Franke, Cocron, et al., 2012; Franke, Rauh, et al., 2015, 2016; Nilsson, 2011a; Rauh et al., 2015a; Rauh, Franke, & Krems, 2017). On the other side, system features like support through advanced information systems, the availability of fast charging stations en route, or advanced range displays have been argued as viable options to further reduce range stress (e.g. Jung, Sirkin, Gür, & Steinert, 2015; Neaimeh, Hill, Hübner, & Blythe, 2013; Philipsen, Schmidt, & Ziefle, 2015; Rauh, Franke, & Krems, 2015b; Yilmaz & Krein, 2013; Zhang et al., 2012). Overall, the research on user-range interaction shows that several factors can lead to a suboptimal utilisation of range resources and result in experienced discomfort or even range stress. Addressing these factors is important for developing sustainable electric mobility systems to optimise the system's design and to develop effective green ergonomics interventions.

CHARGING BEHAVIOUR IN BEV USAGE

Aside from a sustainability-oriented utilisation of the available range resources, an optimal replenishment of mobility resources (i.e. energy resources) is also crucial. EVs can only achieve a maximum sustainability effect if charged on excess energy from renewable resources and consequently not represent a risk to grid stability (Chan, Jian, & Tu, 2014; Jochem, Babrowski, & Fichtner, 2015). Supporting appropriate charging behaviour is therefore a central task in a user-centred design of sustainable electric mobility systems. Initial research has shown that users adopt diverse charging routines when using a BEV (Daina, Polak, & Sivakumar, 2015; Labeye, Hugot, Brusque, & Regan, 2016), and often charge despite having substantial remaining range (Carroll, 2010; Franke, Neumann, et al., 2012). A reason for this pattern can be user diversity in system interaction styles.

Individual Differences in Charging Styles

Similar to the notion that individual differences in driving behaviour can be explained by differences in driving styles (Elander, West, & French, 1993; Lajunen & Özkan, 2011), it has been suggested that at least part of the variance in charging behaviour can be characterised by stable differences in charging styles (Franke & Krems, 2013b). Indeed, BEV drivers typically have considerable leeway regarding their charging decisions because for many mobility profiles, there is an abundance of charging opportunities, hence no need to charge at every available charging opportunity. Yet of course there can also be cases where this is not true because of limited charging opportunities (e.g. no charging opportunity at home) or very long daily travel distances (e.g. Franke et al., 2014).

Focusing on the facet of user interaction with range resources and building upon previous research on mobile phone charging (Rahmati, Qian, & Zhong, 2007; Rahmati & Zhong, 2009), it has been proposed that users' charging decisions are either primarily based on the perceived charging necessity (if range is subjectively becoming too low) or on the availability of convenient charging opportunities. This means that users differ in the extent to which they actively interact with (e.g. monitor) the battery resource. This charging style dimension was thus labelled as user-battery interaction style (UBIS), where users with a higher UBIS interact more actively with the battery resources and primarily recharge based on charging necessity.

Research has found UBIS to be a relatively stable user characteristic (Franke & Krems, 2013b). Furthermore, results of this study have shown UBIS and comfortable range to account for approximately 35% of the variance in typical charging behaviour, that is, the charge level where users typically initiate a charging process. Related to this, users with a higher UBIS also make more efficient use of the battery resources. A higher UBIS can, however, also have negative consequences from a sustainability perspective.

Human Factors Issues in Green Charging

A higher UBIS correlates with a lower efficiency of utilising excess energy from wind (Franke & Krems, 2013b). In this specific study, a smart charging (i.e. green charging) system was implemented that aims to shift the power input to the so-called green windows with a high availability of excess energy from wind. As drivers with a higher UBIS only use charging opportunities when necessary, the system was less flexible to shift the charging process to green windows, as it had to charge more in a shorter time. The finding demonstrates that smart charging systems should also take drivers' charging styles into account to achieve sustainable charging patterns.

In recent years, research on optimised charging has rapidly increased. Smart charging systems represent a promising tool to support the integration of renewable energies into the grid and to help in grid balancing (e.g. Amoroso & Cappuccino, 2012). High user involvement via providing information about, for instance, departure times and required range at departure is supposed to maximise the potential of such systems (Isaksson & Fagerholt, 2012). User-centred research provided evidence that smart charging systems are assimilable in fleets (Pettersson, 2013) and in private contexts (e.g. Schmalfuß et al., 2015). Although experiences from field study research indicated that BEV drivers evaluate a smart charging system with high user involvement as rather acceptable and can successfully integrate it into daily behavioural routines, costs for the user (such as reduced flexibility and a higher effort to plan charging processes) need to be addressed. It appears that appropriate incentives and a positive outcome of users' personal cost-benefit analysis are mandatory for widespread acceptance (Schmalfuß et al., 2015, 2016; Schmalfuß, Kreußlein, et al., 2017). Furthermore, a well-designed human-machine interface is required to adjust charging settings and obtain performance feedback regarding successful participation with smart charging and its potential rewards (Schmalfuß et al., 2016).

ECODRIVING IN EV USAGE

Apart from making full use of available range resources and recharging in an optimal way, an overarching topic of sustainable electrified road transport that concerns all types of EVs pertains to actual energy efficiency achieved in usage. There is considerable variance in the energy efficiency that drivers can achieve with EVs (see, e.g. Birrell, Taylor, McGordon, Son, & Jennings, 2014; Franke, Arend, McIlroy, & Stanton, 2016; Knowles, Scott, & Baglee, 2012; Neumann, Franke, Cocron, Bühler, & Krems, 2015), making this topic key to achieving an optimal sustainability effect. Moreover, research suggests that drivers have a particularly high impact on energy consumption in EVs compared to conventional combustion vehicles (McIlroy, Stanton, & Harvey, 2014; Neumann et al., 2015; Walsh, Carroll, Eastlake, & Blythe, 2010). Ecodriving is the term that has evolved to describe this driver impact, that is, all the influences users have on the real-world energy efficiency of a road vehicle (see, e.g. Barkenbus, 2010; Jamson, Brouwer & Seewald, 2015; Sivak & Schoettle, 2012; Stillwater & Kurani, 2013; Young, Birrell, & Stanton, 2011). Ecodriving therefore does not only involve specific behavioural strategies while driving (i.e. operational ecodriving strategies; Sivak & Schoettle, 2012) but also constitutes behaviours like optimising tyre pressure or route choice (i.e. strategic and tactical ecodriving measures; Sivak & Schoettle, 2012). Nevertheless, in the following, we focus on operational ecodriving strategies applied while driving. This is because the highly dynamic and partly complex behavioural adaptation that drivers need to implement such strategies entails particular human factor challenges.

A Human Factors Perspective on Energy Flows in EVs

While certain ecodriving strategies are transferable from conventional vehicles (such as avoiding high speeds to reduce air resistance), EVs have specific energy dynamics (e.g. powertrain efficiency characteristics; Kuriyama, Yamamoto, & Miyatake, 2010) that make behavioural adaptation necessary to achieve high energy efficiency. Most important in this respect, from a psychological point of view, is bidirectional energy flow. While there is only one relevant direction of energy flow in conventional combustion vehicles (i.e. energy storage to wheels), there is a bidirectional flow in EVs (from energy storage to wheels and back) because the regenerative braking system directly recharges the battery in deceleration manoeuvres. Typically, this system requires behavioural adaptation with regard to general driving behaviour (i.e. general longitudinal control, e.g. Cocron et al., 2013; Helmbrecht, Olaverri-Monreal, Bengler, Vilimek, & Keinath, 2014) and is also highly relevant to ecodriving behaviour (e.g. Labeye et al. 2016; Schmitz, Maag, Jagiellowicz, & Hanig, 2013).

A key question from the ergonomics perspective is how to best configure regenerative braking, that is, how strong it should already act once the driver releases the foot from the accelerator pedal, or if it should only become activated once the braking pedal has been touched. Essentially this is a question of which powertrain state should represent the 'neutral state' (i.e. no foot on pedals) in EV driving. In terms of drivers' comfort, there is some evidence of a preference for 'one-pedal driving' (Cocron et al., 2013; Helmbrecht et al., 2014; Schmitz et al., 2013); this involves a strong electric deceleration on the accelerator pedal that enables driving without using the brake pedal in many driving situations. This design can also help to decrease the energy-wasting usage of the mechanical brakes (Schmitz et al., 2013) because it reduces drivers' reaction and motor time for switching from acceleration to deceleration (particularly if drivers are attentive to anticipating the longitudinal dynamics of preceding traffic) and acts as a kind of assistance system that pushes drivers towards deceleration as a default action in ambiguous driving situations. However, this is still not the most sustainable solution with regard to energy efficiency. Regenerative braking still entails high energy losses (with only a fraction of kinetic energy being recovered in the batteries) with this system layout tending to cause additional (i.e. avoidable) deceleration and acceleration events (i.e. from signal detection perspective the system induces 'false alarms' in deceleration decisions).

The most energy-efficient state in any EV is, however, a neutral glide in which the vehicle is coasting in the absence of energy transformations. Indeed, findings on BEV driver behaviour in relation to actual energy consumption support this notion (Neumann et al., 2015). This research showed that both more accelerations and a higher amount of recaptured energy are related to lower energy efficiency. It appears essential therefore to make 'neutral glide' a default state by designing pedal configurations accordingly to produce affordances that push (i.e. nudge; Lehner, Mont, & Heiskanen, 2016) drivers towards switching to neutral glide as often as possible, instead of switching to electric deceleration in each non-acceleration manoeuvre (see also Arend & Franke, 2016). For such a system design to be as effective as possible, however, it is also important that drivers develop accurate mental models and appropriate behavioural strategies.

Drivers' Adoption of Ecodriving Strategies

These already relatively complex considerations regarding only one specific aspect of the energy user-interface design (the designing pedal configurations) indicate that achieving optimal energy-efficient behaviour in EV driving can be challenging for users and users need to go through considerable adaptation processes. Research has found such adaptation effects in ecodriving-related variables, like a higher amount of reported ecodriving strategies demonstrated by drivers having acquired practical experience driving a BEV (Neumann et al., 2015). Yet an unguided learning process does not necessarily lead to optimal adaptation.

Hybrid electric vehicles (HEVs) constitute the most complex electric powertrain because of the highly dynamic interplay between the powertrain components (the combustion and electric engine versus battery), and the central role of the bidirectional energy flow (this being highly dynamic and prominent from the drivers' perspective). A central question arising from these complexities is what specifically makes ecodriving in HEVs challenging for drivers and how should we design ecodriving support systems to resolve these challenges? One of the first human factors studies in this field showed that ecodriving motivation and technical system knowledge could account for substantial variance in achieved energy efficiency in HEV drivers who had at least some interest in ecodriving (Franke, Arend, et al., 2016). Moreover, there was some indication that particularly drivers with relatively low technical system knowledge developed detrimental ecodriving strategies (e.g. actively using electric energy for driving manoeuvres like acceleration) and false mental models (such as being of the opinion that active utilisation of electric energy is energy efficient). In a similar fashion, a qualitative study of PHEV drivers' ecodriving experiences (Stillwater & Kurani, 2013) concluded that drivers might need clear information concerning the vehicle operation to avoid confusion and correctly learn ecodriving methods. Hence, there is considerable evidence suggesting that drivers need specific support via an appropriate energy interface and optimal feedback to acquire optimal ecodriving strategies in HEV/PHEV driving.

An Outlook on Future Research Topics

When taking into account the available published research on user interaction with EVs from the perspective of human factors and ergonomics, we conclude that research is still at a relatively early stage. It has to be acknowledged here, that this is partly also due to the fact that a considerable part of human factors research in this dynamic field is conducted within companies and is not published. With regard to public available knowledge, particularly more research is required that examines actual user interaction with EVs to identify further strategies for user-interface improvements to support a user-system interaction that is beneficial from a sustainability standpoint.

Firstly, with regard to acceptance, factors that influence BEV acceptance are diverse. Perhaps most relevant, in terms of human factors, is the factor of experience. Further research should continue to explore which specific kind of experience is helpful, and which false conceptions particularly contribute to reduced BEV purchase intentions. Similarly, the field of optimal range setups, from a customer and sustainability perspective (e.g. Weiss et al., 2016), deserves further attention, as does the relationship of user needs and the sustainability potential of EREVs (e.g. Derollepot et al., 2014).

Secondly, in reference to range, research has achieved an initial understanding of user-interaction patterns, such as the recognition that comfortable range is a central variable for users' interaction with range resources. Research should particularly focus on tests of intervention strategies (e.g. user-interface design, tutor systems, gamification approaches) specifically designed to increase benchmark variables such as comfortable range (i.e. range stress resistance) and provide users with a more usable range given a certain technical range setup. In addition to this, users' coping strategies need to be better understood in order to develop assistance systems that help users to, for instance, continuously adapt their driving behaviour to finish long-distance trips with a certain available range; this would enable drivers, for example, to calculate the tolerable average energy consumption for certain trip segments.

Thirdly, supporting green charging behaviour is a central challenge in realising EVs' sustainable potential with regard to the more efficient utilisation of renewable energy. Further research is required on this subject to advance knowledge on effective incentive strategies and user-interface designs that may also need to be adaptive to a user's preferred charging style. Thereby, the level of automation and user involvement are key elements that substantially influence the system design.

Fourthly, initial research shows particular challenges exist for optimal ecodriving in EVs. Research needs to continue to examine ways to reduce drivers' workload in optimally controlling energy flows and supporting the development of effective ecodriving strategies. Particularly because of the highly dynamic nature of ecodriving, it would be beneficial for research to further explore the potential of multimodal representation of energy dynamics (e.g. haptic feedback opposed to only visual feedback to reduce detrimental distraction effects) and optimal system states (see also, e.g. Birrell, Young, & Weldon, 2013). Research should also be concerned with the topic of meaningful aggregation and representation of energy flows, such as finding optimal averaging intervals for displaying average energy consumption and augmenting energy consumption information with environmental context information like the elevation profile. Finally, ecodriving support systems should also help to balance different driving goals like time, safety, and energy efficiency.

In conclusion, human factors and ergonomics for sustainable development should take a central role in guiding the advancement of electric mobility systems. While electrification is a key measure to increase the sustainability of road transport, the user factor ultimately determines the actual sustainability effect achieved. Considerable adaptation processes are required to optimally transfer from conventional to electric mobility, and system design should facilitate this to as great a degree as possible. Moreover, in extended daily usage, it appears essential to improve the usability of user-energy interaction—that is, enhanced user-resource interaction. To this end, a comprehensive understanding of user-interaction patterns and related user experience qualities is of paramount importance. The field of green ergonomics offers useful tools to contribute to this challenge. Finally, because of the focus of the present chapter, there are of course further key factors regarding the sustainability of EVs that are not addressed such as the various facets of life-cycle sustainability and strategies for battery recycling at end of life (e.g. Hawkins et al., 2012). These also have to be considered in the big picture of sustainable electric vehicles.

REFERENCES

Achtnicht, M., Bühler, G., & Hermeling, C. (2012). The impact of fuel availability on demand for alternative-fuel vehicles. *Transportation Research Part D: Transport and Environment, 17,* 262–269.

Ajzen, I. (1991). The theory of planned behavior. *Organizational Behavior and Human Decision Processes, 50,* 179–211.

Almasy, S. (2010, October 20). *The new fear: Electric car 'RA'.* Retrieved February 14, 2017, from http://edition.cnn.com/2010/US/10/18/ev.charging.stations/index.html

Amoroso, F. A., & Cappuccino, G. (2012). Advantages of efficiency-aware smart charging strategies for PEVs. *Energy Conversion and Management, 54,* 1–6.

Arend, M. G., & Franke, T. (2016). The role of interaction patterns with hybrid electric vehicle eco-features for drivers' eco-driving performance. *Human Factors: The Journal of the Human Factors and Ergonomics Society.* https://doi.org/10.1177/0018720816670819

Bakker, S., & Trip, J. J. (2013). Policy options to support the adoption of electric vehicles in the urban environment. *Transportation Research Part D: Transport and Environment, 25,* 18–23.

Barkenbus, J. N. (2010). Eco-driving: An overlooked climate change initiative. *Energy Policy, 38,* 762–769.

Barth, M., Jugert, P., & Fritsche, I. (2016). Still underdetected – Social norms and collective efficacy predict the acceptance of electric vehicles in Germany. *Transportation Research Part F: Traffic Psychology and Behaviour, 37,* 64–77.

BBC. (2013, March 7). *Electric car maker on 'RA' among potential buyers.* Retrieved February 14, 2017, from http://www.bbc.co.uk/news/business-21695901

Birrell, S. A., Taylor, J., McGordon, A., Son, J., & Jennings, P. (2014). Analysis of three independent real-world driving studies: A data driven and expert analysis approach to determining parameters affecting fuel economy. *Transportation Research Part D: Transport and Environment, 33,* 74–86.

Birrell, S. A., Young, M. S., & Weldon, A. M. (2013). Vibrotactile pedals: Provision of haptic feedback to support economical driving. *Ergonomics, 56,* 282–292.

Bühler, F., Cocron, P., Neumann, I., Franke, T., & Krems, J. F. (2014). Is EV experience related to EV acceptance? Results from a German field study. *Transportation Research Part F: Traffic Psychology and Behaviour, 25*(A), 34–49.

Bühler, F., Franke, T., Schleinitz, K., Cocron, P., Neumann, I., Ischebeck, M., & Krems, J. F. (2014). Driving an EV with no opportunity to charge at home – Is this acceptable? In D. de Waard, K. Brookhuis, R. Wiczorek, F. di Nocera, R. Brouwer, P. Barham, C. Weikert, A. Kluge, W. Gerbino, & A. Toffetti (Eds.), *Proceedings of the human factors and ergonomics society Europe chapter 2013 annual conference*, pp. 369–379. Retrieved February 14, 2017, from http://www.hfes-europe.org/largefiles/proceedingshfeseurope2013.pdf

Bunzeck, I., Feenstra, C. F. J., & Paukovic, M. (2011). *Preferences of potential users of electric cars related to charging – A survey in eight EU countries.* Retrieved February 14, 2017, from http://www.d-incert.nl/wp-content/uploads/2011/05/rapportage_ECN.pdf

Burgess, M., King, N., Harris, M., & Lewis, E. (2013). Electric vehicle drivers' reported interactions with the public: Driving stereotype change? *Transportation Research Part F: Traffic Psychology and Behaviour, 17*, 33–44.

Capros, R., Tasios, N., De Vita, A., Mantzos, L., & Paroussos, L. (2012). Transformations of the energy system in the context of the decarbonisation of the EU economy in the time horizon to 2050. *Energy Strategy Reviews, 1*, 85–96.

Carroll, S. (2010). *The smart move trial: Description and initial results.* London: Cenex. Retrieved February 14, 2017, from http://www.cenex.co.uk/wp-content/uploads/2013/06/2010-03-23-Smart-move-trial-report-v3-Compatibility-mode-11.pdf

Chan, C. C., Jian, L., & Tu, D. (2014). Smart charging of electric vehicles – Integration of energy and information. *IET Electrical Systems in Transportation, 4*, 89–96.

Clark, D. A., & Beck, A. T. (2011). *Cognitive therapy of anxiety disorders – Science and practice.* New York: The Guilford Press.

Cocron, P., Bühler, F., Franke, T., Neumann, I., Dielmann, B., & Krems, J. F. (2013). Energy recapture through deceleration – Regenerative braking in electric vehicles from a user perspective. *Ergonomics, 56*, 1203–1215.

Cocron, P., & Krems, J. F. (2013). Silent driving in urban traffic – Results of two field studies on electric vehicles. *Accident Analysis and Prevention, 58*, 122–131.

Dagsvik, J. K., Wennemo, T., Wetterwald, D. G., & Aaberge, R. (2002). Potential demand for alternative fuel vehicles. *Transportation Research Part B: Methodological, 36*, 361–384.

Daina, N., Polak, J. W., & Sivakumar, A. (2015). Patent and latent predictors of electric vehicle charging behavior. *Transportation Research Record: Journal of the Transportation Research Board, 2502*, 116–123.

Daziano, R. A. (2013). Conditional-logit Bayes estimators for consumer valuation of electric vehicle driving range. *Resource and Energy Economics, 35*, 429–450.

Derollepot, R., Weiß, C., Kolli, Z., Franke, T., Trigui, R., Chlond, B., et al. (2014). *Optimizing components size of an extended range electric vehicle according to the use specifications*. In IFSTTAR proceedings of the 5th transport research arena (TRA) 2014, Paris, France. Retrieved February 14, 2017, from http://www.evrest-project.org/pdf/TRA2014_Fpaper_18329.pdf

Dimitropoulos, A., Rietveld, P., & van Ommeren, J. N. (2013). Consumer valuation of changes in driving range: A meta-analysis. *Transportation Research Part A: Policy and Practice, 55*, 27–45.

Duarte, G. O., Gonçalves, G. A., & Farias, T. L. (2016). Analysis of fuel consumption and pollutant emissions of regulated and alternative driving cycles based on real-world measurements. *Transportation Research Part D: Transport and Environment, 44*, 43–54.

Eberle, U., & von Helmolt, R. (2010). Sustainable transportation based on electric vehicle concepts: A brief overview. *Energy & Environmental Science, 3*, 689–699.

Egbue, O., & Long, S. (2012). Barriers to widespread adoption of electric vehicles: An analysis of consumer attitudes and perceptions. *Energy Policy, 48*, 717–729.

Elander, J., West, R., & French, D. (1993). Behavioral correlates of individual differences in road-traffic crash risk: An examination of methods and findings. *Psychological Bulletin, 113*, 279–294.

Ewing, G. O., & Sarigöllü, E. (1998). Car fuel-type choice under travel demand management and economic incentives. *Transportation Research Part D: Transport and Environment, 3*, 429–444.

Franke, T., Arend, M. G., McIlroy, R. C., & Stanton, N. A. (2016). Ecodriving in hybrid electric vehicles – Exploring challenges for user-energy interaction. *Applied Ergonomics, 55*, 33–45.

Franke, T., Cocron, P., Bühler, F., Neumann, I., & Krems, J. F. (2012). Adapting to the range of an electric vehicle: The relation of experience to subjectively available mobility resources. In P. Valero Mora, J. F. Pace, & L. Mendoza (Eds.), *Proceedings of the European conference on human centred design for intelligent transport systems, Valencia, Spain, June 14–15 2012* (pp. 95–103). Lyon, France: Humanist Publications.

Franke, T., Günther, M., Trantow, M., Krems, J. F., Vilimek, R., & Keinath, A. (2014). Examining user-range interaction in battery electric vehicles – A field study approach. In N. Stanton, S. Landry, G. Di Bucchianico, & A. Vallicelli (Eds.), *Advances in human aspects of transportation part II. AHFE conference* (pp. 334–344.) AHFE Conference.

Franke, T., Günther, M., Trantow, M., Rauh, N., & Krems, J. F. (2015). Range comfort zone of electric vehicle users – Concept and assessment. *IET Intelligent Transport Systems, 9*, 740–745.

Franke, T., & Krems, J. F. (2013a). Interacting with limited mobility resources: Psychological range levels in electric vehicle use. *Transportation Research Part A: Policy and Practice, 48*, 109–122.

Franke, T., & Krems, J. F. (2013b). Understanding charging behaviour of electric vehicle users. *Transportation Research Part F: Traffic Psychology and Behaviour, 21*, 75–89.

Franke, T., & Krems, J. F. (2013c). What drives range preferences in electric vehicle users? *Transport Policy, 30*, 56–62.

Franke, T., Neumann, I., Bühler, F., Cocron, P., & Krems, J. F. (2012). Experiencing range in an electric vehicle – Understanding psychological barriers. *Applied Psychology, 61*, 368–391.

Franke, T., Rauh, N., Günther, M., Trantow, M., & Krems, J. F. (2016). Which factors can protect against range stress in everyday usage of battery electric vehicles? Towards enhancing sustainability of electric mobility systems. *Human Factors, 58*, 13–26.

Franke, T., Rauh, N., & Krems, J. F. (2015). Individual differences in BEV drivers' range stress during first encounter of a critical range situation. *Applied Ergonomics, 57*, 28–35.

Franke, T., Schneidereit, T., Günther, M., & Krems, J. F. (2015). *Solving the range challenge? Range needs versus range preferences for battery electric vehicles with range extender.* In Proceedings of the EVS28, Goyang, Korea. Retrieved February 14, 2017, from http://www.evs28.org/event_file/event_file/1/pfile/Franke_et_al._RangeExtender_EVS28_0195_2.pdf

Franke, T., Trantow, M., Günther, M., Krems, J. F., Zott, V., & Keinath, A. (2015). Advancing electric vehicle range displays for enhanced user experience – The relevance of trust and adaptability. In *Proceedings of the 7th international conference on automotive user interfaces and interactive vehicular applications* (pp. 249–256). New York: ACM.

Fuller, R. (2005). Towards a general theory of driver behaviour. *Accident Analysis & Prevention, 37*, 461–472.

Giffi, C., Vitale, J. Jr., Drew, M., Kuboshima, Y., & Sase, M. (2011). *Unplugged: Electric vehicle realities versus consumer expectations.* Retrieved February 14, 2017, from https://www2.deloitte.com/content/dam/Deloitte/global/Documents/Manufacturing/gx_us_auto_DTTGlobalAutoSurvey_ElectricVehicles_100411.pdf

Hancock, P. A., & Warm, J. S. (1989). A dynamic model of stress and sustained attention. *Human Factors, 31*, 519–537.

Hawkins, T. R., Gausen, O. M., & Strømman, A. H. (2012). Environmental impacts of hybrid and electric vehicles – A review. *International Journal of Life Cycle Assessment, 17*, 997–1014.

Helmbrecht, M., Olaverri-Monreal, C., Bengler, K., Vilimek, R., & Keinath, A. (2014). How electric vehicles affect driving behavioral patterns. *IEEE Intelligent Transportation Systems, 6*(3), 22–32.

Hidrue, M. K., Parsons, G. R., Kempton, W., & Gardner, M. P. (2011). Willingness to pay for electric vehicles and their attributes. *Resource and Energy Economics, 33,* 686–705.

Higgins, A., Paevere, P., Gardner, J., & Quezada, G. (2012). Combining choice modelling and multi-criteria analysis for technology diffusion: An application to the uptake of electric vehicles. *Technological Forecasting and Social Change, 79,* 1399–1412.

International Energy Agency. (2016). *Global EV outlook 2016: Beyond one million electric cars.* Retrieved from https://www.iea.org/publications/freepublications/publication/Global_EV_Outlook_2016.pdf

Isaksson, O., & Fagerholt, A. (2012). Smart charging for electric vehicles. *Ericsson Business Review, 3,* 32–36.

Jamson, S., Brouwer, R., & Seewald, P. (2015). Technologies to support green driving [special issue]. *Transportation Research Part C: Emerging Technologies, 58*(D), 629–782.

Jensen, A. F., Cherchi, E., & Mabit, S. L. (2013). On the stability of preferences and attitudes before and after experiencing an electric vehicle. *Transportation Research Part D: Transport and Environment, 25,* 24–32.

Jochem, P., Babrowski, S., & Fichtner, W. (2015). Assessing CO_2 emissions of electric vehicles in Germany in 2030. *Transportation Research A: Policy and Practice, 78,* 68–83.

Jung, M. F., Sirkin, D., Gür, T. M., & Steinert, M. (2015). Displayed uncertainty improves driving experience and behavior: The case of range anxiety in an electric car. In *Proceedings of the 33rd annual ACM conference on human factors in computing systems 2015* (pp. 2201–2210). New York: ACM.

Klöckner, C. A. (2014). The dynamics of purchasing an electric vehicle – A prospective longitudinal study of the decision-making process. *Transportation Research Part F: Traffic Psychology and Behaviour, 24,* 103–116.

Knowles, M., Scott, H., & Baglee, D. (2012). The effect of driving style on electric vehicle performance, economy and perception. *International Journal of Electric and Hybrid Vehicles, 4,* 228–247.

Kuriyama, M., Yamamoto, S., & Miyatake, M. (2010). *Theoretical study on eco-driving technique for an electric vehicle with dynamic programming.* In 2010 international conference on electrical machines and systems, Incheon, South Korea, pp. 2026–2030.

Kuwano, M., Tsukai, M., & Matsubara, T. (2012). *Analysis on promoting factors of electric vehicles with social conformity.* Paper presented at the 13th international conference on travel behaviour research, Toronto, Canada.

Labeye, E., Hugot, M., Brusque, C., & Regan, M. A. (2016). The electric vehicle: A new driving experience involving specific skills and rules. *Transportation Research Part F: Traffic Psychology and Behaviour, 37,* 27–40.

Lajunen, T., & Özkan, T. (2011). Self-report instruments and methods. In B. E. Porter (Ed.), *Handbook of traffic psychology* (pp. 43–59). London: Elsevier.

Larson, P. D., Viáfara, J., Parsons, R. V., & Elias, A. (2014). Consumer attitudes about electric cars: Pricing analysis and policy implications. *Transportation Research Part A: Policy and Practice, 69,* 299–314.

Lazarus, R. S. (1995). Psychological stress in the workplace. In R. Crandall & P. L. Perrewé (Eds.), *Occupational stress: A handbook* (pp. 3–14). Philadelphia: Taylor & Francis.

Lazarus, R. S., & Folkman, S. (1984). *Stress, appraisal, and coping.* New York: Springer.

Lehner, M., Mont, O., & Heiskanen, E. (2016). Nudging – A promising tool for sustainable consumption behaviour? *Journal of Cleaner Production, 134*(A), 166–177.

Lieven, T., Mühlmeier, S., Henkel, S., & Waller, J. F. (2011). Who will buy electric cars? An empirical study in Germany. *Transportation Research Part D: Transport and Environment, 16,* 236–243.

Luettringhaus, H., & Nilsson, M. (2012). *ELVIRE approaches to mitigate EV driver's range anxiety – Technical paper by ELVIRE – European research project on ICT for electric vehicles.* In Proceedings of 19th ITS world congress, Vienna, Austria.

Lundström, A., & Bogdan, C. (2012). *COPE1 – Taking control over EV range.* In Adjunct proceedings of the 4th international conference on automotive user interfaces and interactive vehicular applications (AutomotiveUI'12), Portsmouth, NH, pp. 17–18.

McCollum, D., Krey, V., Kolp, P., Nagai, Y., & Riahi, K. (2014). Transport electrification: A key element for energy system transformation and climate stabilization. *Climatic Change, 123,* 651–664.

McIlroy, R. C., Stanton, N. A., & Harvey, C. (2014). Getting drivers to do the right thing: A review of the potential for safely reducing energy consumption through design. *IET Intelligent Transport Systems, 8,* 388–397.

McManus, M. C. (2012). Environmental consequences of the use of batteries in low carbon systems: The impact of battery production. *Applied Energy, 93,* 288–295.

Moons, I., & De Pelsmacker, P. (2012). Emotions as determinants of electric car usage intention. *Journal of Marketing Management, 28,* 195–237.

Nayum, A., Klöckner, C. A., & Mehmetoglu, M. (2016). Comparison of socio-psychological characteristics of conventional and battery electric car buyers. *Travel Behaviour and Society, 3,* 8–20.

Neaimeh, M., Hill, G. A., Hübner, Y., & Blythe, P. T. (2013). Routing systems to extend the driving range of electric vehicles. *IET Intelligent Transport Systems, 7,* 327–336.

Neubauer, J., & Wood, E. (2014). The impact of range anxiety and home, work-place, and public charging infrastructure on simulated battery electric vehicle lifetime utility. *Journal of Power Sources, 257,* 12–20.

Neumann, I., Franke, T., Cocron, P., Bühler, F., & Krems, J. F. (2015). Eco-driving strategies in battery electric vehicle use – How do drivers adapt over time? *IET Intelligent Transport Systems, 9,* 746–753.

Nilsson, M. (2011a). *Electric vehicles – An interview study investigating the phenomenon of range anxiety.* Retrieved February 14, 2017, from http://e-mobility-nsr.eu/fileadmin/user_upload/downloads/info-pool/report_result_interview_elvire.pdf

Nilsson, M. (2011b). *Electric vehicle: The phenomenon of range anxiety.* Retrieved from http://e-mobility-nsr.eu/fileadmin/user_upload/downloads/info-pool/the_phenomenon_of_range_anxiety_elvire.pdf

Noppers, E. H., Keizer, K., Bolderdijk, J. W., & Steg, L. (2014). The adoption of sustainable innovations: Driven by symbolic and environmental motives. *Global Environmental Change, 25,* 52–62.

Ozaki, R., & Sevastyanova, K. (2011). Going hybrid: An analysis of consumer purchase motivations. *Energy Policy, 39,* 2217–2227.

Pathak, S. K., Sood, V., Singh, Y., & Channiwala, S. A. (2016). Real world vehicle emissions: Their correlation with driving parameters. *Transportation Research Part D: Transport and Environment, 44,* 157–176.

Peters, A., & Dütschke, E. (2014). How do consumers perceive electric vehicles? A comparison of German consumer groups. *Journal of Environmental Policy & Planning, 16,* 359–377.

Pettersson, S. (2013). *ELVIIS – Final report to Göteborg Energi Forskningsstiftelse,* ICT Sweden Viktoria. Retrieved February 14, 2017, from https://www.viktoria.se/sites/default/files/pub/www.viktoria.se/upload/publications/elviis_final.pdf

Philipsen, R., Schmidt, T., & Ziefle, M. (2015). A charging place to be – Users' evaluation criteria for the positioning of fast-charging infrastructure for electro mobility. *Procedia Manufacturing, 3,* 2792–2799.

Pietzcker, R. C., Longden, T., Chen, W., Fu, S., Kriegler, E., Kyle, P., et al. (2014). Long-term transport energy demand and climate policy: Alternative visions on transport decarbonization in energy-economy models. *Energy, 64,* 95–108.

Rahmati, A., Qian, A., & Zhong, L. (2007). *Understanding human-battery interaction on mobile phones.* In Proceedings of the 9th international conference on Human computer interaction with mobile devices and services, Innsbruck, Austria, pp. 265–272.

Rahmati, A., & Zhong, L. (2009). Human-battery interaction on mobile phones. *Pervasive and Mobile Computing, 5,* 465–477.

Rauh, N., Franke, T., & Krems, J. F. (2015a). Understanding the impact of electric vehicle driving experience on range anxiety. *Human Factors, 57,* 177–187.

Rauh, N., Franke, T., & Krems, J. F. (2015b). User experience with electric vehicles while driving in a critical range situation – A qualitative approach. *IET Intelligent Transport Systems, 9,* 734–739.

Rauh, N., Franke, T., & Krems, J. F. (2017). First-time experience of critical range situations in BEV use and the positive effect of coping information. *Transportation Research Part F, 44*, 30–41.

Rezvani, Z., Jansson, J., & Bodin, J. (2015). Advances in consumer electric vehicle adoption research: A review and research agenda. *Transportation Research Part D: Transport and Environment, 34*, 122–136.

Riga, D. (2015). *Hybrid electric vehicles: Driving towards sustainability.* Unpublished master thesis, University of the Witwatersrand, South Africa.

Rogers, E. M. (2010). *Diffusion of innovations.* New York: Simon and Schuster.

Schmalfuß, F., Kreußlein, M., Döbelt, S., Kämpfe, B., Wüstemann, R., Mair, C., et al. (2016). *Smart charging systems as one solution to overcome grid stability problems in the future? A field study for examining the BEV user acceptance of a smart charging system.* In European conference on human centred design intelligent transport system, pp. 114–124. Retrieved February 14, 2017, from http://conference2016.humanist-vce.eu/Proceedings/session-4.pdf

Schmalfuß, F., Kreußlein, M., Mair, C., Döbelt, S., Heller, C., Wüstemann, R., et al. (2017). Smart charging in daily routine – Expectations, experiences and preferences of potential users. In J. Liebl (Ed.), *Grid integration of electric mobility – 1st international ATZ conference 2016.* Wiesbaden, Germany: Springer Vieweg.

Schmalfuß, F., Mair, C., Döbelt, S., Kämpfe, B., Wüstemann, R., Krems, J. F., et al. (2015). User responses to a smart charging system in Germany: Battery electric vehicle driver motivation, attitudes and acceptance. *Energy Research & Social Science, 9*, 60–71.

Schmalfuß, F., Mühl, K., & Krems, J. F. (2017). Direct experience with battery electric vehicles (BEVs) matters when evaluating vehicle attributes, attitude and purchase intention. *Transportation Research Part F: Traffic Psychology and Behaviour, 46*(A), 47–69.

Schmitz, M., Maag, C., Jagiellowicz, M., & Hanig, M. (2013). Impact of a combined accelerator-brake pedal solution on efficient driving. *IET Intelligent Transport Systems, 7*, 203–209.

Schuitema, G., Anable, J., Skippon, S., & Kinnear, N. (2013). The role of instrumental, hedonic and symbolic attributes in the intention to adopt electric vehicles. *Transportation Research Part A: Policy and Practice, 48*, 39–49.

Schwartz, S. H., & Howard, J. A. (1981). A normative decision-making model of altruism. In J. P. Rushton & R. M. Sorrentino (Eds.), *Altruism and helping behavior* (pp. 189–211). Hillsdale, MI: Lawrence Erlbaum.

Seeking Alpha. (2013, August 1). *Beyond RA: The EV's next challenge.* Retrieved February 14, 2017, from http://seekingalpha.com/article/1595562-beyond-range-anxiety-the-evs-next-challenge?

Sierzchula, W., Bakker, S., Maat, K., & van Wee, B. (2014). The influence of financial incentives and other socio-economic factors on electric vehicle adoption. *Energy Policy, 68*, 183–194.

Sivak, M., & Schoettle, B. (2012). Eco-driving: Strategic, tactical, and operational decisions of the driver that influence vehicle fuel economy. *Transport Policy, 22*, 96–99.

Stillwater, T., & Kurani, K. S. (2013). Drivers discuss ecodriving feedback: Goal setting, framing, and anchoring motivate new behaviors. *Transportation Research Part F: Traffic Psychology and Behaviour, 19*, 85–96.

Summala, H. (2007). Towards understanding motivational and emotional factors in driver behaviour: Comfort through satisficing. In C. Cacciabue (Ed.), *Modelling driver behaviour in automotive environments* (pp. 189–207). London: Springer.

Tate, E. D., Harpster, M. O., & Savagian, P. J. (2008). The electrification of the automobile: From conventional hybrid, to plug-in hybrids, to extended-range electric vehicles. *SAE International Journal of Passenger Cars – Electronic and Electrical Systems, 1*(1), 156–166.

Thatcher, A. (2013). Green ergonomics: Definition and scope. *Ergonomics, 56*, 389–398.

Turrentine, T., Garas, D., Lentz, A., & Woodjack, J. (2011). *The UC Davis MINI E consumer study*. Report no. UCD-ITS-RR-11-05. Davis, CA: University of California.

Venkatesh, V., Morris, M. G., Davis, G. B., & Davis, F. D. (2003). User acceptance of information technology: Toward a unified view. *MIS Quarterly*, 425–478.

Vilimek, R., Keinath, A., & Schwalm, M. (2012). The MINI E field study – Similarities and differences in international everyday EV driving. In N. A. Stanton (Ed.), *Advances in human aspects of road and rail transportation* (pp. 363–372). Boca Raton, FL: CRC Press.

Walsh, C., Carroll, S., Eastlake, A., & Blythe, P. (2010). *Electric vehicle driving style and duty variation performance study*. Retrieved February 14, 2017, from http://www.cenex.co.uk/wp-content/uploads/2013/06/Electric-vehicle-driver-and-duty-variation-performance-study1.pdf

Weiss, C., Mallig, N., Heilig, M., Schneidereit, T., Franke, T., & Vortisch, P. (2016). *How much range is required? A model based analysis of potential battery electric vehicle usage*. In Transportation Research Board (Ed.), TRB 95th annual meeting compendium of papers.

Wikström, M., Hansson, L., & Alvfors, P. (2014). Socio-technical experiences from electric vehicle utilisation in commercial fleets. *Applied Energy, 123*, 82–93.

Yilmaz, M., & Krein, P. T. (2013). Review of battery charger topologies, charging power levels, and infrastructure for plug-in electric and hybrid vehicles. *IEEE Transactions on Power Electronics, 28*, 2151–2169.

Young, M. S., Birrell, S. A., & Stanton, N. A. (2011). Safe driving in a green world: A review of driver performance benchmarks and technologies to support "smart" driving. *Applied Ergonomics, 42*, 533–539.

Yuan, X., Li, L., Gou, H., & Dong, T. (2015). Energy and environmental impact of battery electric vehicle range in China. *Applied Energy, 157,* 75–84.

Zhang, Y., Wang, W., Kobayashi, Y., & Shirai, K. (2012). Remaining driving range estimation of electric vehicle. In *IEEE international electric vehicle conference* (pp. 95–101). Red Hook, NY: Curran Associates.

Ziegler, A. (2012). Individual characteristics and stated preferences for alternative energy sources and propulsion technologies in vehicles: A discrete choice analysis for Germany. *Transportation Research Part A: Policy and Practice, 46,* 1372–1385.

CHAPTER 7

HFE in Biophilic Design: Human Connections with Nature

Ryan Lumber, Miles Richardson, and Jo-Anne Albertsen

INTRODUCTION

There are three main interactions in human factors and ergonomics (HFE) practice, using a systems approach and the outcomes of well-being and productivity. HFE has the potential to design work environments to optimise worker performance and well-being, which themselves are linked outcomes. However, the potential for HFE design to achieve this has so far been underexploited, with calls to better integrate HFE practices for all stakeholders (Dul et al., 2012). This is exemplified by the discipline and many models of workplace well-being tending to ignore the benefits of nature to human well-being (Richardson et al., 2017) and by extension, productivity too. Further, as a discipline, ergonomics should engage with global challenges, such as sustainability, climate change, and the loss of biodiversity, particularly since the latter is linked to reductions in well-being (e.g. Moray, 1993; Thatcher, 2013). As research and policy agendas

R. Lumber
Faculty of Health and Life Sciences, De Montfort University, Leicester, UK

M. Richardson (✉) • J.-A. Albertsen
College of Life and Natural Sciences, The University of Derby, Derby, UK

© The Author(s) 2018   161
A. Thatcher, P. H. P. Yeow (eds.), *Ergonomics and Human Factors for a Sustainable Future*, https://doi.org/10.1007/978-981-10-8072-2_7

recognise the human need for nature, ergonomics should extend the scope of the solutions it offers, bringing nature into workplaces for well-being, restoration, and the wider benefits to the natural environment.

Studying the effects of nature in the workplace is important, because many people spend at least one third of their time at work (Chang & Chen, 2005). Various research has found that exposure to nature, or depictions thereof, in the workplace decreases stress and depression, restores concentration and attention, and increases prosocial behaviour, creativity, comfort, job satisfaction, and the health and well-being of employees (Richardson et al., 2017). Further, a connectedness to nature brings additional benefits such as pro-environmental behaviour. HFE is paying insufficient attention to these nature-based solutions for workplace issues (Richardson et al., 2017); guidance on creating healthy workplaces (e.g. Day, Kelloway, & Hurrell, 2014) and reviews and models of workplace well-being (e.g. Danna & Griffin, 1999; Wilson, Dejoy, Vandenberg, Richardson, & McGrath, 2004) often overlook the well-being benefits of nature. Likewise, despite exposure to nature being an easy and inexpensive intervention (Trau, Keenan, Goforth, & Large, 2015), workplace health promotion is often limited to traditional approaches such as exercise (e.g. Kuoppala, Lamminpää, & Husman, 2008). Similarly, the benefits of the natural environment for well-being and restoration of performance are not promoted in key ergonomics texts (e.g. Salvendy, 2012). Given the importance of the work environment on health and well-being, there is a need for HFE practitioners to use a full range of solutions to promote these—including the use of nature. One area where nature is being brought into the workplace is through the application of biophilic design principles to buildings (e.g. Ryan, Browning, Clancy, Andrews, & Kallianpurkar, 2014), but there is still a need for nature-focused solutions to become part of ergonomics practice.

Green ergonomics is a branch of HFE that considers how the human-nature connection might promote human well-being and productivity and thus has a role to play in biophilic design (Thatcher, 2013). Based upon the biophilia hypothesis, biophilic design aims to include natural elements and systems in the built environment (Gillis & Gatersleben, 2015). It forms an integral part of restorative environmental design (Kellert, 2005) and assists in creating a favourable interior while promoting environmental sustainability, human well-being, and productivity (Kellert & Calabrese, 2015). As a holistic approach that steers architectural and interior design choices (Augustin, 2009), biophilic design interventions in places where people live and work secure the long-lasting inclusion of nature in expanding and crowded urban environments (Kellert, Heerwagen, & Mador, 2008). This chapter introduces the concepts of the human-nature relationship and nature connectedness before moving onto the benefits of nature for health, well-being, and restoration, with a specific focus on ergonomics in the workplace. Concepts of biophilic design and ways of bringing nature into the workplace are also considered.

THE HUMAN-NATURE RELATIONSHIP

The human species spread from the savannahs of Africa hundreds of thousands of years ago where the landscape offered both survival opportunities and threats to survival, leading to preferences for certain aspects of nature and aversion to others (Kahn, 2011; Wilson, 2002). Such preferences endure: even today, savannah-like environments often receive high preference ratings, and other natural environments such as coastal areas or inland bodies of water are also rated as preferred natural spaces (Hinds & Sparks, 2011). Humanity evolved to make sense of the natural world to which we belong; whether through hunting or farming, our cognitive processes and emotional states have been shaped through our interactions with nature (Gullone, 2000). It has been argued that humanity has an innate tendency to have an affiliation for nature, natural life, or life-like processes, surmised as the biophilia hypothesis (Gullone, 2000; Kahn, 1997; Kellert & Wilson, 1993). While the term biophilia was originally coined by Erich Fromm, who used it to describe a passionate love for life, Fromm's focus was mainly human-to-human interactions, which at times also included non-human life. The biophilia hypothesis relates to the human desire to connect with life in all its diversity, and while a love for life is part of this, biophilia in this sense is expressed additionally through other aspects (Kellert, 2003) as part of the nine values (see Table 7.1). Having an affiliation with nature is theorised to stem from an evolutionary history spent searching for survival-enhancing settings (Frumkin, 2001; Kellert & Wilson, 1993; Windhager, Atzwanger, Bookstein, & Schaefer, 2011). As urban living has occurred relatively recently in humanity's evolutionary history, the embedded learning rules derived from nature are unlikely to have been erased from our biology completely (Nisbet, Zelenski, & Murphy, 2011). Biophilia is composed of nine succinct values that cover a range of ways in which individuals relate to, or interact with, nature (Kellert, 1993) and are often unconsciously manifested in cognitions,

Table 7.1   The nine values of biophilia based on Kellert and Wilson (1993)

Value	Definition	Function
Utilitarian	Using natural material for a practical use	Sustaining one's physical life and security
Naturalistic	The feeling of pleasure from natural contact	Development of cognitive, motor, and outdoor skills and development
Ecologistic-scientific	Studying scientifically the interconnectedness of nature and natural systems	Observing nature, increasing knowledge, and understanding through scientific means
Aesthetic	The visual appeal of nature's physical beauty	Sense of security, inspiration, and contentment
Symbolic	Using nature-based language and metaphors to express ideas	Mental development, communicating with others (including nature)
Humanistic	A love for and emotional bond with nature	Co-operation, bonding, and companionship
Moralistic	Making judgements based upon ethical concern, along with revering nature	Moral reasoning, meaning of life, affiliation with nature/others
Dominionistic	Control/dominance over nature	The use of technology/mechanical skill, physicality, and control
Negativistic	Aversion and/or fear of nature	Physical protection/security

emotive responses, artistry, and ethics (Kahn 1997, 2011). It is suggested that an active participation through one or more of the values allows for an innate learning of nature (Gullone, 2000); thus, biophilia is a predisposition for certain natural settings driven by a hardwired biological process (Wilson, 2002). As a result, biophilia is suggested to be a biocultural model that occurs through inherited prepared learning (Wilson, 2002) that has been maintained through reliance on and affiliation towards nature, leading to greater survival and evolutionary fitness (Wilson, 1993); biophilia may therefore be crucial to optimum human functioning both affectively and psychologically (Kellert, 1997, as cited in Nisbet et al., 2011). Tentative evidence exists for innate biophilia, as children between the ages of eight and eleven are more likely to prefer savannah-like landscapes, with older children preferring savannah landscapes and their home environment equally (Wilson, 2002). It has been noted that the transmission of biophilic tendencies through genetic heritability is questionable and is far more likely a result of experiential learning (Simaika & Samways, 2010). The expression of biophilic tendencies may be optimised through a combination of learning, culture, and direct experiences with nature

(Hinds & Sparks, 2011). Empirical support for the hypothesis has been mixed (Kahn, 1997) with evidence supporting the hypothesis drawn from studies into restoration and preferences for nature. Biophilia is a complicated web of learning processes that persist throughout generations even when we are separated from natural environments (Wilson, 1993). Hence, mismatches arise between our adaptive evolutionary attributes and the modern built environment (Fitzgerald & Danner, 2012). These mismatches can cause stress and a variety of health problems.

NATURE CONNECTION

It is estimated that around 50% of the world's population currently reside in urban environments, with this trend set to increase (Lin, Fuller, Bush, Gaston, & Shanahan, 2014). Historically, shifts in economics, from the traditional to those built upon industry, mechanisation, and increasing land ownership for the few, forced rural populations to migrate to urban environments, resulting in rapid urbanisation (Castles, De Haas, & Miller, 2014). As a result, the connections with nature that were possible in rural environments may have been lost, given that opportunities to engage in a meaningful way with the natural world in urban locations are diminished (Pyle, 2003). However, given the innate, evolutionary basis for humanity's need to form relationships with the wider natural world, a reconnection with nature was (and is) still possible, which may be best achieved through nature connection. Nature connection is a multidimensional construct and is composed of a number of factors. Nature connection is the sense that an individual is part of a larger natural community (Leopold, 1966; Mayer, Frantz, Bruehlman-Senecal, & Dolliver, 2009) leading to nature and the self becoming one (Schultz, 2001) through personal and social influences as part of an environmental identity (Clayton, 2003, 2012). This connection is composed of cognitive (Schultz, 2001), affective (Mayer & Frantz, 2004), learnt, experiential (Nisbet, Zelenski, & Murphy, 2009), and personality (Kals, Schumacher, & Montada, 1999) factors that together create an individual, subjective relationship with nature.

Humanity is a social species, one which places an importance upon social connectedness with others as a direct result of our biology and evolutionary history (Cacioppo & Patrick, 2008). It was essential for our ancestors that a psychological connection with other group members was formed, a connection that was based upon similarity (Lakin, Jefferis,

Cheng, & Chartrand, 2003). It is theorised that humanity's capacity for co-operation emerged from this similarity and our emotional bonds with others, which in turn produced compassionate helping which could offset any destructive tendencies humanity possesses (Gilbert, 2014). The need to form social connections with others was (and continues to be) important for our health and genetic legacy, something that could be threatened by social isolation. In modern Western societies, social isolation has been found to have a negative effect on cognitive functioning which can be alleviated through connecting with other human beings and non-human life such as pets (Cacioppo & Hawkely, 2009), or through the restorative effect of viewing nature (Kaplan, 1995). Re-connecting with nature may have an evolutionary basis in humanity's need and predisposition to form connected relationships with others, both human and non-human life. The ability to connect not only with other members of the human species but wider nature is an important one for modern small band hunter-gatherer communities; the sharing of resources, knowledge, and company is prevalent, and a partnership and embeddedness with the wider natural world are lived and practised (Narvaez, 2013). Unfortunately, the prevailing view held by modern, Westernised societies is that humanity is set apart from (Vining, Merrick, & Price, 2008) and even above nature (Maller et al., 2009), which has been outlined as being one of the principal causes of environmentally harmful behaviour (Haila, 1999). An expansion of an individual's concept of self to include nature is necessary in order to become connected to nature (Mayer & Frantz, 2004; Schultz, 2001). It is thought that extending the concept of self to include nature creates a feeling of kinship (Olivos, Aragones, & Amerigo, 2011) and commonality with all life (Fox, 1990) as nature and the self are perceived as one and the same (Light, 2000). A nature connection therefore creates a sense of belonging to the wider natural world as part of a larger community of nature (Mayer et al., 2009). More recently, nature connection has been found to be predicted by engagement with activities related to the humanistic, moralistic, symbolic, naturalistic, and aesthetic values of biophilia (Lumber, Richardson, & Sheffield, 2017). The need for a connection to nature is a Western notion, and for indigenous cultures such as the Inuit, the natural landscape forms a crucial part of their cultural identity (Russell et al., 2013). Thus, the concept of self as including a relationship with nature is an ancient one, as traditional indigenous belief systems often see the Earth and self as one and the same, with an individual's identity entwined with the fate of the wider environment (Macy, 2007).

Experience and culture are important as they inform an eco-identity, which is intertwined with an active engagement with nature (Russell et al., 2013).

In addition to the concept of self, an emotional attachment or affiliation with nature is also important for nature connection (Davis, Green, & Reed, 2009) as an emotional attachment to nature may be crucial to the formation of connectedness and feeling part of the natural world (Mayer & Frantz, 2004). Individuals living in close proximity with nature tend to report feelings of inner calm and happiness, while gardens, beaches, parks, and rivers elicit self-reported sensations of fun and relaxation (Hinds & Sparks, 2011). While the emotional response to nature can be either positive, negative or a mixture of the two, there is some evidence that the positive emotional response to nature is more widespread than negative or neutral emotions (Hinds & Sparks, 2011). Having a personal relationship with nature may lead to environmentally protective behaviours as it utilises the affective connection in eliciting pro-environmental attitudes to local nature (Nisbet et al., 2009). This affective connection or emotional affiliation towards nature is thought to comprise of four aspects of natural affection—love, freedom, security, and being part of nature (Muller, Kals, & Pansa, 2009). Becoming emotionally connected with nature is thought to occur through positive interactions during childhood (Hinds & Sparks, 2008; Muller et al., 2009) where the memory of the connecting experience becomes imprinted upon the individual (Hawkes & Acott, 2013). The emotional attachment to nature formed through childhood engagement with nature is enduring and contributes to a desire to have contact with nature in adulthood (Muller et al., 2009). While an emotional attachment to nature comprises (in part) a connection to nature, being connected goes beyond simply a love for nature (Frantz, Mayer, Norton, & Rock, 2005). This emotional connection goes beyond a surface love for pleasing nature, through an understanding of the interconnectedness and value of natural life (Nisbet et al., 2009) that is similar to the concept of deep ecology, whereby the richness and diversity of nature is valued regardless of its potential for human use, as humanity and nature are inter-related and part of the same community of life (Drengson & Devall, 2010).

The Benefits of Being Connected with Nature

Having introduced the concept of nature connectedness and its benefits, a focus on ergonomics and the workplace can be introduced. There is a growing evidence base for the physical and mental well-being benefits of engaging with nature. Given that nature connection predicts an engagement with nature (Nisbet & Zelenski, 2013), the following subsections discuss the links between connectedness and engagement with nature that in turn lead to physical and mental well-being benefits. As HFE's fundamental aim is well-being, the wider benefits of nature should be part of the core knowledge of all those in the profession. As the ergonomics knowledge base—and wider models of workplace well-being—does not tend to consider the health benefits of nature, it is important to introduce the breadth of benefits as they show how bringing nature into the workplace could be advantageous for well-being, performance, and absenteeism.

Finally, the link between nature connection and pro-environmentalism should be highlighted in the wider context of sustainability. Given that the harm caused to nature by humanity is thought in part to result from the perception that humanity and nature are separate (McPhie & Clarke, 2015), nature connection could potentially counter this disconnection and, in turn, lead to pro-environmental behaviours (Tam, 2013).

Mental Well-Being and Nature Connection

Nature has long been associated with positive outcomes for humanity, historically, anecdotally, and in more recent empirical work on well-being (Russell et al., 2013). Nature connection is associated with overall benefits to well-being with an effect size similar to the established factors of income and education in social psychology (Capaldi, Dopko, & Zelenski, 2014). Research has found that nature connection is positively related to satisfaction with life (Mayer & Frantz, 2004; Russell et al., 2013), happiness (Zelenski & Nisbet, 2012), perspective taking (Russell et al., 2013) as well as social and psychological well-being (Howell, Dopko, Passmore, & Buro, 2011), personal growth, vitality, and meaning in life (Nisbet et al., 2011) and decreases in trait and state anxiety (Martyn & Brymer, 2014). Even brief exposure to nature can have positive benefits (McMahon & Estes, 2015; Nisbet & Zelenski, 2011). Clearly, satisfied and happy workers with vitality and less anxiety are important to a successful working environment. Further, improvements to cognition and mood can be gained through contact with nature (Capaldi et al., 2014); this effect has also been found in individuals residing in urban environments after spending time in nature (Russell et al., 2013). Even simply viewing nature has been associated with a reduction in stress and an increase in overall well-being (Russell et al., 2013), with these benefits explained via a restoration of mental resources derived from viewing aspects of nature that captures attention while creating the sensation of being away from the stressor (Kaplan, 1995; McDonald, Kirk, & Kinns, 2015; Wyles, Pahl, Thomas, & Thompson, 2016). However, actual natural environments have a greater overall effect on well-being than virtual nature (McMahon & Estes, 2015). Therefore, modest changes to workplace design to facilitate access to, and a view of, nature can potentially have a positive impact, with interventions (e.g. noting three good things in nature; Richardson & Sheffield, 2017) that increase nature connection also being worthwhile.

Physical Health

Natural environments also have benefits for physical health (Russell et al., 2013). For example, a review by Logan and Selhub (2012) of the benefits to health that exposure to nature may bring found that contemplating for 20 minutes while in nature created the feeling of relaxation that in turn lowered cortisol (through a reduction of activity in the amygdala) with the suggestion that this may improve immune functioning. Nature connection may also play a role as, coupled with positive personality traits, it is thought to offer coping options to meet the challenges of stress that may in turn provide resilience to disease due to an improved functioning of the immune system (Cervinka, Roderer, & Hefler, 2012). A recent review supports this possibility, with the suggestion that engaging with nature leads to greater immune functioning, acting as a central pathway to further physical health outcomes (Kuo, 2015); therefore, the benefit to mental well-being and physical health from nature connectedness is a dual process, given that the benefit to well-being from feeling connected has a direct effect on physical health. The link between nature and health has also been demonstrated in urban work environments, where lower stress levels and lower reported health complaints were related to contact with indoor nature (Largo-Wight, Chen, Dodd, & Weiler, 2011). Such findings have implications for break-taking in the workplace, and the use of outdoor space within the working day where possible.

Nature and Restoration

A benefit of particular note to the HFE profession is restoration. Stress can occur in the absence of mental fatigue and vice versa, with an over-arousal due to stress leading to an over-loading of attentional capacity (Berto, 2014). Mental fatigue arises from continuously focusing and holding one's attention on a task. A decline in concentration can result in difficulties such as minor and catastrophic errors, irritability, and annoyance with colleagues (Kaplan, 1993). As such, mental fatigue can be considered an after-effect of stress while also leading to an increased vulnerability to further stress (Berto, 2014). Engaging with nature, either directly or indirectly, has been shown to provide psychological benefits through restoration following attentional fatigue or stress (Kaplan, 1995). An environment is considered restorative if it assists in returning an individual to a more favourable, previously held state (White et al., 2010).

The two main theoretical frameworks explaining the restorative benefits of nature are Ulrich's (1983) stress recovery theory (SRT) and Kaplan and Kaplan's (1987) attention restoration theory (ART). As a psycho-evolutionary theory, SRT argues that people are physiologically and psychologically suited to natural rather than urban milieus, because human evolution occurred in natural environments (Berto, 2014). SRT focuses on nature and how it can aid in reducing the physiological arousal caused by stress (Ratcliffe, Gatersleben, & Sowden, 2013). ART focuses on mental fatigue, and humans are viewed as having an instinctive tendency to observe and react favourably to natural elements such as flora (Berto, 2014). Lottrup, Grahn, and Stigsdotter (2013) describe observing natural features as *soft fascination*, which supports mental restoration as attention is briefly rested from a task. The theory is well supported through research and is considered to provide the best explanation for restoration of attention and the subsequent well-being benefits it creates (Russell et al., 2013).

Research related to these theories agree that (a) an individual's selection of environment to engage with is influenced by their need for restoration, and (b) natural environments are more likely to be restorative than man-made ones (Berto, 2014). For example, Beil and Hanes (2013) found that salivary alpha-amylase levels (an indicator of stress) increased after participants were exposed to a predominantly built environment. Another study found that contact with nature outdoors was strongly associated with decreased stress and improved health (Largo-Wight et al., 2011). Finding restoration through engaging with nature has been linked to better cognitive performance in proofreading tasks (Kaplan, 1995). Further, Lee, Williams, Sargent, Williams, and Johnson (2015) found that attention can be restored by a 40 second view of a green roof, with Chow and Lau (2015) finding that people have greater persistence in logic and reasoning tasks after being exposed to photos of nature to restore their 'inner strength.' These findings can inform advice given regarding break-taking behaviour.

IMPLICATIONS FOR ERGONOMICS AND HUMAN FACTORS

From a systems perspective, there is a need to bring nature into the workplace in a methodical manner. As has been indicated above, breaks from work are likely to be more restorative if they include exposure to natural elements and this can inform health promotion activities. The wider benefits and concept of nature connection itself provide a holistic systems perspective on what makes a good life and good workplace. This wider interconnectedness can inform work, for example, in areas such as situation awareness and accidents. When a systems-orientated perspective to the relationship between an individual and nature is taken (e.g. Bateson, 1972), our goal-directed consciousness can be seen as only a partial window on our wider systemic and dynamic relationship with the environment. This perspective could facilitate a move away from the goal-directed focus of people and work, allowing a more holistic relationship with our environment to develop (Bateson, 1972). Within HFE, the value of such a general integrative perspective has been acknowledged by those who highlight that it is difficult to establish where the system ends and environment begins (Dekker, 2013).

Green Ergonomics and Green Building Design

Green ergonomics focuses on developing human systems that harmonise with the natural environment in a sustainable manner (Thatcher, 2013). Sustainability proposes that current generations should not jeopardise the future needs of others. For something to be sustainable it should have the possibility to be maintained and it should proceed continuously (Martin, Legg, & Brown, 2013). Green ergonomics gives consideration to (a) how systems can aid the protection and recovery of natural resources, and (b) how the human-nature connection may assist people's performance and general well-being (Thatcher, 2013). It plays a role in areas such as product and job design (Hanson, 2013; Hedge, Rollings, & Robinson, 2010), and it can contribute to enhancing people's well-being in the built environment (Thatcher, 2013).

Green (or sustainable) building design focuses on conserving energy, utilising renewable resources, producing optimal indoor environmental quality, and constructing in a resource-efficient and durable manner (Martin et al., 2013). The Leadership in Energy and Environmental Design (LEED) Green Building Rating System, of the US Green Building Council (USGBC), is a system that promotes the development of sustainable buildings (Hedge, 2008). A study found that as much as 83% of occupants in a LEED building preferred this structure to their previous non-LEED building (Hedge & Dorsey, 2013). The USGBC also acknowledges that successful green design has to support the creation of optimal workplaces. Their building rating system therefore includes an ergonomics perspective (Hedge, 2008). However, not every environmentally sustainable design is guaranteed to contribute to human well-being (e.g. increased daylight might cause unwanted glare) (Thatcher & Milner, 2014). A multidisciplinary and collaborative design approach may, however, assist in creating buildings that are both green and satisfactory to occupants (Fiore, Phillips, & Sellers, 2014).

Sustainable design, as described above, tends to focus on preventing environmental degradation (also called *low environmental impact design*). However, it does not fully address the diminishing human contact with nature. Low environmental impact design and biophilic design should ideally be combined to form *restorative environmental design* that has more lasting sustainability (Kellert et al., 2008). The following discussion examines biophilia and biophilic design, and its role in human well-being and the workplace, in greater depth.

Contact with Nature Through Biophilic Design

Biophilic design is defined by Gillis and Gatersleben (2015, p. 948) as '...a design philosophy that encourages the use of natural systems and processes in the design of the built environment.' The following six biophilic design elements have been identified by Kellert et al. (2008): evolved human-nature relationships, place-based relationships, environmental features, natural patterns and processes, natural shapes and forms, and light and space. Further illustrating the variety found in nature, over 70 corresponding features within these six elements (e.g. water, views, botanical motifs, biomimicry, sunlight, space, and security) have been outlined (Kellert et al., 2008). These elements and features can be incorporated into the built environment in a subtle or more direct manner, and may appear in the building's exterior and/or interior, as ornamental features, or in the outdoor landscape (Kellert, 2005).

Architectural components like windows, doors, balconies, courtyards, and patios help to connect the indoors with the outside environment (Kellert & Calabrese, 2015). Windows especially play a major role in biophilic design. Not only do operable windows allow occupants to benefit from natural ventilation, but they also provide an opportunity for views and for natural daylight to enter the interior (Bluyssen, 2009). However, natural ventilation from open windows cannot be utilised when an airtight building is required by mechanical ventilation systems, when outside noise and pollution exists, or when heat recovery is necessary (Bluyssen, 2009). Even though people may have access to the outdoor environment through some of the abovementioned architectural components at their workplace, studies show that most office workers do not venture outdoors because they consider themselves to be too engaged at work and because the working culture does not encourage it (Lottrup et al., 2013). Yet, research by Largo-Wight et al. (2011) found that physical access to nature in the workplace had more advantages than solely visual access, and Lottrup et al. (2013) found that contact with outdoor nature had a greater correlation with well-being and a decline in stress than indirect contact with nature. Current workplace health programmes are inclined to neglect the role of nature in well-being (Richardson et al., 2017), even though such programmes and HFE interventions can promote contact with nature at work and encourage workers to engage in more physical activity during the day, such as walking in a park at lunchtime, or taking a short break in the workplace garden. These types of activities offer benefits to both the individual (e.g. lowering stress) and the organisation (e.g. increased workforce productivity) (Lottrup et al., 2013) and should be a serious consideration when aiming to create a positive work environment.

Culture and tradition are important factors in interior design. They influenced the size, type, and arrangement of plants in a Danish office, and guided plant choice for specific occasions (e.g. Easter) (Thomsen, Sønderstrup-Andersen, & Müller, 2011). People, culture, and the physical environment should therefore be considered holistically, because people and their culture are influenced by the environment, and vice versa (Reddy, Chakrabarti, & Karmakar, 2012). In workplaces where living elements of nature are inappropriate, representations thereof (e.g. images, sculptures, and computer simulations) may be used instead (Gillis & Gatersleben, 2015). However, not all natural elements or representations of nature are restorative (Gillis & Gatersleben, 2015). For example, Thomsen et al. (2011) found that several people were allergic to certain flowering plants. Biophilic design should therefore consider people's different responses to nature, and acknowledge that some individuals may not even like it (Gillis & Gatersleben, 2015).

While people have control over the design of their homes (e.g. adding extra windows for light), they have very little or no control over such design in the workplace (Fitzgerald & Danner, 2012). Scientific literature or other standards can inform the most optimal workplace interior design, but it does not guarantee satisfying occupants' requirements as this is largely dictated by individual preferences (Reddy et al., 2012). Workplace identity can be fostered by allowing employees to contribute to interior design (e.g. by choosing artwork) (Freeman, 2011). Additionally, workers' input into the design of their work environment can increase job satisfaction, comfort, and productivity (Knight & Haslam, 2010). However, if complete responsibility for design options is given to employees, it may lead to conflict and a neglect of the important technical aspects of indoor environmental quality (Reddy et al., 2012). There has to be a balance between the desires of occupants and the technical feasibility of the building. Nonetheless, providing employees with the opportunity for contact with nature is an inexpensive intervention that appears to offer significant benefits for employees and companies alike (Kaplan, 1993).

The design of the workplace is an important facet of work system design, and, because systems occur within a context, the components of this context affect general system functioning—including the comfort, safety, and efficiency of workers (Salvendy, 2012). As biophilic design features are concerned with the physical environment, a HFE approach to addressing these issues can be positioned under the domain of environmental ergonomics. Such an approach can aid in eliminating or reducing the occurrence of adverse environmental conditions (e.g. glare from sunlight) (Proctor & Van Zandt, 2008). Nevertheless, irrespective of all preventative efforts, some environmental problems may only emerge once the various constituent parts of the workplace interact (Proctor & Van Zandt, 2008). HFE practitioners should therefore consider the whole environment and not solely its individual components in isolation (Parsons, 2000).

Biophilic Design Features

The following discussion explores a few of the key biophilic design features that are particularly relevant in the workplace. However, it should be noted that not all of these features may fit every work environment. For example, living plants are not practical in a setting that requires sterile conditions (e.g. a medical laboratory). It is thus imperative that designers and other workplace stakeholders understand the context and environment—including the occupants, tasks, and location—for which they are proposing biophilic design (Gillis & Gatersleben, 2015). Subjective (e.g. rating scales) and more objective methods (e.g. measurement of task performance) can be used to evaluate the effects of the environment on workers (Parsons, 2000). Furthermore, studies that seek to understand people's preferences for natural elements are also crucial for guiding design choices that fulfil occupants' expectations and needs.

Colour
Colour is considered to have a compelling, emotive, and perplexing effect in interior design (Reddy et al., 2012). People may experience physical discomfort (e.g. headaches) from conflicting colours (Reddy et al., 2012), or they may experience a pleasurable effect from appropriate colour schemes that contribute positively to aesthetics (Richardson, Hallam, & Lumber, 2015). Even though people perceive colours similarly (Salvendy, 2012), preferences for interior colour schemes differ between countries and cultures (Augustin, 2009). Men and women appear to respond differently to colour, with women being more sensitive to the effects of a colour's brightness and saturation than men (Augustin, 2009). Colour choice therefore requires careful thought in buildings that accommodate numerous occupants and complicated functions (e.g. a hospital) (Reddy et al., 2012). While colour is commonly addressed in HFE issues pertaining to screen and display colour, and colour coding for warnings (Proctor & Van Zandt, 2008), Meerwein, Rodeck, and Mahnke (2007) noted that ergonomics pays insufficient attention to the value of colour in the interior workplace environment. An ergonomics approach to colour design can support workers' psychological and physiological requirements by taking into account how the interplay between illumination and colour coordination impacts, for example, concentration, visual pleasure, and fatigue (Meerwein et al., 2007).

To reflect the natural world, and to indirectly reinforce the human-nature connection, Kellert and Calabrese (2015) recommend the use of earthy tones found in sand, rock, and vegetation instead of bright and clashing colours. While there is no consensus about the most suitable colour scheme for an interior environment, studies have found that workers preferred offices that were white, beige, or neutral in colour (Schatz & Bowers, 2005). Anshel (2005) also recommends that walls should be light colours and ceilings should be white. Jalil, Yunus, and Said's (2012) review of colour research found, however, that most studies focused on a very narrow selection of colours (i.e. red, blue, white, and green). Additionally, the authors also uncovered conflicting results between studies examining the effects of colour on people's arousal levels—red had the most arousal effects in one study, while blue had the most arousal effects in another (Jalil et al., 2012). It is thus inadvisable to generalise the effect that colour may have on workers (Meerwein et al., 2007), because confounding factors such as colour trends, age, social norms, and workplace conditions may all influence people's perceptions and experiences of colour (Schatz & Bowers, 2005).

Despite the abovementioned contradiction, various other colour studies have concluded that wall colour impacts performance (Schatz & Bowers, 2005). The colour of walls should therefore complement the tasks that are performed in the space—colours should be calm enough to facilitate concentration, but energising enough to prevent drowsiness (Augustin, 2009). Additionally, bright reflective surfaces (e.g. white table-tops) that result in glare should be avoided (Anshel, 2005), and light-dark contrasts (e.g. black furniture against white walls) and distracting patterns of colour should be omitted from the direct visual field (Meerwein et al., 2007). In short, determinants such as the quality of light, other existing colours, the colour's distance from view, and the surrounding environment can all influence colour vision and thereby an individual's experience of their immediate environment (Elliot, 2015).

Daylight
Suitable lighting is essential for successful task performance, safety, comfort, and general enjoyment of the interior space. It has been found that exposure to sunlight in the workplace promotes relaxation, improves job satisfaction, and reduces employees' intentions to resign (Fitzgerald & Danner, 2012) with sunlight acting as a buffer to workplace stress (An, Colarelli, O'Brien, & Boyajian, 2016). Conversely, a lack of exposure to light—whether in the form of natural daylight or artificial light—causes an increase in melatonin levels, which in turn results in drowsiness and a depressed mood (Kellert et al., 2008). However, too much daylight can result in overheating and glare that impacts task execution negatively (Edwards & Torcellini, 2002). Research has also found that workers often reported that sunlight made electronic and video monitors hard to read or completely unreadable (Salvendy, 2012).

The intensity of natural light changes as the day progresses, and ideally a combination of daylight and artificial light should be used to create the most favourable lighting conditions in the workplace (Augustin, 2009). By measuring daylight and light from other sources, the illumination levels proposed by visual ergonomics can be applied to guide potential lighting adjustments and ensure that the amount and quality of light meet the needs of the workforce and the tasks at hand (Proctor & Van Zandt, 2008). Bluyssen (2009) states that a minimum illumination level of 200 lux is needed in areas usually occupied by people, but these levels can be increased or decreased to create the desired ambience or fit task requirements. Light shelves and adjustable window coverings (e.g. blinds) can be used to control the amount of daylight entering the interior space (Salvendy, 2012). Optimal levels of natural light have been shown to have a positive impact on a company's finances; the productivity of workers is enhanced due to increased health and well-being, and energy consumption is reduced as a result of the decreased need for artificial lighting (Edwards & Torcellini, 2002).

Plants
The benefits of plants for humans and the built environment is a widely researched topic (Freeman, 2011). A qualitative study by Thomsen et al. (2011) found that employees disliked artificial plants and regarded natural plants as contributing positively to the workplace and to their well-being, because they were living things. This finding reinforces the biophilia hypothesis. Positive affect was reported when the plants flowered or flourished (Thomsen et al., 2011), and similar findings (e.g. references to budding plants) emerged as the second most frequent theme in Richardson et al.'s (2015) research on the qualitative accounts of what things people find to be 'good' in nature. This may indicate that the continuous changes in nature constitute a notable feature that captivates human attention.

Plants also contribute invisible properties to the environment, such as fragrances (Grinde & Patil, 2009), and the risk of sick building syndrome is reduced as plants filter the air and decrease indoor air pollution (Fitzgerald & Danner, 2012). HFE evaluations of workplace air quality can include surveys that examine the health status and indoor air quality problems experienced by building occupants either prior to, or alongside, using special instruments (e.g. gas detector tubes) to test for indoor air pollutants (Hedge, 2005). These kinds of evaluation techniques can provide the practitioner with information that may be useful when developing a strategy to improve indoor air quality. Identifying pollutants also guides the choice of plant species for an indoor environment, as some plants were found to be better at purifying the air from certain pollutants than others—for example, *Osmunda japonica*, also known as the Japanese royal fern, was found to have a very high efficiency rate for removing formaldehyde from the air (Claudio, 2011). However, Kays (as cited in Claudio, 2011) pointed out that some indoor plants (including the pots, pesticides, microorganisms that they carry, and the medium in which they are grown) can possibly add volatile organic compounds to the air, making the exact function of plants in improving indoor air quality uncertain.

Nonetheless, Bringslimark, Hartig, and Patil (2007) found that having many plants in view resulted in decreased sick leave and stress and increased productivity in three Norwegian offices. However, in Thomsen et al.'s (2011) study, the condition of the plants sometimes distracted employees from their work, decreasing performance accordingly. Plants in a poor condition contributed to negative feelings (e.g. irritation) and affected overall well-being adversely (Thomsen et al., 2011). This finding highlights the unfavourable influence that indoor nature can have on a building's occupants if it is of an inferior quality or type. Maintenance of the plants was used to aid relaxation, undertake physical activity, and have a break from work (Thomsen et al., 2011). Work breaks are essential for ensuring that workers retain a balanced mental and physical workload, as both a high and a low workload can have negative effects on task performance and the overall work experience (Salvendy, 2012). For example, visual fatigue and repetitive strain injuries can be avoided or minimised by having frequent breaks away from computer workstations (Stranks, 2007). Stress reduction appears to be greater when people are exposed to plants during a recess than while executing a task. Therefore, plants in areas away from actual work (e.g. a cafeteria) may be more beneficial than plants in and around workstations (Bringslimark et al., 2007). When it is not desired or practical to have potted plants with foliage that branches out into a space, facade or interior wall greening with climbing or creeping plants (e.g. ivy and other vine plants) can add vegetation to the building itself (Kellert, 2005).

Views of Nature
Views of nature do not have to be expansive or of a long duration in order to be effective. Research by Kaplan (1993) found that employees with views from windows that included nature reported better health, higher job, and life satisfaction, and they were more patient and tolerant than those without views of natural elements. Window views also help to reduce

eyestrain, because the blend of long- and short-range views helps the eye to regain its focus (Edwards & Torcellini, 2002). This indicates that visual comfort is not solely affected by optimal lighting conditions for task execution, but that views are also necessary for such comfort (Bluyssen, 2009).

Chang and Chen (2005) conducted an experimental study where various combinations of office environments were simulated using computer software. Electroencephalography, electromyography, and blood volume pulse sensors were used to measure the participants' physiological responses to the simulated environments, and the results showed that people's physiological conditions were affected by views (e.g. blood volume pulse (an indication of psychological arousal) was lowest with window views of nature). Similar to Kaplan's (1993) findings, views of nature had more impact on physiological measures than views of the built environment (Chang & Chen, 2005). The psychological data also revealed that anxiety levels were reduced by window views and the presence of indoor plants. Both the psychological and physiological data showed that window views have more impact than indoor plants (Chang & Chen, 2005). This might explain why views of nature through windows often outweigh indoor plants in the workplace (Augustin, 2009). In dense urban areas where views of natural landscapes are absent, green roofs (i.e. roofs upon which vegetation is grown) have also been found to provide restoration to nearby building occupants (Gillis & Gatersleben, 2015).

When viewing waterscapes, people may remember or mentally reproduce the positive affect attached to being immersed in water; research has found that immersion in water reduces psychological and physiological stress (White et al., 2010). Yet, there is a lack of literature on the psychological benefits of water (Gillis & Gatersleben, 2015). When individuals rated 120 photographs of natural and built environments, aquatic scenes with greenery were rated higher than scenes only containing water for restorativeness and preference, which suggests that the combination of land and water may be especially appreciated (White et al., 2010). Similar to the findings concerning the quality of plants, the quality of water also influences its perceived restorative capacity. Clean water is perceived to be more restorative than brown or dirty water (Gillis & Gatersleben, 2015).

Workers' tasks and personal traits affect how window views are perceived and appreciated. For example, work requiring minimal concentration calls for a more dynamic view than work requiring higher levels of concentration (Bluyssen, 2009). Horticulturists and landscape designers should therefore consider the plants and landscapes around buildings from the perspective of the indoor occupants (Chang & Chen, 2005).

Artwork and Representations of Nature

Artwork and representations of nature contribute to the aesthetics of an interior. They can also be used to compensate for a lack of natural views in windowless workspaces, or to introduce natural themes when limited space or workplace conditions do not permit the inclusion of details like ornamental plants. Research shows that the three most appropriate types of images to use in healthcare settings, which are fitting for the workplace as well, are artwork depicting the countryside with meadows and trees, cultivated gardens, and tranquil water scenes (Augustin, 2009). It has been shown that images of nature in the workplace aid in reducing anger and stress and, together with plants, may possibly enhance job satisfaction, identification with the organisation, psychological and physical comfort, as well as productivity (Knight & Haslam, 2010). Abstract artwork—including abstract representations of nature—should be carefully chosen though, because it can evoke negative feelings that are not particularly restorative (Augustin, 2009).

Despite research indicating that the presence of a dog fosters unity, collaboration, and trust amongst group members working together on a task (Fitzgerald & Danner, 2012), it is often challenging and impractical to include living animals in the built environment (Kellert & Calabrese, 2015). Even aesthetically pleasing and relaxing aquariums with live fish may be difficult to install and maintain (Augustin, 2009). Hence, sculptures and pictures of animals can be added to the interior instead. However, research has found that patients who viewed images of nature that are considered stimulating (e.g. zebras staring directly at the viewer) had a higher systolic blood pressure than patients who viewed images of tranquil scenes (e.g. waterscapes) (Ulrich, 1991). Augustin (2009) also suggests that one should avoid displaying close-up images of animals in healthcare settings. Although further studies of people's psychological and physiological responses to depictions of animals in the work environment are needed, it may be tentatively assumed that designers should generally refrain from using any imagery or design features that might increase workers' stress levels by evoking feelings of threat or danger (e.g. a wall mural of animals exhibiting stalking behaviour).

Artificial products and finishes designed to look like natural materials (e.g. synthetic laminated flooring with a wood design, and wallpaper illustrating vegetation and foliage) are useful for creating an indoor environment that is more reminiscent of nature (Freeman, 2011). However,

Kellert et al. (2008) recommend that natural materials (e.g. stone tiles) should be used in design whenever it is possible, because people generally seem to prefer natural materials over imitations, and natural materials do not emit harmful substances.

More suggestive and indirect representations of nature, in the form of architectural features, patterns, and motifs, can be added to the interior and exterior of a building (Kellert et al., 2008). For example, a staircase banister with a decorative plant motif or a building exterior boasting tree-like column supports can enhance the aesthetics of a building while simultaneously exposing people to impressions of elements found in the natural world.

Sound

Sound is always present, and it directly affects the comfort, safety, health, and performance of building occupants (Reddy et al., 2012). Unwanted sounds are considered to be noise that causes a disturbance and prevents people from hearing the sounds that they would like to hear (Wang, Bakker, de Groot, & Wörtche, 2014). Although an individual's response to sound depends on factors such as hearing, age, and the context in which it is experienced, exposure to excessive noise can cause anxiety, stress, high blood pressure, and other physical ailments (Reddy et al., 2012). Research indicates that mechanical sounds are usually regarded as unpleasant, while natural sounds are not (Alvarsson, Stefan Wiens, & Nilsson, 2010).

Ratcliffe et al. (2013) refer to research where natural sounds were found to restore mental fatigue in office settings, and where listening to a bird song accompanied by classical music decreased participants' negative emotions. These studies show that sounds from nature might be instrumental in recovering from stress and adverse emotions. Bird song was perceived as a pleasant distraction that encouraged a break from task demands (Ratcliffe et al., 2013). Likewise, results from Alvarsson et al.'s (2010) study indicate that nature sounds aid the quicker recovery of the sympathetic nervous system after exposure to a stressful mental task. This outcome may be tied to the positive emotions that nature sounds evoke. However, not all bird songs or calls are pleasant. Some bird calls may represent a threat—based on the intensity, pitch, and coarseness of its acoustic qualities—and thereby stimulate arousal and hinder recovery from stress (Ratcliffe et al., 2013). White noise created from natural sounds—such as the sound of the ocean or a waterfall—masks undesirable sounds and appears to increase concentration in environments with other audible distractions (Augustin, 2009). Also, the leaves of indoor and outdoor plants at the workplace may help to reduce noise by partially absorbing sound waves (Wang et al., 2014).

While the above provides examples of the benefits of nature and the ways in which it can be merged with the built environment, it is still crucial that those involved in the planning and development of property look for ways in which to erect a building in a natural environment—instead of merely eliminating nature in order to make way for construction (Kellert et al., 2008). Ideally, biophilic design should encourage the long-term well-being and sustainability of the natural environment and not simply apply interim strategies that modify the physical conditions of a building or landscape (Kellert & Calabrese, 2015).

CONCLUSIONS

A large body of literature demonstrates that nature provides many benefits for humans. People who are exposed to nature or feel connected with it experience higher levels of well-being. Nature has a restorative effect on both depleted attention capacity and a stressed autonomic nervous system. In general, a straightforward message for the practitioner exists: exposure to nature is beneficial to well-being. While there is a need for research to identify best practice and most effective ways to introduce nature into the workplace, there is sufficient evidence of the benefit of nature in the work environment, and examples of this are happening in numerous and inventive ways. As these nature-based solutions come to the fore, an understanding of how to accommodate nature within working environments and working patterns needs to be developed. Biophilic design does not have to be the sole responsibility of, for example, designers and building managers. HFE professionals and others concerned with creating healthy workplaces can educate workers about the benefits of the human-nature connection, encourage behaviour change that prompts workers to engage in more frequent contact with nature at work, and also suggest changes to the physical environment that would increase their exposure to nature in the workplace. In short, ergonomists should understand the value of nature and include this knowledge in behavioural and work environment interventions in order to make the most of nature's beneficial effects, and help foster a positive attitude towards nature as we work towards a sustainable future.

REFERENCES

Alvarsson, J. J., Stefan Wiens, S., & Nilsson, M. E. (2010). Stress recovery during exposure to nature sound and environmental noise. *International Journal of Environmental Research and Public Health, 7*, 1036–1046.

An, M., Colarelli, S. M., O'Brien, K., & Boyajian, M. E. (2016). Why we need more nature at work: Effects of natural elements and sunlight on employee mental health and work attitudes. *PLoS One, 11*, e0155614. https://doi.org/10.1371/journal.pone.0155614

Anshel, J. (Ed.). (2005). *Visual ergonomics handbook*. Hoboken, NJ: CRC Press.

Augustin, S. (2009). *Place advantage: Applied psychology for interior architecture*. Hoboken, NJ: Wiley.

Bateson, G. (1972). *Steps to an ecology of mind. Collected essays in anthropology, psychiatry, evolution and epistemology*. Northvale, NJ: Jason Aronson.

Beil, K., & Hanes, D. (2013). The influence of urban natural and built environments on physiological and psychological measures of stress – A pilot study. *International Journal of Environmental Research and Public Health, 10*, 1250–1267.

Berto, R. (2014). The role of nature in coping with psycho-physiological stress: A literature review on restorativeness. *Behavioural Sciences, 4*, 394–409.

Bluyssen, P. M. (2009). *The indoor environment handbook: How to make buildings healthy and comfortable*. London/Sterling, VA: Earthscan.

Bringslimark, T., Hartig, T., & Patil, G. G. (2007). Psychological benefits of indoor plants in workplaces: Putting experimental results into context. *HortScience, 42*, 581–587.

Cacioppo, J. T., & Hawkely, L. C. (2009). Perceived social isolation and cognition. *Trends in Cognitive Science, 13*, 447–454. https://doi.org/10.1016/j.tics.2009.06.005

Cacioppo, J. T., & Patrick, W. (2008). *Loneliness: Human nature and the need for social connection*. New York: W. W. Norton Company.

Capaldi, C. A., Dopko, R. L., & Zelenski, J. M. (2014). The relationship between nature connectedness and happiness: A meta-analysis. *Frontiers in Psychology, 5*, 1–28.

Castles, S., De Haas, H., & Miller, M. J. (2014). *The age of migration international population movements in the modern world* (Fifth ed.). Basingstoke, UK: Palgrave Macmillan.

Cervinka, R., Roderer, K., & Hefler, E. (2012). Are nature lovers happy? On various indicators of well-being and connectedness with nature. *Journal of Health Psychology, 17*, 379–388.

Chang, C. Y., & Chen, P. K. (2005). Human response to window views and indoor plants in the workplace. *Hortscience, 40*(5), 1354–1359.

Chow, J. T., & Lau, S. (2015). Nature gives us strength: Exposure to nature counteracts ego depletion. *The Journal of Social Psychology, 155*, 70–85.

Claudio, L. (2011). Planting healthier indoor air. *Environmental Health Perspectives, 119*, 426–427.

Clayton, S. (2003). Environmental identity: A conceptual and operational definition. In S. Clayton & S. Opotow (Eds.), *Identity and the natural environment* (pp. 45–65). Boston: MIT Press.

Clayton, S. (2012). *Environment and identity. The Oxford handbook of environmental and conservation psychology*. New York: Oxford University Press.

Danna, K., & Griffin, R. W. (1999). Health and well-being in the workplace: A review and synthesis of the literature. *Journal of Management, 25*, 357–384.

Davis, J. L., Green, J. D., & Reed, A. (2009). Interdependence with the environment: Commitment, interconnectedness, and environmental behaviour. *Journal of Environmental Psychology, 29*, 173–180. https://doi.org/10.1016/j.jenvp.2008.11.001

Day, A., Kelloway, E. K., & Hurrell, J. J. (Eds.). (2014). *Workplace well-being: How to build psychologically healthy workplaces*. Chichester, UK: Wiley.

Dekker, S. W. (2013). On the epistemology and ethics of communicating a cartesian consciousness. *Safety Science, 56*, 96–99.

Drengson, A., & Devall, B. (2010). The deep ecology movement: Origins, development & future prospects. *The Trumpeter, 26*, 48–69.

Dul, J., Bruder, R., Buckle, P., Carayon, P., Falzon, P., Marras, W. S., et al. (2012). A strategy for human factors/ergonomics: Developing the discipline and profession. *Ergonomics, 55*, 377–395.

Edwards, L., & Torcellini, P. (2002). *A literature review of the effects of natural light on building occupants*. Oak Ridge, TN: U.S. Department of Energy.

Elliot, A. J. (2015). Color and psychological functioning: A review of theoretical and empirical work. *Frontiers in Psychology, 6*, 1–8.

Fiore, S. M., Phillips, E., & Sellers, B. C. (2014). A transdisciplinary perspective on hedonomic sustainability design. *Ergonomics in Design: The Quarterly of Human Factors Applications, 22*, 22–29.

Fitzgerald, C. J., & Danner, K. M. (2012). Evolution in the office: How evolutionary psychology can increase employee health, happiness, and productivity. *Evolutionary Psychology, 10*, 770–781.

Fox, W. (1990). Transpersonal ecology: "Psychologising" ecophilosophy. *The Journal of Transpersonal Psychology, 22*, 59–96.

Frantz, C., Mayer, F. S., Norton, C., & Rock, M. (2005). There is no "I" in nature: The influence of self-awareness on connectedness to nature. *Journal of Environmental Psychology, 25*, 427–436.

Freeman, K. (2011). *Nature-inspired interior landscaping: How to promote well-being in buildings by using the principles of biophilia in interior landscape design*. Retrieved February 27, 2015, from http://www.ambius.co.uk/about-ambius/biophilia/biophilia.pdf

Frumkin, H. (2001). Beyond toxicity. Human health and the natural environment. *American Journal of Preventative Medicine, 20*, 234–240.

Gilbert, P. (2014). The origins and nature of compassion focused therapy. *British Journal of Clinical Psychology, 53,* 6–41.

Gillis, K., & Gatersleben, B. (2015). A review of psychological literature on the health and wellbeing benefits of biophilic design. *Buildings, 5,* 948–963. https://doi.org/10.3390/buildings5030948

Grinde, B., & Patil, G. G. (2009). Biophilia: Does visual contact with nature impact on health and well-being? *International Journal of Environmental Research and Public Health, 6,* 2332–2343.

Gullone, E. (2000). The biophilia hypothesis and life in the 21st century: Increasing mental health or increasing pathology? *Journal of Happiness Studies, 1,* 293–321.

Haila, Y. (1999). Biodiversity and the divide between culture and nature. *Biodiversity and Conservation, 8,* 165–181.

Hanson, M. A. (2013). Green ergonomics: Challenges and opportunities. *Ergonomics, 56,* 399–408.

Hawkes, F. M., & Acott, T. G. (2013). People, environment and place: The function and significance of human hybrid relationships at an allotment in south east England. *Local Environment: The International Journal of Justice and Sustainability, 18,* 1117–1133.

Hedge, A. (2005). Indoor air quality: Chemical exposures. In N. Stanton, A. Hedge, K. Brookhuis, E. Salas, & H. Hendrick (Eds.), *Handbook of human factors and ergonomics methods* (pp. 64-1–64-11). Boca Raton, FL: CRC Press.

Hedge, A. (2008). The sprouting of "green" ergonomics. *Human Factors and Ergonomics Society (HFES), 51*(12), 1–3.

Hedge, A., & Dorsey, J. A. (2013). Green buildings need good ergonomics. *Ergonomics, 56,* 492–506.

Hedge, A., Rollings, K., & Robinson, J. (2010). "Green" ergonomics: Advocating for the human element in buildings. *Proceedings of the Human Factors and Ergonomics Society Annual Meeting, 54,* 693–697.

Hinds, J., & Sparks, P. (2008). Engaging with the natural environment: The role of affective connection and identity. *Journal of Environmental Psychology, 28,* 109–120.

Hinds, J., & Sparks, P. (2011). The affective quality of human-natural environment relationships. *Evolutionary Psychology, 9,* 451–469.

Howell, A. J., Dopko, R. L., Passmore, H. A., & Buro, K. (2011). Nature connectedness: Associations with well-being and mindfulness. *Personality and Individual Differences, 51,* 166–171.

Jalil, N. A., Yunus, R. M., & Said, N. S. (2012). Environmental colour impact upon human behaviour: A review. *Procedia – Social and Behavioral Sciences, 35,* 54–62.

Kahn, P. H. (1997). Developmental psychology and the biophilia hypothesis: Children's affiliation with nature. *Developmental Review, 17,* 1–61.

Kahn, P. H. (2011). *Technological nature: Adaptation and the future of human life.* Cambridge, MA: MIT Press.

Kals, E., Schumacher, D., & Montada, L. (1999). Emotional affinity toward nature as a motivational basis to protect nature. *Environment and Behaviour, 31,* 178–202.

Kaplan, R. (1993). The role of nature in the context of the workplace. *Landscape and Urban Planning, 26,* 193–201.

Kaplan, S. (1987). Aesthetics, affect, and cognition: Environmental preferences from an evolutionary perspective. *Environment and Behaviour, 19,* 3–32.

Kaplan, S. (1995). The restorative benefits of nature: Toward an integrative framework. *Journal of Environmental Psychology, 15,* 169–182.

Kellert, S. R. (1993). The biological basis for human values of nature. In S. R. Kellert & E. O. Wilson (Eds.), *The biophilia hypothesis.* Washington, DC: Island Press.

Kellert, S. R. (2003). *Kinship to mastery: Biophilia in human evolution and development.* Washington, DC: Island Press.

Kellert, S. R. (2005). *Building for life: Designing and understanding the human-nature connection.* Washington, DC: Island Press.

Kellert, S. R., & Calabrese, E. F. (2015). *The practice of biophilic design.* Retrieved September 16, 2015, from http://www.biophilicdesign.com

Kellert, S. R., Heerwagen, J., & Mador, M. (Eds.). (2008). *Biophilic design: The theory, science, and practice of bringing buildings to life.* Hoboken, NJ: Wiley.

Kellert, S. R., & Wilson, E. O. (Eds.). (1993). *The biophilia hypothesis.* Washington, DC: Island Press.

Knight, C., & Haslam, S. A. (2010). The relative merits of lean, enriched, and empowered offices: An experimental examination of the impact of workspace management strategies on well-being and productivity. *Journal of Experimental Psychology: Applied, 16,* 158–172.

Kuo, M. (2015). How might contact with nature promote human health? Promising mechanisms and a possible central pathway. *Frontiers in Psychology, 6,* 1093.

Kuoppala, J., Lamminpää, A., & Husman, P. (2008). Work health promotion, job well-being, and sickness absences: A systematic review and meta-analysis. *Journal of Occupational and Environmental Medicine, 50*(11), 1216–1227.

Lakin, J. L., Jefferis, V. E., Cheng, C. M., & Chartrand, T. L. (2003). The chameleon effect as social glue: Evidence for the evolutionary significance of nonconscious mimicry. *Journal of Nonverbal Behavior, 27,* 145–162.

Largo-Wight, E., Chen, W. W., Dodd, V., & Weiler, R. (2011). Healthy workplaces: The effects of nature contact at work on employee stress and health. *Public Health Reports, 126*(1), 124–130.

Lee, K. E., Williams, K. J., Sargent, L. D., Williams, N. S., & Johnson, K. A. (2015). 40-second green roof views sustain attention: The role of micro-breaks in attention restoration. *Journal of Environmental Psychology, 42,* 82–189.

Leopold, A. (1966). *A sand country almanac: With essays on conservation from Round River.* New York: Ballantine Books.

Light, A. (2000). What is an ecological identity? *Environmental Politics, 9,* 37–41.

Lin, B. B., Fuller, R. A., Bush, R., Gaston, K. J., & Shanahan, D. F. (2014). Opportunity or orientation? Who uses urban parks and why. *PLoS One, 9,* 1–7.

Logan, A. C., & Selhub, E. M. (2012). Vis medicatrix naturae: Does nature "minister to the mind"? *BioPsychoSocial Medicine, 6,* 11.

Lottrup, L., Grahn, P., & Stigsdotter, U. K. (2013). Workplace greenery and perceived level of stress: Benefits of access to a green outdoor environment at the workplace. *Landscape and Urban Planning, 110,* 5–11.

Lumber, R., Richardson, M., & Sheffield, D. (2017). Beyond knowing nature: Contact, emotion, meaning, compassion, and beauty are pathways to nature connection. *PLoS ONE, 12,* e0177186. https://doi.org/10.1371/journal.pone.0177186

Macy, J. (2007). *World as lover, world as self: Courage for global justice and ecological renewal.* California, CA: Parallax Press.

Maller, C., Townsend, M., Leger, L. S., Henderson-Wilson, C., Pryor, A., Prosser, L., et al. (2009). Healthy parks, healthy people: The health benefits of contact with nature in a park context. *The George Wright Forum, 26,* 51–83.

Martin, K., Legg, S., & Brown, C. (2013). Designing for sustainability: Ergonomics – Carpe diem. *Ergonomics, 56,* 365–388.

Martyn, P., & Brymer, E. (2014). The relationship between nature relatedness and anxiety. *Journal of Health Psychology,* Advance online publication, 1–10. https://doi.org/10.1177/1359105314555169.

Mayer, F. S., & Frantz, C. M. (2004). The connectedness to nature scale: A measure of individuals' feeling in community with nature. *Journal of Environmental Psychology, 24,* 503–515.

Mayer, F. S., Frantz, C. M., Bruehlman-Senecal, E., & Dolliver, K. (2009). Why is nature beneficial? The role of connectedness to nature. *Environment and Behaviour, 41,* 607–643.

McDonald, S., Kirk, D., & Kinns, N. B. (2015). *Nature bot: Experiencing nature in the built environment.* Proceedings of the 2015 ACM SIGCHI conference on creativity and cognition, pp. 173–176. Retrieved June 30, 2015, from http://dl.acm.org/citation.cfm?id=2764547

McMahon, E. A., & Estes, D. (2015). The effect of contact with natural environments on positive and negative affect: A meta-analysis. *The Journal of Positive Psychology, 10*(6), 507–519.

McPhie, J., & Clarke, D. A. G. (2015). A walk in the park: Considering practice for outdoor environmental education through an imminent take on the material turn. *Journal of Environmental Education, 46,* 230–250.

Meerwein, G., Rodeck, B., & Mahnke, F. H. (2007). *Color: Communication in architectural space.* Basel, Switzerland: Birkhauser Verlag.

Moray, N. (1993). Technosophy and humane factors. *Ergonomics in Design, 1*(4), 33–39.

Muller, M. M., Kals, E., & Pansa, R. (2009). Adolescents' emotional affinity to nature: A cross-societal study. *The Journal of Developmental Processes, 4*(1), 59–69.

Narvaez, D. (2013). Development and socialization within an evolutionary context: Growing up to become "a good and useful human being". In D. Fry (Ed.), *War, peace and human nature: The convergence of evolutionary and cultural views* (pp. 643–672). New York: Oxford University Press.

Nisbet, E. K., & Zelenski, J. M. (2011). Underestimating nearby nature: Affective forecasting errors obscure the happy path to sustainability. *Psychological Science, 22*, 1101–1106.

Nisbet, E. K., & Zelenski, J. M. (2013). The NR-6: A new brief measure of nature relatedness. *Frontiers in Psychology, 4*, 813.

Nisbet, E. K., Zelenski, J. M., & Murphy, S. A. (2009). The nature relatedness scale: Linking individuals' connection with nature to environmental concern and behaviour. *Environment and Behaviour, 41*, 715–740.

Nisbet, E. K., Zelenski, J. M., & Murphy, S. A. (2011). Happiness is in our nature: Exploring nature relatedness as a contributor to subjective well-being. *Journal of Happiness Studies, 12*, 303–322.

Olivos, P., Aragones, J. I., & Amerigo, M. (2011). The connectedness to nature scale and its relationship with environmental beliefs and identity. *International Journal of Hispanic Psychology, 4*, 5–19.

Parsons, K. C. (2000). Environmental ergonomics: A review of principles, methods and models. *Applied Ergonomics, 31*, 581–594.

Proctor, R. W., & Van Zandt, T. (2008). *Human factors in simple and complex systems* (2nd ed.). Boca Raton, FL: Taylor & Francis Group.

Pyle, R. M. (2003). Nature matrix: Reconnecting people with nature. *Oryx, 37*, 206–214.

Ratcliffe, E., Gatersleben, B., & Sowden, P. T. (2013). Bird sounds and their contributions to perceived attention restoration and stress recovery. *Journal of Environmental Psychology, 36*, 221–228.

Reddy, S. W., Chakrabarti, D., & Karmakar, S. (2012). Emotion and interior space design: An ergonomic perspective. *Work, 41*, 1072–1078.

Richardson, M., Hallam, J., & Lumber, R. (2015). One thousand good things in nature: Aspects of nearby nature associated with improved connection to nature. *Environmental Values, 24*, 603–619.

Richardson, M., & Sheffield, D. (2017). Three good things in nature: Noticing nearby nature brings sustained increases in connection with nature. *Psyecology, 8*, 1–32.

Richardson, R., Maspero, M., Golightly, D., Sheffield, D., Staples, V., & Lumber, R. (2017). Nature: A new paradigm for well-being and ergonomics. *Ergonomics, 60*(2), 292–305.

Russell, R., Guerry, A. D., Balvanera, P., Gould, R. K., Basurto, X., Chan, K. M.-A., et al. (2013). Humans and nature: How knowing and experiencing nature affect well-being. *Annual Review of Environment and Resources, 38*, 6.1–6.30. https://doi.org/10.1146/annurev-environ-012312-110838

Ryan, C. O., Browning, W. D., Clancy, J. O., Andrews, S. L., & Kallianpurkar, N. B. (2014). Biophilic design patterns: Emerging nature-based parameters for health and well-being in the built environment. *International Journal of Architectural Research: ArchNet-IJAR, 8*, 62–76.

Salvendy, G. (2012). *Handbook of human factors and ergonomics* (4th ed.). Hoboken, NJ: Wiley.

Schatz, S., & Bowers, C. (2005). 10 questions on room color: Answers for workplace designers. *Ergonomics in Design, 13*(4), 21–27.

Schultz, P. W. (2001). The structure of environmental concern: Concern for self, other people, and the biosphere. *Journal of Environmental Psychology, 21*, 327–339. https://doi.org/10.1006/jevp.2001.0227

Simaika, J. P., & Samways, M. J. (2010). Biophilia as a universal ethic for conserving biodiversity. *Conservation Biology: The Journal of the Society for Conservation Biology, 24*, 903–906.

Stranks, J. W. (2007). *Human factors and behavioural safety* (1st ed.). Burlington, ON: Elsevier.

Tam, K. P. (2013). Concepts and measures related to connection to nature: Similarities and differences. *Journal of Environmental Psychology, 34*, 64–78.

Thatcher, A. (2013). Green ergonomics: Definition and scope. *Ergonomics, 56*, 389–398.

Thatcher, A., & Milner, K. (2014). Green ergonomics and green buildings. *Ergonomics in Design: The Quarterly of Human Factors Applications, 22*, 5–12.

Thomsen, J. D., Sønderstrup-Andersen, H. K. H., & Müller, R. (2011). People-plant relationships in an office workplace: Perceived benefits for the workplace and employees. *Hortscience, 46*, 744–752.

Trau, D., Keenan, K. A., Goforth, M., & Large, V. (2015). Nature contacts: Employee wellness in healthcare. *HERD: Health Environments Research & Design Journal, 9*, 47–62.

Ulrich, R. S. (1983). Aesthetic and affective response to natural environment. In I. Altman & J. Wohlwill (Eds.), *Human behavior and environment volume 6: Behavior and natural environment* (pp. 85–125). New York: Plenum.

Ulrich, R. S. (1991). Effects of interior design on wellness: Theory and recent scientific research. *Journal of Health Care Interior Design, 3*, 97–109.

Vining, J., Merrick, M. S., & Price, E. A. (2008). The distinction between humans and nature: Human perceptions of connectedness to nature and elements of the natural and unnatural. *Research in Human Ecology, 15*, 1–11.

Wang, Y., Bakker, F., de Groot, R., & Wörtche, H. (2014). Effect of ecosystem services provided by urban green infrastructure on indoor environment: A literature review. *Building and Environment, 77*, 88–100.

White, M., Smith, A., Humphryes, K., Pahl, S., Snelling, D., & Depledge, M. (2010). Blue space: The importance of water for preference, affect, and restorativeness ratings of natural and built scenes. *Journal of Environmental Psychology, 30,* 482–493.

Wilson, E. O. (1993). Biophilia and the conservation ethic. In S. R. Kellert & E. O. Wilson (Eds.), *The biophilia hypothesis* (pp. 31–41). Washington, DC: Island Press.

Wilson, E. O. (2002). *The future of life.* New York: Alfred A. Knopf.

Wilson, M. G., Dejoy, D. M., Vandenberg, R. J., Richardson, H. A., & McGrath, A. L. (2004). Work characteristics and employee health and well-being: Test of a model of healthy work organization. *Journal of Occupational and Organizational Psychology, 77,* 565–588.

Windhager, S., Atzwanger, K., Bookstein, F. L., & Schaefer, K. (2011). Fish in a mall aquarium: An ethological investigation of biophilia. *Landscape and Urban Planning, 99,* 23–30.

Wyles, K. J., Pahl, S., Thomas, K., & Thompson, R. C. (2016). Factors that can undermine the psychological benefits of coastal environments exploring the effect of tidal state, presence, and type of litter. [Electronic version]. *Environment and Behaviour, 48,* 1095–1126.

Zelenski, J. M., & Nisbet, E. K. (2012). Happiness and feeling connected: The distinct role of nature relatedness. [Electronic version]. *Environment and Behaviour.* https://doi.org/10.1177/0013916512451901

CHAPTER 8

Building Sustainable Organisations: Contributions of Activity-Centred Ergonomics and the Psychodynamics of Work

Claudio Marcelo Brunoro, Ivan Bolis,
and Laerte Idal Sznelwar

INTRODUCTION—SUSTAINABLE ORGANISATIONS: A COMPLEX SYSTEM-OF-SYSTEMS APPROACH AND THE INDIVIDUAL LEVEL

As discussed in previous chapters, ergonomics can contribute in various ways to different scopes of sustainability (e.g. responsible consumption, interactions with sustainable devices, human connections with nature, mobility, green buildings, sustainable product cycles, sustainable supply chains, and green ergonomics). Concepts of sustainability are incorporated in different ways in society (e.g. the sustainable development concept) and in organisations (i.e. the corporate sustainability concept). In this regard, one of the main challenges for sustainable development is related to promoting an integration of these different scopes within a perspective of complexity, as proposed by Thatcher and Yeow (2016) in the complex system-of-systems approach (Fig. 8.1).

C. M. Brunoro (✉) • I. Bolis • L. I. Sznelwar
Production Engineering Department, Polytechnic School, University of São
Paulo, São Paulo, Brazil

© The Author(s) 2018   191
A. Thatcher, P. H. P. Yeow (eds.), *Ergonomics and Human Factors for a
Sustainable Future,* https://doi.org/10.1007/978-981-10-8072-2_8

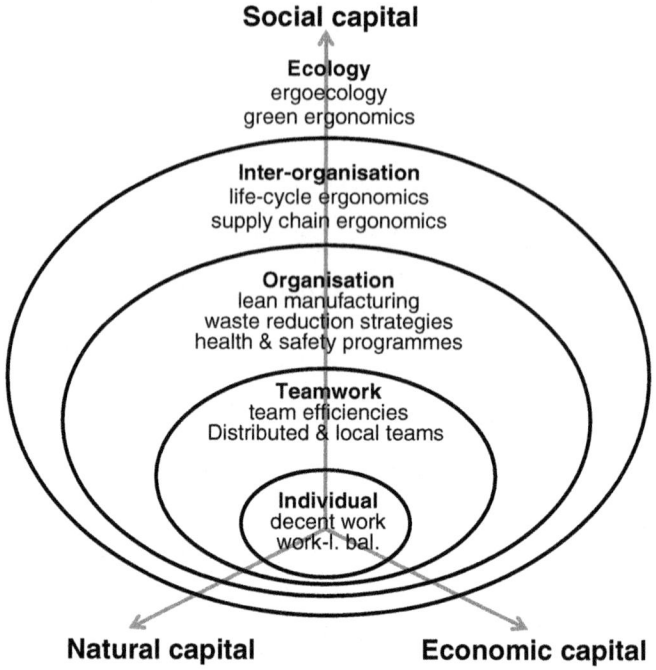

Fig. 8.1 A complex system-of-systems approach (Thatcher & Yeow, 2016)

From this perspective, an organisation aligned with principles of sustainability (corporate sustainability) should be seen as a system (or subsystem) that simultaneously interferes in and undergoes interference from other internal subsystems, particularly at the level of teamwork and of people. Moreover, if the system is particularly considered as an open system, external interferences also modulate how decision makers act. Our focus is related to the construction of a sustainable perspective by different actors in an organisation; specially the relevant issues pertaining to the internal social dimension that cannot be ignored. So, how do corporate actors perceive relevant issues about the relationships between work activity and corporate sustainability? Have they already been implemented? It is important to identify relationships between work and corporate sustainability, not only in order to ensure favourable conditions for companies to achieve their goals, but also from the standpoint of sustainability for each person and the company's workforce as a group.

In this sense, activity-centred ergonomics (ACE) and the psychodynamics of work (PDW) are particularly relevant in order to contribute to this discussion. As both ACE and PDW are not actually incorporated into a corporate sustainability perspective by actors inside organisations, the relationship with corporate sustainability (CS) has to be built. Therefore, the objective of this chapter is to point out how ideas, premises, and approaches based on ACE and PDW can contribute to creating an organisation with a sustainability perspective. Both approaches consider that work activity is fundamental not only for the constitution of the individual human being but also for the society, since work systems are essential for civilising processes, that is, the possibility to live with others (Arendt, 1959). So, what do sustainability principles adopted by companies imply for people in terms of transforming work in a broader sense, especially in terms of its relationship with an individual's (mental) health?

Moreover, if an organisation declared itself as sustainable, based on a system-of-systems approach, people's work must be revalued and transformed based on the context of sustainability. However, as will be shown in this chapter, internal social corporate sustainability politics are generally not related to work activities themselves (e.g. voluntary work) or what happens through limited and isolated politics. For instance, it is commonly organisations that consider the benefits of politics and the programmes aimed at welfare as part of corporate social sustainability. Despite being useful, they are usually indirect and mitigatory and not related to what workers are actually doing. These actions focus mainly on the effects caused to the worker rather than the factors that modulate a given situation (i.e. the way the work is conceived, organised, and evaluated, and the content of the tasks, which ACE and PDW state).

Third, work from a sustainability perspective requires full consideration of potential externalities that may be generated, since the individuals involved in a productive system are not only workers but also members of society. Whatever happens at the micro level (e.g. the individual and the population of workers in the organisation) has direct and indirect impacts on the macro level (i.e. society), and vice versa. This is an important aspect to promote the development of individuals (and society). This also involves the affective qualities of an activity that is performed well and the comfort conditions that enable a person to act conveniently. It can also become significant for organisations that unsustainable situations can engender high turnover rates, loss of quality in production, accidents, and illnesses.

Before answering these questions, it is essential to examine the current concepts of (social) corporate sustainability and its relationship with the theme of work. After, we will proceed by highlighting ergonomics and PDW connections with corporate sustainability that have already established. Following, Brazilian case studies will be presented in order to illustrate corporate perceptions and how social corporate sustainability is implemented. Last, a discussion will follow based on the ACE and PDW approaches.

WORK AND CORPORATE SUSTAINABILITY

The concept of (social) corporate sustainability is not limited to one definition; it is somehow polysemic, since different authors use this concept with diverse meanings. Some of its definitions and their authors are listed below (Table 8.1).

In general, the main elements mentioned by these authors involve (Morioka & Carvalho, 2016):

1. Goals, synergies, and economic, environmental, and social trade-offs: the relevance of considering at least important dimensions of sustainability and their interdependence.
2. Corporate stakeholders, including investors, employees, customers, the natural environment, community, the state, and so on: the relevance of considering all stakeholders.
3. Short-, medium-, and long-term impacts for current and future generations: the relevance of considering temporality for all decisions.

There was also unanimity concerning the importance of internal workplace practices which highlight health and safety polices and working conditions, especially for physical aspects. This is in line with the assumptions for sustainable work system propositions (Docherty, Kira, & Shami, 2009; Fischer & Zink, 2012; Zink, 2014).

Moreover, when organisational propaganda states an alignment with the guidelines of corporate sustainability, what can be clearly identified is the introduction of proposals related to the Global Reporting Initiative (GRI, 2013), ISO26000 (ISO, 2010) and the Global Compact (UNGC, 2014). Concerning social issues, those approaches presumably include work-related issues, such as employment and labour relations, working

Table 8.1 Definitions of (social) corporate sustainability

Author	Definition
Dyllick and Hockerts (2002)	Meet the needs of direct and indirect stakeholders (such as shareholders, employees, clients, advocacy groups, communities, etc.) without compromising the ability of future stakeholders to meet their own needs
Gladwin, Krause, and Kennelly (1995, p. 42)	A socially sustainable corporation is one that does not cause direct or indirect net loss of social capital, human rights, and human capital and that meets the basic needs of its own employees and of the communities in which it operates
Zadek, Pruzan, and Evans (1997, p. 13)	Socially sustainable corporations are those that are considered fair and trustworthy by all their stakeholders
Dyllick and Hockerts (2002)	Socially sustainable companies add value to the communities in which they operate, improving the human capital of their members, and promoting the social capital of those communities. They generate social capital such that their stakeholders can understand their motivations and broadly agree with the company's value system
Savitz and Weber (2007)	A sustainable company conducts its business with a view to naturally generating a flow of benefits for all its stakeholders, including employees, customers, business partners, the community in which it operates, and, of course, its shareholders
Figge and Hahn (2004)	Definition of corporate sustainability: the sustainability of a company is judged based on its economic, environmental, and social performance
Salzman, Ionescu-Somers, and Steger (2005, p. 27)	Corporate sustainability management should be defined as a strategic and for-profit response to environmental and social issues resulting from the organisation's primary and secondary activities
Baumgartner and Ebner (2010, pp. 79–80)	Corporate social sustainability is a company's awareness of the responsibility of its own actions, as well as an authentic commitment (especially long-term) to all its business and other activities, with the goal of remaining in the market successfully for a long period of time
Van Marrewijk and Werre (2003)	Corporate sustainability refers to the activities of a company—voluntary by definition—that demonstrate the inclusion of social and environmental concerns in its business operations and in its interactions with stakeholders. This is the broad—some would say "vague"—definition of corporate sustainability
Hart and Milstein (2004)	A sustainable company, therefore, is one that contributes to sustainable development by simultaneously generating economic, social, and environmental benefits—known as the three pillars of sustainable development
Bansal (2005)	Sustainable corporate development is based on three principles: economic integrity, social equity, and environmental integrity

conditions and social protection, social dialogue, health and safety at work, human development and training at work, and those pertaining to human rights.

There are large spaces to publicise the adherence to sustainability policies. In general, the descriptions include actions in each dimension of the Triple Bottom Line statements, including the issue of work as an aspect pertaining to the internal social dimension. An analysis of corporate websites (Bolis, Brunoro, & Sznelwar, 2014) and of sustainability reports (Bolis, Morioka, Brunoro, & Sznelwar, 2013) revealed that statements are strongly focused on work-related issues. However, data contained in those reports is generic, lacking information about the impact on workers' activities resulting from changes aimed at sustainability. Moreover, both studies found evidence of actions solely developed in operations management (i.e. top-down actions).

On the other hand, it is widely recognised that workers and their expertise are important for the development of sustainability-related actions (as described later in Brazilian cases) (Bolis, Brunoro, & Sznelwar, 2016). The vast majority of the reports include statements that comply with labour rules and regulations pertaining to the minimum requirements of working conditions, as is the case in considerations involving decent work.

However, the challenges to sustainability in terms of people who work in those companies must go further, and this means an understanding of actual work and its consequences. These include not only the incidence and prevalence of musculoskeletal disorders and other work-related disorders, diseases, and accidents but also the increasing rates of psychological disorders accompanied by sick leave, as well as high turnover rates, which are related to the work conditions and work organisation. Professional development and the possibilities of building one's health are central for this debate, since sustainable work would not be related to the issue of illness. It is not simply a matter of prevention, but rather, of health promoting and professional development for both the individual and the collective.

That's why multiple scales of analysis is an important strategic part of corporate sustainability in order to conceive, design, and manage production (or work) systems in accordance with the concepts of sustainability (Fischer & Zink, 2012; Zink, 2014). According to Docherty et al. (2009), the purpose of a sustainable work system should be to regenerate the resources it uses—human and natural resources—returning them to society preserved or improved (i.e. developed), and the development of one type of resource should not come at the expense of the exploitation of the other resources, thus incurring the responsibility for external costs (Kira & Eijnatten, 2009).

It is also important to distinguish between human and other resources. If natural resources can be preserved, human beings are the only "resource" that can be developed (Hubault & Tertre, 2008). If working systems afford conditions for professional development and for being healthier, it would be possible to state that the systems are more sustainable in terms of social issues. Since there is not a neutral working system (Dejours, 2009), if the scenario doesn't afford conditions for workers' development, the results are opposed, like accidents, diseases, and other negative externalities.

Within the scope of corporate sustainability, a relevant question is "How can we design and maintain economically viable work systems that contribute positively to human, social, and ecological sustainability?" (Docherty et al., 2009, p. 3). The next section addresses this question in light of ACE and PDW.

CONNECTIONS BETWEEN ERGONOMICS AND CORPORATE SUSTAINABILITY

Overall Connections

Ergonomics provides supporting evidence for issues pertaining to the internal social dimension of corporate sustainability, particularly those involved in decisions that impact the organisation and content of work, identifying not only the instigators of an integrated long-term view of performance and health that goes beyond legal issues but also benefiting both management and workers and, hence, society (Zink, 2006).

All corporate sustainability's main elements are considered from the ergonomics perspective and what this may involve (Bolis et al., 2014):

- The organisation's performance and workers' well-being: mainly related to economic, environmental, and social synergies (and trade-offs) and the relevance of considering all stakeholders. It is of the utmost importance to consider whether environmentally or economically sustainable production processes have any impact on health and professional development of workers, and also what the workers actually do to provide conditions to help to implement sustainable processes at companies. Contributions of ergonomics are highly significant with respect to these issues (Hanson, 2013; Ryan & Wilson, 2013). Its contributions are relevant even in situations where the company's focus is concentrated solely on the field of environmental sustainability (Thatcher, 2013). One could even include the impact of the work environment on creativity (Lukersmith & Burgess-Limerick, 2013). Moreover, even in workplaces with less adequate physical and environmental characteristics, there are workers who suffer increasingly from mental disorders (Harnois & Gabriel, 2000; Houtman & Kompier, 2011).

- Improving working conditions to render them practicable for the individual throughout their life: mainly related to the relevance of considering all stakeholders from a short-, medium-, and long-term perspective. Ergonomics deals with this by stating the need for work to be adapted to human characteristics. The causes of the most prevalent issues need to be addressed, avoiding mitigating actions. Health and wellness programmes are clearly necessary and positive, for example, programmes that encourage the practice of physical exercise (Bridger, Brasher, & Bennett, 2013). Therefore, when dealing with broader (macro) aspects of work, the aim of ergonomics is not to limit its scope of action (Dul et al., 2012). From this standpoint, job content and other organisational aspects are very relevant, mainly to ensure suitable conditions for the development of quality work, and hence, for professional development and the construction of health.
- Support for changes: mainly related to short-, medium-, and long-term perspective and the relevance of considering all stakeholders. Greater consideration of workers' opinions in planning and implementation of sustainability-related projects can contribute to improve organisational performance (Metzner & Fischer, 2010).

Thus, within a complex system-of-systems approach (Thatcher & Yeow, 2016) and also the framework for the design of sustainable work systems (Zink, 2014), a reality of work from the perspective of sustainability should recognise and avoid the pressures and constraints that undermine the health of a team, an area, or even an organisation. The next subsection provides both ACE's and PDW's relevant elements aligned with this perspective.

ACTIVITY-CENTRED ERGONOMICS AND PSYCHODYNAMICS OF WORK (RELEVANT ELEMENTS)

ACE (Daniellou, 2005; Daniellou & Rabardel, 2005), through ergonomic work analysis (Guérin, Laville, Daniellou, Duraffourg, & Kerguelen, 2001; Wisner, 1995a, 1995b), meets a dual objective (i.e. the concerns of the organisation and workers), considering health in a broader sense as a dynamic process (Dejours, 1986).

Health is mainly related to constraints of an organisational scope (Westgaard & Winkel, 2011). This point poses a challenge for ergonomics to expose reality, since the nature of these relationships is becoming increasingly qualitative and subtle (Kleine & Hauff, 2009). However, many of these relationships are often invisible and usually not considered by management. At most, they are dealt with as externalities (Hubault & Tertre, 2008).

As Duarte, Béguin, Pueyo, & Lima (2015) point out, it is highly essential that the design of sustainable work systems focuses particularly on the political and technical dimensions of project management in order to consider work activity within sustainable development. For ACE, it is essential that the company develop a strategic management that considers work activity as central to the company itself, because by taking into account what its workers actually do, regardless of their position in the hierarchy, it can promote actions that result in effective change in work.

In this case, the engagement of the workers involved is essential not only to ensure a better understanding of the activity but also to develop better projects (Daniellou, 2005; Daniellou & Rabardel, 2005; Guérin et al., 2001; Wisner, 1995a, 1995b). It should be kept in mind that this is not simply a strategy to gain people's greater involvement, but especially to place on the agenda of discussions, knowledge that the company's top management is usually unaware of. In this context, in order to ensure a corporate sustainability perspective, it is also very relevant to identify whether an environmentally or economically "sustainable process" causes an impact on workers.

It is necessary to identify aspects directly related to the content, conditions, and organisation of work, which must be taken into account. This means that we cannot address only one of the aspects that make up people's activities, such as those of a physical, cognitive, or psychological nature; it is also necessary to understand the different dimensions of work activities and the actual situation of production.

PDW, a clinical discipline, is the discipline that primarily studies what is related with the subjective mobilisation that occurs through interrelations established in and with work (Molinier, 2013). Work plays a central role in people's lives, since the ability to act is connected to the individual's desire to feel useful. This is an element that fundamentally presupposes the collective and that enables the development of processes of identification with work and of developing intelligence (Dejours, 2004a). Thus, the main themes that stand out are those pertaining to self-realisation, to reinforcement of the process of identification, and considering the individual as a protagonist (Sznelwar, 2015) in a given work situation, as well as collective issues, hierarchical relations of coordination, and of cooperation.

In this context, work is not limited solely to producing, but also represents opportunities for self-transformation. These require an atmosphere of trust, promoted by a relationship that emphasises genuine cooperation, as opposed to a lesser and merely utilitarian version (Dejours, 2009). An organisation of work that enables the mobilisation of intelligence and the dynamics of recognition and (symbolic) retribution provides opportunities for transforming suffering into pleasure, for coping with what is real, and for acting on a perspective of achieving self-realisation as well as emancipation (Dejours & Gernet, 2011, p. 42).

The common background shared by these two approaches (ACE and PDW) is the proposal of conceiving an interesting work (i.e. work that makes sense in a given profession and respects its values and traditions). In this regard, the challenge lies in designing work that makes sense. There is no definitive answer, but there are elements to be considered and that should be incorporated to render work interesting, as opposed to work that is alienating and has a strong potential to lead to suffering and illness. Hence, it is important to focus on and develop organisational mechanisms that enable the construction of trust and genuine cooperation (among individuals), instead of those based on fear and reluctant cooperation (Dejours, 2004b; Molinier, 2001).

The next section presents some Brazilian case studies. The objectives of this research were: (1) to capture corporate perceptions about internal social corporate sustainability and the relationships between work activity and corporate sustainability and (2) to analyse the findings according to both ACE's and PDW's relevant elements.

WORK-RELATED ISSUES AND CORPORATE SUSTAINABILITY: CASES OF BRAZILIAN ORGANISATIONS

General Aspects

In order to illustrate corporate perceptions about internal social corporate sustainability and its analyses based on ACE and PDW, this section presents main results from ten case studies conducted in Brazilian multinational companies considered benchmark in corporate sustainability (Brunoro, Bolis, Sznelwar, & Kawasaki, 2014). Corporate perception was captured based on corporate official documents (e.g. sustainability reports and corporate websites) and in-depth interviews (semi-structured interviews and open-ended questions). Table 8.2 describes the industrial sector of each company that was interviewed and the respective positions of the respondents.

Data were gathered according to complex system-of-systems approach (macro, medium, and individual level and their interconnected and interdependent links) and analysed in light of both ACE's and PDW's relevant elements. Worker's health (in a broader sense) was the principal internal social corporate sustainability issue to be examined.

Macro Level (Companies' Vision, Mission, and Values)

The visions interviewees described for "sustainable work" were holistic and convergent. It should favour a work-life balance, make sense, and promote health and well-being. Thus, they argue that work should be an

Table 8.2 Number and position of respondents in each area of the companies

Company	Industrial sector	Position and number of respondents from each area		
		Sustainability	Engineering	HR
1	Chemical	Director (1)	Director (1)	Director (1)
2	Food	Manager (1)	–	Manager (1)
3	Transport	Manager (1)	Manager (2)	–
4	Finance	Manager (1)	–	Manager (3)
5	Auditing	Director (1)	–	Director (1)
6	Consumer goods	–	–	Manager (1)
7	Food	–	–	Manager (1)
8	Consumer goods	Analyst (1)	–	Manager (1)
9	Packaging	Analyst (1)	–	–
10	Food	–	–	Manager (1)

action that is positive for both the company and the worker, in a win-win relationship. The centrality of the human being also emerges, geared primarily to relationships and trust.

For instance, this paragraph extracted from a corporate website expresses the most common issues addressed (direct or indirect related to worker's health):

> *Building a stimulating and creative work environment in which people feel respected in their individuality, recognized for their contributions, encouraged to accept challenges, and seeking the new is a daily exercise that involves all of us. We have the ongoing challenge of building and cultivating ethical and transparent relationships and creating a safe and healthy environment where dynamism, pleasure, trust and cooperation are always present. With the commitment of all, with the living of our principles, we will improve our relationships and our way of working and we will live happier. We establish open and honest dialogue with our managers, our teams and our peers. We encourage diversity in our staff and respect individuality. We guarantee equal opportunities and fair treatment for all. We value teamwork, recognizing and rewarding each one's contribution based on the achievement of goals and competencies. We seek to have compensation systems that allow a fair distribution of company results. We seek the individual and team training for the full exercise of our functions. [...] We give transparency to the criteria used in evaluating the activities and we try to know how our work is evaluated. We do not engage in abusive, inappropriate or offensive behaviour in the work environment, whether verbal, physical or gestural. Sexual and moral harassment are unacceptable behaviours. [...]. We are committed to the permanent pursuit of quality of life, considering professional achievement, social and family integration, good physical and mental health. We are all responsible for this achievement.*

However, little was mentioned about the predominance of organisational aspects (influencing work organisation or work content) in this regard. In general, the individual issues (time of life, family history, individual values, etc.) and the leader's responsibility to maintain balance in the work environment overshadowed the consideration of the implications for the work organisation.

At the same time, not all work-related issues are considered as priorities in a corporate sustainability strategy. In general, the sustainability strategies identified were introduced to shift production more broadly towards an overall policy of (limited) sustainability, involving mainly economic issues and the preservation of natural resources.

Moreover, changes in labour and pension laws have driven companies to reduce accident and illness rates by investing in health and safety programmes so as to avoid paying heavier fines or higher tax contributions. Nevertheless, even companies considered exemplary in terms of their organisational climate have to face worrisome data concerning employee turnover rates, sick leave due to mental disorders, and job dissatisfaction.

Medium Level (Internal Social Corporate Sustainability Practices)

When social sustainability is officially considered from the company's perspective, the primary understanding is usually focused on the external audience. Various examples of this can be cited (e.g. programmes for community well-being, social foundations, sponsorship of arts and culture programmes, and supply contracts). The latter are more stringent in terms of ensuring decent jobs throughout the supply chain.

Practices addressing internal audiences (considered as internal social sustainability actions) were also enumerated: voluntary work, workplace exercises, safety training programmes, home office programmes, psychological assistance, rehabilitation and return-to-work programmes, channels to report harassment complaints, and a wide range of benefits and life quality programmes (e.g. quit smoking programmes, private health and retirement plans, kindergartens, meal and transportation allowance, etc.).

Regarding the issue of healthier workplaces, health and safety actions were identified, particularly those involving physical aspects. As for psychological (or mental) issues, actions aimed at health, well-being, and quality of life were identified (e.g. meditation rooms, yoga classes and stress management programmes, as well as channels of individual attention).

Companies studied are organised functionally, with each of the departments acting almost independently of the others, although the importance and the need to act sustainably were shared premises. In short, each department had a specific "function" and the issue of sustainability was the responsibility of a specific area. The absence of a vision and of core actions is even more evident when the functional area responsible for sustainability does not include work-related issues, especially as they pertain to health. The interviewees stated that these issues are the responsibility of the health and safety sector, which also addresses work-related issues without connecting to the production sectors, focusing its actions more on the prevention of illness and accidents instead of on a perspective of health construction and professional development.

The need for individual commitment was emphasised, because even though all the conditions in the company may be favourable, people's behaviour (in relationships) and work postures (physical) must be in keeping with these conditions so as to avoid rendering them unfavourable, including for the individual themselves. However, the prevailing discourse still ascribes problems of health and safety to people's behaviour, as though the issue of health was secondary for the worker, because their behaviour is inconsistent with its preservation, or even with its promotion. Therefore, many initiatives involve training aimed at changing how people behave in situations that could endanger their own health, rather than being aimed at effective work transformation processes. Organisational issues are not mentioned as inducers of these situations when these situations are explicitly posited as possible causes of such misalignments. Even then, the responsibility is relegated to individual decisions, as if the organisation provides the proper conditions, but people do not know how to take advantage of what is offered and they are therefore the ones most responsible for any health problems, or even for failures in production. Hence, it is understood that it is up to people themselves to decide whether or not they are willing to deal with these issues. It should also be noted that individual issues such as values and goals, family history, the time of life, and the leader-led relationship are crucial factors in triggering what they call "stress" or personal problems. In this regard, both physical and psychological issues would be resolved through the initiative of individuals, and it would be up to the companies to support and guide them, without considering work per se.

Individual Level (Worker's Health)

There is a mental model of the successful worker. An ideal professional is one who possesses the necessary characteristics to cope with intense, varied, and unpredictable demands. He must be proactive, multi-skilled, resilient, and be able to deal with all sorts of unforeseen challenges. At some companies, due to a strong commitment to operations safety, boldness is a personal trait to be watched for, but even in such cases people are expected to take on an increasing number of responsibilities.

The situation can be summarised by an apparent paradox: while tasks are complex and heterogeneous, ideal professionals are fundamentally homogeneous. Some interviewees described the organisational climate as a pressure cooker, in which those who show poor performance were at higher risk of being penalised, transferred, or fired. These issues are only dealt with in a more systemic and integrated way in an analysis of indicators for the preparation of annual reports. From the point of view discussed here, this causes distortions, since the results did not effectively show a concern with sustainable work, given that the indicators do not portray the reality of what the workers experience in their daily lives. Moreover, sustainability can add significantly to their workload. For instance, at one company, the sustainability discourse was so rich and refined that the interviewee confessed to a general feeling of silent frustration among employees, as everything they could actually do fell short of the company's sustainability expectations. At another company, the concern for internal transparency was so high that individual performance results were openly shown, and criteria for career promotions were clearly defined. The interviewee claimed that this situation made people feel as if they were living in a showcase and locked into a "corporate game" as their actions were constantly watched and had to be carefully weighed. These examples show that, in a scenario of pressure for aggressive and "sustainable" results, even the best-intentioned initiatives can eventually raise workers' psychological suffering to severe or even pathological levels.

It should be highlighted that the organisation of work and production must be highly flexible in terms of task variability (i.e. content, schedule, and location), yet fundamentally stringent in the negotiation of targets, lead times, or priorities. Given the corporations' resistance to expand their workforce—which would unburden workers in terms of their required performance—pressure for productivity culminates in severe psychological suffering that, under the name of "stress", is often interpreted as a physiological response which should be managed individually. The costs of human resources are allegedly high and render hiring unfeasible, but this argument should be weighed against the extreme income inequality among workers in the same company.

Interviewees envisioned an ideal not only for the professional but also for the worker-company dyad. For its part, an exemplary corporation should offer the worker a wide range of benefits and life quality programmes as well as training, structure, and technology support. The worker, in turn, should always give the company their best in terms of skills, flexibility, and moral commitment. Such investments therefore stimulate individual capacities and at the same time legitimate an unfair individualising approach to complex demands. Moreover, strict individual alignment of the worker to corporate values is explicitly expected, indicating that little room is allowed for human variability. As long as these points remain unaddressed, social sustainability practices will probably remain essentially palliative in terms of a positive attitude towards the worker's health and professional development.

CONCLUSIONS

Our analysis indicates that complex demands are usually interpreted through the individualising lens of corporate culture, when in fact they should also be addressed collectively at the level of organisations. Returning to the complex system-of-systems approach, moving to a sustainability perspective requires analysis not only restricted to each level (in the hierarchy of systems) but with a full understanding that all levels are interconnected and interdependent. The system-of-systems must be coherent at all levels. This requires coherent frameworks capable of analysis accordingly and explicit links to the inherent complexity (Dekker, Hancock, & Wilkin, 2013; Zink, & Fischer, 2013). Both ACE and PDW frameworks propose to interconnect how the organisational aspects can be inductors for the individual level, in this case, for workers' health.

Both ACE and PDW frameworks can contribute to work sustainability (e.g. by understanding the gap between prescribed and actual work), considering human limitations and variability, engaging workers in organisational decisions, and creating opportunities to discuss constraints and impasses without fear of being penalised, thereby favouring the construction of a collective identity and a mutually supportive workplace environment. These are examples of practices that are not palliative, unlike corporate social responsibility practices of the past. Instead, they enable workers to act upon the causes of physical and mental suffering (Abrahão, Sznelwar, Silvino, Sarmet, & Pinho, 2009; Daniellou, 2004; Dejours, 1986).

Lastly, inspired by Dejours (2009), we propose that the concepts related to sustainable development include the possibility of building (or rebuilding) interpersonal relationships in the workplace (i.e. solidarity, trust, and cooperation). This proposal is justified, given that this collective construction has positive impacts on society and on the development of culture. In other words, despite the presence of elements conducive to sustainability, of both an individual nature—which cannot and should not be ignored—and a collective nature, within a context of sustainability, the question rests on the opportunity provided by the forms of work organisation for the mobilisation of intelligences in a context based on trust and genuine cooperation, in order to enable the construction of health and professional development.

REFERENCES

Abrahão, J., Sznelwar, L. I., Silvino, A., Sarmet, M., & Pinho, D. (2009). *Introdução À Ergonomia: Da Prática À Teoria*. São Paulo, Brazil: Blucher.

Arendt, H. (1959). *The human condition. A study of the central dilemmas facing modern man*. Chicago: University of Chicago Press.

Bansal, P. (2005). Evolving sustainably: A longitudinal study of corporate sustainable development. *Strategic Management Journal, 26*, 197–218.

Baumgartner, R., & Ebner, D. (2010). Corporate sustainability strategies: Sustainability profiles and maturity levels. *Sustainable Development, 18*, 76–89.

Bridger, R. S., Brasher, K., Bennett, A. (2013). Sustaining person-environment fit with a changing workforce. *Ergonomics, 56*(3), 565–577. Retrieved August 15, 2013, from http://www.ncbi.nlm.nih.gov/pubmed/22928675

Bolis, I., Brunoro, C. M., & Sznelwar, L. I. (2014). Mapping the relationships between work and sustainability and the opportunities for ergonomic action. *Applied Ergonomics, 45*, 1225–1239.

Bolis, I., Brunoro, C. M., & Sznelwar, L. I. (2016). Work for sustainability: Case studies of Brazilian companies. *Applied Ergonomics, 57*, 72–79.

Bolis, I., Morioka, S. N., Brunoro, C. M., & Sznelwar, L. I. (2013, September). Sustainability policies and Corporate Social Responsibility (CSR) Ergonomics contribution regarding work in companies. *Proceedings of the Human Factors and Ergonomics Society Annual Meeting, 57*(1), 1080–1084. Sage, CA/Los Angeles: SAGE Publications.

Brunoro, C. M., Bolis, I., Sznelwar, L., & Kawasaki, B. (2014). *Work in a sustainability perspective: Corporates' perception and ergonomics*. In 11th International Symposium on Human Factors in Organisational Design and Management (ODAM), 2014, Copenhagen, Denmark. Ergonomic challenges in the new economy. IEA Press.

Daniellou, F. (2004). Introdução. Questões epistemológicas acerca da ergonomia. In F. Daniellou (Ed.), *Aergonomia em busca de seus princípios: debates epistemológicos* (pp. 1–18). Edgard Blücher: São Paulo.

Daniellou, F. (2005). The French-speaking ergonomists' approach to work activity: Cross-influences of field intervention and conceptual models. *Theoretical Issues in Ergonomics Science, 6*(5), 409–427.

Daniellou, F., & Rabardel, P. (2005). Activity-oriented approaches to ergonomics: Some traditions and communities. *Theoretical Issues in Ergonomics Science, 6*(5), 353–357.

Dejours, C. (1986). Por um novo conceito de saúde. *Revista brasileira de saúde ocupacional, 14*(54), 7–11.

Dejours, C. (2004a). A Metodologia Em Psicodinâmica Do Trabalho. In S. Lancman & L. I. Sznelwar (Eds.), *Christophe Dejours: Da Psicopatologia À Psicodinâmica Do Trabalho* (pp. 105–126). Brasília, Brazil: Paralelo 15.

Dejours, C. (2004b). Sofrimento E Prazer No Trabalho: A Abordagem Pela Psicopatologia Do Trabalho. In S. Lancman & L. I. Sznelwar (Eds.), *Christophe Dejours: Da Psicopatologia À Psicodinâmica Do Trabalho* (pp. 141–156). Brasília, Brazil: Paralelo 15.

Dejours, C. (2009). *Travail vivant. 2: Travail et émancipation.* Paris: Payot.

Dejours, C., & Gernet, I. (2011). Trabalho, Subjetividade E Confiança. In L. I. Sznelwar (Ed.), *Saúde Dos Bancários* (1st ed., pp. 33–44). São Paulo, Brazil: Publisher Brasil & Editora Gráfica Atitude.

Dekker, S. W. A., Hancock, P. A., & Wilkin, P. (2013). Ergonomics and sustainability: Towards an embrace of complexity and emergence. *Ergonomics, 56*(3), 357–364.

Docherty, P., Kira, M., & Shami, A. B. (Rami) (2009). *Creating sustainable work systems – Developing social sustainability* (2nd ed.). London: Routledge.

Duarte, F., Béguin, P., Pueyo, V., & Lima, F. (2015). Work activities within sustainable development. *Production, 25*(2), 257–265.

Dul, J., Bruder, R., Buckle, P., Carayon, P., Falzon, F., Marras, W. S., et al. (2012). A strategy for human factors/ergonomics: Developing the discipline and profession. *Ergonomics, 55*(4), 377–395.

Dyllick, T., & Hockerts, K. (2002). Beyond the business case for corporate sustainability. *Business Strategy and the Environment, 11*(2), 130–141.

Figge, F., & Hahn, T. (2004). Sustainable value added—Measuring corporate contributions to sustainability beyond eco-efficiency. *Ecological Economics, 48*(2), 173–187.

Fischer, K., & Zink, K. J. (2012). Defining elements of sustainable work Systems-A System-oriented approach. *Work* (Reading, Mass.), *41*(Suppl 1), 3900–3905.

Gladwin, T. N., Krause, T.-S., & Kennelly, J. J. (1995). Beyond eco-efficiency: Towards socially sustainable business. *Sustainable Development, 3*, 35–43.

GRI. (2013). Global Reporting Initiatives – Sustainability Reporting Guidelines (G3.1). Retrieved from: https://www.globalreporting.org/resourcelibrary/G3.1-Guidelines-Incl-Technical-Protocol.pdf

Guérin, F., Laville, A., Daniellou, F., Duraffourg, J., Kerguelen, A., et al. (2001). *Compreender O Trabalho Para Transformá-Lo: A Prática Da Ergonomia.* São Paulo, Brazil: Blucher.

Hanson, M. (2013). Green ergonomics: Challenges and opportunities. *Ergonomics, 53*(3), 399–408.

Harnois, G., & Gabriel, P. (2000). *Mental health and work: Impact, issues and good practices.* Geneva, Switzerland: [S.L.] WHO.

Hart, S. L., & Milstein, M. B. (2004). Criando Valor Sustentável. *Rae Executivo, 3*(2), 65–79.

Houtman, I. L. D., & Kompier, M. A. J. (2011). Work and mental health. In J. M. Stellman (Ed.), *Encyclopedia of occupational health and safety.* Geneva, Switzerland: International Labor Organization.

Hubault, F., & Du Tertre, C. (2008). Le Travail D'évaluation. In F. Hubault (Ed.), *Évaluation Du Travail, Travail D'évaluation* (pp. 95–114). Toulouse, France: Octarès.

ISO. (2010). *Guidance on social responsibility: Draft.* International Standard ISO/Dis26000. Geneva, Switzerland: International Organization For Standardization.

Kira, M., & Eijnatten, F. M. Van. (2009). Sustainability by work: Individual and social sustainability in work organizations. In: P. Docherty, M. Kira, A. B. Shami (Rami) (Eds.), *Creating sustainable work systems* (2nd ed., pp. 233–246). New York: Routledge.

Kleine, A., & Hauff, M. (2009, October 9). Sustainability-driven implementation of corporate social responsibility: Application of the integrative sustainability triangle. *Journal of Business Ethics, 85*(S3), 517–533.

Lukersmith, S., & Burgess-Limerick, R. (2013). The perceived importance and the presence of creative potential in the health professional's work environment. *Ergonomics, 56*(6), 922–934.

Metzner, R. J., & Fischer, F. M. (2010). Fatigue and workability in Brazilian textile companies in different corporate social responsibility score groups. *International Journal of Industrial Ergonomics, 40*(3), 289–294.

Molinier, P. (2001). Souffrance Et Théorie De L'Action. *Travailler, 7,* 131–146.

Molinier, P. (2013). *O Trabalho E A Psique – Uma Introdução À Psicodinâmica Do Trabalho.* Brasília, Brazil: Paralelo 15.

Morioka, S. N., & de Carvalho, M. M. (2016). A systematic literature review towards a conceptual framework for integrating sustainability performance into business. *Journal of Cleaner Production, 136,* 134–146.

Ryan, B., & Wilson, J. R. (2013). Ergonomics in the development and implementation of organisational strategy for sustainability. *Ergonomics, 56*(3), 541–555.

Salzmann, O., Ionescu-Somers, A., & Steger, U. (2005). The business case for corporate sustainability. *European Management Journal, 23*(1), 27–36.

Savitz, A. W., & Weber, K. (2007). *A Empresa Sustentável* (1st ed.). Rio De Janeiro, Brazil: Elsevier.

Sznelwar, L. I. (2015). *Quando trabalhar é ser protagonista e o protagonismo do trabalho.* São Paulo: Editora Blucher.

Thatcher, A. (2013). Green ergonomics: Definition and scope. *Ergonomics, 56*(3), 389–398.

Thatcher, A., & Yeow, P. H. P. (2016). A sustainable system of systems approach: A new HFE paradigm. *Ergonomics, 59*(2), 167–178.

UNGC. (2014). United Nation Global Compact. Retrieved from: http://www.unglobalcompact.org/

Van Marrewijk, M., & Werre, M. (2003). Multiple levels of corporate sustainability. *Journal of Business Ethics, 44,* 107–119.

Westgaard, R. H., & Winkel, J. (2011). Occupational musculoskeletal and mental health: Significance of rationalization and opportunities to create sustainable production systems – A systematic review. *Applied Ergonomics, 42*(2), 261–296.

Wisner, A. (1995a). Situated cognition and action: Implications for ergonomic work analysis and anthropotechnology. *Ergonomics, 38,* 1542–1557.

Wisner, A. (1995b). Understanding problem-building: Ergonomics work analysis. *Ergonomics, 38*(3), 595–605.

Zadek, S., Pruzan, P., & Evans, R. (1997). *Building corporate accountability – Emerging practices in social and ethical accounting, auditing and reporting.* London: Earthscan.

Zink, K. J. (2006). Human factors, management and society. *Theoretical Issues in Ergonomics Science, 7*(4), 437–445.

Zink, K. J. (2014). Designing sustainable work systems: The need for a systems approach. *Applied Ergonomics, 45*(1), 126–132.

Zink, K. J., & Fischer, K. (2013). Do we need sustainability as a new approach in human factors and ergonomics? *Ergonomics, 56*(3), 348–356.

CHAPTER 9

Green Buildings: The Role of HFE

Erminia Attaianese

INTRODUCTION

Building and construction are receiving rising attention worldwide for the way they may influence sustainable development, since it has been demonstrated that the amounts of natural resources involved during a building's life cycle have a significant impact on the environment (US EPA, 2009). In fact, it has been shown that the great deal of energy and water consumption, ozone layer depletion, carbon dioxide emissions, raw materials, and waste products are comparable or larger to those of the transportation and industry sectors (US EPA, 2009). Considering the critical impact on the natural environment and the consequent economic costs, the identification and adoption of green methods and strategies are needed in order to support the design of buildings for addressing the goals of sustainable development. Particularly, a number of green building assessment systems and rating tools have been developed to foster building stakeholders, professionals, and consumers, to request, adopt, and implement sustainable goals in the design of buildings.

Recently, a lot of criticism in current systems has been emerging, probably due to a restricted idea of sustainability, unbalanced toward the green footprint (Berardi, 2013; Zuo & Zhao, 2014). The need to expand

E. Attaianese (✉)
Department of Architecture, University of Naples Federico II, Naples, Italy

© The Author(s) 2018
A. Thatcher, P. H. P. Yeow (eds.), *Ergonomics and Human Factors for a Sustainable Future*, https://doi.org/10.1007/978-981-10-8072-2_9

concepts of building sustainability is emerging, since environmental, social, and economic aspects are inextricably linked to each other, according to general principles of sustainability related to buildings standards (ISO, 2006), even if it is argued that the social aspect of a sustainable building is still rarely investigated. Despite a number of studies demonstrating the rising interest about the socio-technical aspects of energy efficiency in buildings (Cole, Brown, & McKay, 2010; Janda, 2011; Lutzenhiser, 1993), the explicit reference to the human side of building sustainability is only partially assigned to the occupants' role and, moreover, the human side is out-weighed in building assessment methods by environmental aspects such as energy conservation. Therefore, based on a literature overview, the chapter investigates the current role of HFE in the design of green sustainable buildings. A comparison of the main sustainable building rating tools, from the human factors perspective, is provided in order to identify where, how, and what human-related factors are included in the assessment criteria. A discussion about HFE's further contributions to the green building domain is proposed.

Principles of Green Sustainable Buildings

Defining Green Sustainable Buildings

The US Environmental Protection Agency (EPA) defines green building as the "practice of creating structures and using processes that are environmentally responsible and resource-efficient throughout a building's life-cycle from siting to design, construction, operation, maintenance, renovation and deconstruction" (U.S. Environmental Protection Agency, 2009).

The approach based on the saving of natural resources has, for a long time, affected the general understanding about the role of buildings in contributing to reaching sustainability goals, imprinting a distinctive "green" aspect to the design of buildings centered on energy saving, and reducing the environmental footprint. But if on the one hand the environmental imperative has expanded and consolidated, so as to lead to the design of zero energy buildings (GhaffarianHoseini, GhaffarianHoseini, Makaremi, & GhaffarianHoseini, 2012), the "greenness" has been considered a risky attribute, giving particular emphasis to energy questions, without reflecting on the actual impacts of the long-term effects of buildings on the environment (Feifer, 2011).

As stated by the first Conference of the Parties (COP) in Rio in 1992 and inspired by the Brundtland Report (UN, 1987), environmental, economic, and social dimensions of sustainability have been highlighted and taken increasingly into account in a wider concept of sustainable building, described as a subset of sustainable development, which requires a continuous balancing process of all three pillars of sustainability, since popularized into the slogan "people, planet, and profit" (Feifer, 2011). Nevertheless, the attributes of "green" and "sustainable" are often used interchangeably, although the term "sustainable" has a more precise and broader meaning that is limited, often obscured and sometimes distorted, by the abuse of the green profile. So, if "green" is part of being sustainable, but tends to emphasize a design that considers the usefulness of applying passive and active strategies to reduce energy depletion and minimize resource consumption, a sustainable building aims for all of the same green goals, and in addition tends to increase longevity, adaptability, and flexibility of the buildings, accounts for the efficiency of resources spent, addresses safety and security, and includes social and economic issues (Alwaer & Clements-Croome, 2010).

Sustainability is a state that requires that humans carry out their activities in a way that protects the function of the earth's ecosystem as a whole, and referring to building construction, it includes the consideration of its three primary inextricably aspects (ISO, 2006). In accordance with the actual meaning of the term sustainability, expressed in ecology as the dynamic by which biological systems remain different and productive over time, the consideration of the human component enriches the concept of sustainable building, aimed at the creation and responsible management of healthy facilities, designed and built in a resources-efficient manner, using ecological principles (Kibert, 2008). With regard to humans, sustainability is the potential for long-term maintenance of well-being, which has environmental, economic, and social aspects. With regard to buildings the core issues of sustainability are the long-term maintenance and well-being of the users, seen under the aspects of the environmental, economic, and social dimensions (Feifer, 2011).

Including humans in the sustainable building concept, the leading edge of sustainability has been moved beyond the green building strategy, following a trend that goes from a mechanistic view, oriented toward finding technical solutions for energy efficiency and environmental impact mitigation which "distances" humans from buildings, toward an ecological approach, where progressive relationships between human and their built environments are promoted and supported, on the basis of new strategies such as adaptation, resilience, and regeneration (Sorrento, 2012). As a result, in order to encourage people engagement and responsiveness to environmental issues, taking into account the connection between humans and buildings (Cole, 2014), the need for a new concept focused on a high-performance building has grown, in which qualitative data about the actual operational functions, and their association with green strategies, are linked to the physiological, neurocognitive, and psychosocial human response to these buildings (Sorrento, 2012). Thus a high-performance building is called for in order to integrate and optimize, on a life-cycle basis, all major high-performance attributes, including energy conservation, environmental protection, safety, security, durability, accessibility, cost-benefit, productivity, functionality, and operational conditions (Fischer, 2011).

More explicitly focused on humans, the complex systems of the three inter-related basic issues of people (i.e. owners, occupants, users, etc.), products (i.e. materials, fabric, structure, facilities, equipment, automation and controls, services), and processes (i.e. maintenance, performance evaluation, facilities management) are the nucleus of a sustainable intelligent building (Alwaer & Clements-Croome, 2010), responding to the needs of their occupants to be healthy, technologically aware, flexible, and adaptable to cope with change (Clements-Croome, 2013). A building designed to enhance the development of human activities, not necessarily with purely advanced technologies and not necessarily "intelligent" by itself, should be able to empower the occupants' intelligence, for living and working more comfortably and efficiently (Clements-Croome, 2013). A largely shared definition of green sustainable building is still absent, and this has led to difficulties in identifying goals and strategies for implementing sustainability principles in building and construction. To shape a conceptual framework and sketch a new interpretation of sustainability in the built environment, literature reviews have been recently conducted, in order to identify common issues and core elements of sustainable buildings (Akadiri, Chinyio, & Olomolaiye, 2012; Berardi, 2013).

Improving the Triple Bottom Line

The sustainability issues in building design and construction have only seen a theoretical application of the triple bottom line. Economic sustainability relates to the maintenance of high and stable levels of local economic growth and employment through the improvement of project delivery and the increase of profitability and productivity. Environmental sustainability aims at both the effective protection of the environment, by avoiding pollution, protecting and enhancing biodiversity and planning transportation, both the prudent use of natural resources, by improving their efficient use. Social sustainability concerns social progress that recognizes the needs of everyone, working with local communities, and respecting all stakeholders (Akadiri et al., 2012). Natural resource conservation, cost efficiency, and design for human adaptation emerge as the three general objectives for implementing sustainable building design and construction. There is a general consensus about a set of process-oriented principles: complexity, interdisciplinarity, and integration perspectives, with life-cycle and systems approaches, have been highlighted as core dimensions of sustainable building design. Nevertheless those aspects, featured by a time dependence, different application levels, and concerning interaction in different domains, may be seen as causes of uncertainty that may hinder sustainable strategies (Akadiri et al., 2012; Berardi, 2013).

Most of the criticisms are attributed to people-related factors, given the difficulties in understanding people perceptions and their involvement in building sustainability, in establishing common sustainability requirements between people and the different stakeholders, and in taking into account the changing emerging question of quality of life. But, as reported by Berardi (2013), it has been stated that sustainability is socially related, and the participation of people, the consideration of their different expectations, and interpretations of sustainable development are unavoidable, since "If an environmental friendly building can be realized almost everywhere by minimizing its environmental impact, a sustainable building asks for more. It has to consider the impact of the building on the physical and mental health of the occupants; increased social equity, cultural and heritage issues, traditions, human health and social infrastructures" (Berardi, 2013, p. 77).

ERGONOMICS AND HUMAN FACTORS FOR BUILDING SUSTAINABILITY: A LITERATURE REVIEW

In 2013, a special issue of the journal *Ergonomics* reported contemporary research and thinking about sustainability from across ergonomics and human factors, in accordance with two previous initiatives centered on the sustainability issue of the International Ergonomics Association (IEA): the formation of the Human Factors and Sustainable Development

Technical Committee endorsed in 2008 and the 18th triennial congress of the IEA, held in Brazil in 2012, titled "Designing a Sustainable Future" (Haslam & Waterson, 2013). It is interesting to notice that among the articles of the Ergonomics special issue, only limited references are made to ergonomics in building sustainability. Of the 17 commentaries, only one addresses this issue directly (Hedge & Dorsey, 2013) and five partially: by including green building design among other potential challenges and opportunities of the next E/HF profession future (Hanson, 2013) that the built environment may be seen as the place where the application of green ergonomics appears more obvious (Thatcher, 2013) or by presenting specific aspects of the energy issue, reporting examples of user interfaces of domestic appliances (Peffer, Perry, Pritoni, Aragon, & Meier, 2013; Stedmon, Winslow, & Langley, 2013), and an application of sustainable facility management (Lee & Kang, 2013).

Indeed, existing literature in the ergonomic domain, centered on the relationship between HFE and green buildings, is today limited (Martin, Legg, & Brown, 2013), even if the need to improve this linkage is widely shared (Attaianese, 2012, 2014; Brown & Legg, 2011; Flemming, Hilliard, & Jamieson, 2008; Hanson, 2013; Hedge, 2000, 2008; Hedge & Dorsey, 2013; Martin et al., 2013; Sutcliffe, Hooper, & Howell, 2008; Thatcher, 2013). Particularly, Hedge and Dorsey (2013) in their article in the special issue presented a post-occupancy evaluation survey in two certified buildings in which, although health, performance, and satisfaction are reported as generally positive by occupants, some concerns about comfort parameters were described, and a more incisive integration of HFE in green building design and assessment is evoked. Post-occupancy evaluation surveys (POE) represent an important research trend in the relationship between HFE and green buildings, and not only among ergonomics experts, since they let us observe if and how green solutions in buildings affect occupants. Hedge (2000) was among the first, in the HFE domain, to report a literature review about the relationships between workers' health, office lighting, Indoor Air Quality (IAQ), and Sick Building Syndrome (SBS) in offices, explicitly highlighting the necessity to integrate ergonomics and sustainability and foster this integration for designers (Hedge, Rollings & Robinson, 2010). Further, an earlier study by Heerwagen (1998) discussed the connections between design, productivity, and well-being.

Afterward, several other studies have been published, in order to present encouraging outcomes on tenants' productivity in occupational buildings, achieved by improving health and comfort through green practices (Clements-Croome, 2000, 2006; Heerwagen, 1998, 2000; Heerwagen & Zagreus, 2005). Health is an important integrated goal, since ergonomics ensures workability through preventive occupational health and safety, and one of the sustainability goals is sustaining human health; so HFE can contribute to identify positive and negative effects of the design of the built environment on the health of occupants, improving sustainability through the design of a healthy built environment (Leech, Raizenne, & Gusdorf, 2011).

There is strong evidence connecting improvements in work performance to environmental features of buildings (Heerwagen, 2000; Leaman & Bordass, 2007). This work seems to confirm a virtuous cycle linking health, comfort, and building quality, since a better building performance is expected to lead to an increased well-being and thus to better human performance (Clements-Croome, 2014; Hedge & Dorsey, 2013). Comparative assessments, focusing on the differences between green buildings and conventional ones, can be found particularly among HFE references (Hedge, Rollings, & Robinson, 2010; Thatcher & Milner, 2012, 2014). They confirm generally positive effects on occupants of green buildings, but a number of explicit ergonomic concerns have been recently reported (Hedge & Dorsey, 2013; Thatcher & Milner, 2014). Even if occupants of green buildings report more possible sources of satisfaction derived from the built environment (Leaman & Bordass, 2007), it seems to be shown that green systems and practices do not necessarily lead to the well-being of occupants, especially if they do not take into account the human factors issues (Altomonte, Rutherford, & Wilson, 2015; Hedge & Dorsey, 2013; Leaman & Bordass, 2007; Thatcher & Milner, 2014). The complaints are mainly associated with perceived comfort, frequently due to acoustics and lighting problems (Abbaszadeh, Zagreus, Lehrer, and Huizenga 2006; Leaman & Bordass, 2007; Muehleisen, 2010), but a number of other different concerns (i.e. green material toxicity, fire safety, construction workers safety) are reported in technical surveys about green solutions of new buildings (Attaianese, 2014; EASHW, 2013; Tidwell & Murphy, 2010).

On the other hand, direct connections to the natural environment, including daylight and sunlight access, fresh air circulation, and outdoor views of nature, were found to be among the key building features associated with increases in human performance and satisfaction and probably are also important elements also for stimulating creativity (Ceylan et al., 2008;

Dul & Ceylan, 2011 in Thatcher, 2013). Evidence about positive follow-up in the design of schools and healing environments (Clements-Croome, 2014; Heerwagen, 2011) confirms that the pro-nature approach, fostered by green ergonomics, may lead building designs to improve benefits for occupants (and not only in working environments), even if further research to understand what other aspects of how human functioning and well-being at work might be positively influenced by connections with the natural environment is still needed (Thatcher, 2013). Occupants are more comfortable in buildings in which the amount of perceived control over the indoor environment is high. Several studies demonstrate that the perception of comfort is linked to occupants' ability to control solar glare (Barlow & Fiala, 2007; Nicol & Roaf, 2005). This is a relevant element in green buildings in which the need to control temperature, ventilation, and noise is high (Boerstra, Beuker, Loomans, & Hensen, 2013). Indoor comfort is an important topic in building sustainability, looking at the number of studies that focus on this subject. It is at the center of environmental ergonomics (Persons, 2000), given that comfort is now defined as a dynamic condition rather than a uniform and static condition, experienced by humans not only in physiological terms but also in psychological, behavioral, and the social senses (Vink & Hallbeck, 2012). In addition, it is obviously connected to energy efficiency, since it has been demonstrated that the largest amount of energy costs for buildings are spent on improving indoor comfort (Newsham, Mancini, & Birt, 2013). Focusing directly on systems to achieve environmental conditions best fitting their needs (e.g. through opening and closing windows, doors, lights, shading devices, thermostats, vents, and other manual controls), occupants' behaviors significantly influence energy performance of buildings. Thus, the human component has been positioned as an active determinant of energy performance in green building design, and considering occupants' participation, which is essential to design sustainable places, is an important driver for increasing people's attainment of sustainability goals (Cole et al., 2010; Gram-Hanssen, 2014; Janda, 2011).

As two faces of a coin, the comfort and the availability of building controls have been widely reported as crucial & strongly linked to each other (Hauge, Thomsen, & Berker, 2011; Leaman & Bordass, 2007; Nicol & Roaf, 2005). Well-designed feedback tools can increase the likelihood that individuals will conserve energy (Flemming et al., 2007) and ergonomic behavioral interventions may improve sustainability by changing the conduct/behavior of people (Sutcliffe et al., 2008). Buildings and equipment can therefore support adaptive human behavior through effective systems design (Hilliard, 2008). From outside the HFE community, an increasing number of studies on socio-technical aspects of energy efficiency in buildings also concern adaptive comfort technologies (Altomonte et al., 2015).

Altogether, the main discussion topics about ergonomics in green buildings, from an HFE community perspective, relate to the building's final stage, and the effects on occupant perceptions of already built green solutions, designed with no specific reference to the ergonomic approach.

Furthermore, it has been suggested that a wider interpretation of sustainable building design is needed; design that meets the needs of present, without compromising the ability of future generations to meet their own needs. This involves a life-cycle approach to sustainability in building and construction (Charytonowicz, 2007). In this perspective, it is argued that an ergonomic approach can be applied in a wider sense for improving sustainable processes of building design, from the original concept to construction, from operation and maintenance, until the deconstruction stage (Attaianese, 2012). However, few studies focus on these aspects, and a comprehensive ergonomic strategy that promotes building sustainability is still lacking (Alzaed & Boussabaine, 2012; Attaianese & Duca, 2010; Hedge, 2008). Even if health and safety (H&S) of the construction workers has been recognized as integral parts of sustainability building goals (Hinze et al., 2013), not many of the sustainability goals relate to specific concerns about green jobs in the building and construction domain (Attaianese, 2014; Dewlaney, Hallowell, & Fortunato, 2012; Rajendran, Gambatese, & Behm, 2009).

In the HFE community, publications discussing experiences about how ergonomics can support green construction are rare. Recently, a survey carried out in South Africa, about the range of ergonomic problems of construction workers of green buildings and factors influencing them, reveals that construction workers, especially in developing countries, are exposed to many ergonomic and health and safety hazards, causing them ill health, stress, and musculoskeletal disorders (Sass & Smallwood, 2015). The study concludes that, from the construction worker perspective, even though buildings may be rated as green, they may not be "green" from the social sustainability perspective of the construction workers.

It is argued that design plays an important, but underestimated, role in improving construction workers' safety (Gambatese & Hinze, 1999), since the origin of several ergonomic problems seems to be associated with how the buildings are conceptualized and planned, before even being constructed. Indeed green building solutions and onsite construction equipment (Waris, Liew, Kamidi, & Idrus, 2014) may expose workers to additional risks such as increasingly heavy loads due to high-performance blocks and skylights (Attaianese, 2012, 2014), risks of falls and work at heights for the tall spaces and atriums, or the installation of solar panels, eyestrain when installing reflective surfaces, and exposure to harmful and toxic substances (Chen, 2010).

In order to improve the sustainability of construction sites by integrating green ergonomics, an application of a biophilic model of a construction site has been reported. Findings demonstrate that ecological and psychosocial interventions, attempting to reconstitute the natural habitat of the site (i.e. by repositioning original soil to reshape the territory, and preserving existing plants and trees), encouraging a direct interaction between workers and management and nature, may be successfully proposed (Obiozo & Smallwood, 2013).

Since the main goal of ergonomic design is the optimization of human-system interaction (Karwowski, 2006), HFE has been proposed to facilitate the selection of the most appropriate technologies in green building design, for waste reduction and improved building functioning through the optimization of building estate management (Charitonowicz & Leszeks, 2001; Charytonowicz, 2007). Moreover, maintainability of a building is a key performance indicator of the sustainability of the built environment (Chan, 2014) and it may be improved by HFE. Since the maintenance context is a network of interactions among people (as direct and indirect users of the built environment), architectural systems need to be maintained, and it has been shown how socio-technical system interactions can be improved by considering the physical, cognitive, and organizational aspects of maintenance and, at the same time, ensuring safety and well-being of all subjects involved (Attaianese, 2011).

HFE AND RATING TOOLS FOR GREEN SUSTAINABLE BUILDING

Assessing Green Sustainable Performance

In the last few decades, in order to address the growing demand for green buildings worldwide, standards, certifications, and rating systems have been developed in order to endorse and reward new buildings and refurbishments that achieve good levels of sustainable features. Specifically, green building rating or certification systems are multi-criteria assessment systems, where a score is awarded if it can be demonstrated that there is compliance between specific characteristics of the building and a list of green criteria and parameters, and after having compared real performances with referenced ones. The evaluation of sustainability occurs by summing the results of assessed criteria; each criterion has a certain amount of (weighted) available points over the total assessment (Berardi, 2015). Rating systems focus on the project as a whole, and rate relative levels of compliance or performance with specific goals and requirements, considering the building's life cycle: from siting to design, construction, operation, maintenance, renovation, and demolition. While the certification methods vary across these systems, a common objective is that projects awarded or certified within these programs are designed to reduce the overall impact of the built environment on human health and the natural environment (Vierra, 2014).

Although green sustainable building rating tools are voluntary and national, their impact is global, given that many of current systems have been modified from early models that were originally developed in other countries and locally adopted (Reed, Wilkinson, Bilos, & Shulte, 2011). Among the most diffused systems having original models and multiple country applications, we find (1) the Building Research Establishment Environmental Assessment Method or BREEAM, developed in 1990 in the United Kingdom and generally accepted as the world's first green building rating system; (2) the United States' and Canada's Leadership in Energy and Environmental Design (or LEED), (3) Green Globes and EnergyStar; (4) the Japanese Comprehensive Assessment System for Building Environmental Efficiency, or CASBEE; and (5) the SBTool, as an evolution of the GBTool, developed by the International Framework Committee for the Green Building Challenge, which has involved more than 25 countries since 1998 and has now been adopted as the general scheme by several European countries (Fowler & Rauch, 2006; Berardi, 2015). In other cases, rating tools were created modifying or integrating multiple systems, as GreenStar is based in part upon BREEAM and LEED, developed by the Green Building Council of Australia (GBCA) and used in Australia, New Zealand, and South Africa. On the contrary, the recent German "Deutsche Gesellschaft für Nachhaltiges Bauen" (DGNB) Label is an original national system that is increasing its relevance by the support of World Green Building Council (WGBC) as well as the patronage of the International Organization for Standardization (ISO) (Dirlich, 2011).

WELL Certification deserves specific mention, since it is suggested to be the first standard that focuses solely on the health and wellness of building occupants. It is a performance-based standard for measuring, certifying, and monitoring features of the built environment that impact on human health and well-being, through air, water, nourishment, light, fitness, comfort, and the mind. The WELL Building Standard is designed to work harmoniously with the LEED Green Building Rating System and other leading global green building standards (Delos, 2016).

Ergonomics Points in Current Systems

In analyzing the main sustainable building rating tools from a human factors perspective, we need to follow a double track. On the one hand, we need to identify in the current rating systems where explicit credits about ergonomics are included. On the other hand, the hardest one, we need to note the implicit or limited human-related considerations in other credits. Even though for long time, ergonomics credits were completely unknown within sustainable assessment systems (Berardi, 2015), now ergonomics credits are incorporated into some rating tools, even if in a very limited number of cases.

The USGBC LEED rating system was the first rating system to include ergonomics in its rating tool in 2008 (Hedge, 2008). Originally it was available within the "Innovation in Design and Innovation in Operations Credit", but since 2010 the ergonomics credit was included in the USGBC Pilot Credit 44, whereby a built project could get a LEED point if it had applied a full ergonomic strategy for at least 75% of full-time equivalent employees if that strategy included a flexible environment able to address the needs of users and promoted the healthy work, and a comfortable and productive work environment (USGBC, 2008a, 2008b). The strategy was updated again in March 2016. In LEED v4 a possible point is assigned when occupant well-being (human health, sustainability, and performance) is improved through the integration of ergonomics principles, specifically in the design of work spaces for all computer users. The engagement of an ergonomist or health and safety specialist to assist in the development of the ergonomic strategy and a commitment to integrate ergonomics principles into the overall design are both mandatory requirements for the credit assignment. The ergonomist, in conjunction with the client, must develop a description of the ergonomic strategy that will be implemented, including the ergonomic goals, the characteristics of occupants, tasks and tools, the education program, and the process for evaluating and maintaining occupant well-being, to ensure that the ergonomic strategy goals are being met (LEED, 2016b).

Another system that unmistakably includes ergonomics is the current version of Green Star Interiors Rating Tool, launched in 2013 by the Green Building Council of Australia (GBCA) (Green Star, 2016). The tool, released after a significant revision, covered all interior fit-out projects as well as including a number of new credit categories, includes the option to award ergonomics and occupant comfort with an ergonomics credit. The assessment methodology, in accordance with the credit requirements, focuses on furniture and equipment in the areas where employees may work for more than two hours per day. It includes, on the basis of a participative process, the ergonomic evaluation of workstations and chairs, and the provision of user guidance materials and recommendations about office ergonomics. The ergonomics credit must be assessed by a Certified Professional Ergonomist (Aickin & Pollard, 2013).

Similarly, since 2014, the Green Building Council of South Africa (GBCSA) has launched the GreenStar SA Interiors Rating Tool, where credits can be obtained for good ergonomic workstations and workplaces, or for the use of ergonomically certified furniture, fittings, and equipment (Thatcher & Milner, 2014). In order to sustain the certification process, models of survey tools have been proposed for improving the ergonomic design and assessment of occupational spatial layouts (Thatcher & Chunilal, 2015) and for supporting the pre- and post-occupancy evaluations of the interior fit-out's impact on well-being and productivity. The application of this model allowed, in an experimental case, to assign two innovation credits, and today it is considered by GBCSA, as an industrial standard for developing Version 2 of the rating tool (Thatcher & Milner, 2014). In the 2014, GBCSA also launched the GreenStar SA Socio-Economic Category Pilot, in which six of the seven categories include H&S of construction workers. Credits are awarded after design compliance with the health and safety national regulations has been shown, hazard identification of risks assessments has been conducted, and a comprehensive primary health program has been implemented (Sass & Smallwood, 2015).

The WELL Building Standard includes an explicit ergonomics requirement within the comfort category, even though many other requirements focus on HFE issues. In order to reduce physical strain and maximize comfort and safety of the workstation, an ergonomics point is assigned through visual ergonomics features and the flexibility of desks and seats (Delos, 2016).

"Hidden" Human-Related Factors

"Explicit" ergonomics credits are not the only points where HFE can make further contributions in sustainable building rating systems. They could also be identified by looking at those issues affecting environmental performance, such as occupant comfort; the social and economic aspects of sustainability, such as health and safety of construction personnel, users, and operators; and overall efficiency and effectiveness, accessibility and inclusivity, and quality of life, in a broader sense (Attaianese, 2012; Zuo & Zhao, 2014). Literature reviews demonstrate that the leading focus of current green building assessment tools is on environmental aspects, and social factors are largely overlooked (Alayami & Regzui, 2012; Berardi, 2011, 2015; Zuo & Zhao, 2014). However, an analysis of the HFE-related aspects of the green sustainable assessment categories in the international rating systems is currently lacking (Kim, Oh, & Kim, 2013). Based on the information available, a tentative discussion about this issue is provided below.

BREEAM

The BREEAM rating tool bases its assessment method on the sum of points awarded by building design and procurement compliance, with a number of sustainability criteria. It is based on the following seven categories: Management, Health & Well-Being, Energy, Transport, Water, Materials, Land Use, and Ecology and Pollution. Human-related factors can be found, as expected, in the Health & Well-Being category, with credits pertaining to indoor noise, lighting and view issues, ventilation, contaminant levels, and occupant thermal comfort. Since a number of points are related to the observance of fixed standards concerning a "passive" consideration of occupants, the aspect affecting the human component of Health & Well-Being is the availability of personal ambient controls for lighting, glare, ventilation, and thermal zone systems, as well as HVAC and energy systems monitoring (Kordjamshid, 2011). Unfortunately no reference about the usability evaluation of those controls seems to be given, even if a clear understanding of how a building can be sufficiently operated and maintained is requested in a user guide, within the Management category. A hint is given about the inclusion issue, through the credit "Access to amenities" in the Transport category.

LEED

The US LEED rating system is also structured on seven categories of credits including Sustainable Site, Water Efficiency, Energy and Atmosphere, Materials and Resources, Indoor Environmental Quality, Innovation and Design Process, and Regional Priority. Among these categories, five include prerequisites and core credits necessary to gain the certification, and the last two, particularly the Innovation and Design Process, have been added separately, to give exemplary performance, beyond core credits. The ergonomics credit, mentioned above, pertains to this category. Furthermore, human-related factors in LEED can be found in the Indoor Environment Quality category and refer generally to the same aspects of comfort we found in BREEAM (Alayami & Rezgui, 2012). The availability of occupants' personal ambient controls is a relevant topic, given the number of related credits, but no mention can be found about their quality in use. LEED Version 2012 incorporated post-occupancy evaluations of thermal comfort, even if they were reported to be carried out in a deterministic manner (Siew, Balabat, & Carmichael, 2013). In LEED v4, updated in January 2017, additional factors can be observed about outdoor physical accessibility, within the Sustainable Site category, and space flexibility in the Material and Resources category of healthcare buildings (LEED, 2016a). Moreover, the USGBC released one pilot credit aimed at encouraging prevention through design practices (Toole & Gambatese, 2008). The intent of the pilot credit is to support high-performance cost-effective employee safety and health outcomes across the building life cycle through early attention to safety and health hazards (LEED, 2015).

CASBEE

The Japanese CASBEE presents a different assessment model based on the distinction between environmental load and quality of building performance where results are plotted on a graph, with environmental load on one axis and quality on the other. The best buildings will fall in the section representing the lowest environmental load and the highest quality (Cole, 2014; Fowler & Rauch, 2006). The main assessed aspects, detailed into 80 sub-criteria, include Indoor Environment, Quality of Service, Outdoor Environment on-site within the Building Environmental Quality group, Energy, Resources and Materials, and Offsite Environment within the Environmental Load group. The system is very focused on the building energy efficiency rating. As in the previous systems, the Indoor Environment category includes comfort factors, but a number of human-related hidden aspects can also be found in the category of Quality of Service, which is comprised of criteria facilitating building use and operation, such as Functionality and Usability, Flexibility and Adaptability, Durability and Reliability, Controllability of Systems, and Maintenance of Performance (Alayami & Rezgui, 2012).

SBTool

The SBTool rating system provides a multi-factorial assessment of performance that is related to local norms and standards. It covers a wide range of sustainable building issues related to the pre-design, design, construction, and operation stages, to address a number of active and mandatory criteria. Issue areas include:

- Site Location
- Available Services and Site Characteristics
- Site Regeneration and Development
- Urban Design and Infrastructure
- Energy and Resource Consumption
- Environmental Loadings
- Service Quality
- Social, Cultural and Perceptual Aspects
- Cost and Economic Aspects

Even if the systems approach to Comfort credit assessment, within the Indoor Environmental Quality area, is comparable to the rating systems previously analyzed, the SBTool looks particularly significant for the number of hidden human-related criteria within other categories.

In the category Service Quality, 16 active criteria are organized into five sub-categories, explicitly focusing on the assessment of the building quality in use for the occupants and operators.

They include:

- Safety and security (through credits focusing on the maintenance of core building functions during power outages and personal security for building users during normal operations)
- Functionality and efficiency (through credits focusing on spatial efficiency and volumetric efficiency)
- Controllability (through credits focusing on effectiveness of facility management control systems, the capability for partial operation of facility technical systems, and the degree of local control of lighting systems)

- Degree of personal control of technical systems by occupants
- Flexibility and adaptability (through credits focusing on the effectiveness of facility management control systems, the capability for partial operation of facility technical systems, the degree of local control of lighting systems, and the degree of personal control of technical systems by occupants)
- Optimization and maintenance of operating performance (through credits focusing on the ability for building operators or tenants to modify facility technical systems)
- Potential for the horizontal or vertical extension of the structure
- Adaptability constraints imposed by the structure or floor-to-floor heights

In category F: Social, Cultural and Perceptual, eight active criteria and one mandatory criterion are grouped into three sub-categories, partially involving human factors, including topics rarely included in green building rating systems. In fact in the Social Aspects, a direct involvement of HFE can be found in the universal on-site access and in the building credits, while indirect links are in access to direct sunlight from living areas of dwelling units, visual privacy in principal areas of dwelling units, access to private open space from dwelling units, culture and heritage (through credits on the provision of open public space compatible with local cultural values, the impact of the design on existing streetscapes, the use of traditional local materials and techniques), and perceptual access to exterior views from the interior (Larsson, 2015). Using the general SBTool scheme, several countries have proposed national versions of this system, but unfortunately the local adaptation may lead to a lack of significant credits, like in the ITACA Protocol, implemented in 2000 as the Italian version of SBTool, where the assessment criteria are insignificant from an HFE perspective (Asdrubali, Baldinelli, Bianchi, & Sambuco, 2015; UNI/PdR 13.0, 2015; UNI/PdR 13.1, 2015).

GreenStar

GreenStar is a comprehensive voluntary building rating tool, developed by the Green Building Council of Australia (GBCA) and used in Australia, New Zealand, and South Africa. The system covers nine criteria where scores are awarded if targets are met. They include Management, Indoor Environment Quality, Energy, Transport, Water, Materials, Land Use and Ecology, and Emissions.

HFE-related factors can be observed in Indoor Environment Quality (IEQ), which scores target environmental impact along with occupant well-being and performance, by addressing heating, ventilation, and air conditioning (HVAC), lighting, occupant comfort, and pollutants (Siew et al., 2013). IEQ criteria comprise Indoor Air Quality, Hazardous Materials, Lighting Comfort, Daylight & Views, Thermal Comfort, Acoustic Comfort, and Occupant Satisfaction, which comprise four of the eighteen points available for the total IEQ credits. Moreover, in the category Management, two points can be assigned to the credit Building Information (Green Star, 2016), and one point is awarded for an easy-to-use guide, that includes information relevant to users, occupants, and tenants. However, it has been argued that different people will have different understandings of what constitutes "easy-to-use", and they may not arrive at the same understanding (Siew et al., 2013). As described above, an ergonomics credit can be obtained in the current versions of the Green Star Interiors Rating Tool of Australia and South Africa. In addition points may be awarded to workspace efficiency (Thatcher & Chunilal, 2015) and to H&S of construction workers, thanks to the Socio-Economic Category Pilot 2014 of the GBCSA.

DGNB Label

The German DGNB Label applies a holistic approach to building sustainability, basing the assessment on 40 criteria grouped into six quality categories: Environmental Quality, Economic Quality, Socio-cultural and Functional Quality, Technical Quality, Process Quality, and Site Quality. Through a criteria overview (DGNB, 2014), it can be noticed that human-related factors are disseminated in several credits according to the system's intention to equally cover the environmental, economic, and social dimensions of sustainability (Dirlich, 2011). Socio-cultural and Functional Quality comprises three sub-categories including Health, Comfort, and Users' satisfaction; Functionality; and Design Quality. Human-related points may also be found in Thermal Comfort, IAQ, Visual Comfort (including Colour Rendering), Acoustic Comfort (especially in multiple occupant offices), User Control (both in terms of availability of controls and ease of use), Quality of Outdoor Spaces (in terms of qualitative and quantitative evaluations), and Safety and Security, both as a subjective perception of safety and protection against assault and as reduction of damage if an accident occurs. Other human-related indirect aspects can be observed in the Functionality category, including particularly the point about Design for Alland in the Design Quality category, and Layout Quality category. In the Technical Quality category, hidden human credits are included in Fire Safety, including Additional Fire Safety features of the design, structures and technical systems are required; in Adaptability of Technical Systems and in Deconstruction and Disassembly, in which easy to disassemble is included. In addition, it is noteworthy that Flexibility and Adaptability spaces are included in the Economic Quality category of the building.

WELL

The WELL Certification is organized into seven categories of wellness: Air, Water, Nourishment, Light, Fitness, Comfort, and Mind, which are comprised of 102 features intended to address specific aspects of occupant health, comfort, or knowledge. Each feature is ascribed to the human body systems that are intended to benefit from its implementation. This enables project teams to classify the intended benefits of each WELL feature and develop a comprehensive set of strategies. Ergonomics is an essential part of the rating system (Dorsey, 2016), since several HFE points are included in each category, especially in Comfort and Mind (Delos, 2016).

From this overview some considerations emerge. It can be confirmed that, even if HFE credits are included in a very limited number of current rating systems, points affecting human aspects are numerous and may be successfully improved by a comprehensive HFE approach. Comfort and health credits, usually directly linked to the ergonomic domain, are much more indirectly associated with HFE factors than might be expected. The adequacy of the Indoor Environmental Quality to occupant comfort is generally founded on technical standards compliance, according to both energy-saving questions and to a static notion of comfort. These aspects influence the design of buildings, since they lead to the conceptualization of buildings whose post-occupancy conditions are essentially immovable, within a limited variability range. Post-occupancy evaluation surveys, requested in order to demonstrate occupants' satisfaction, are not directly relevant for the use at the design stage of a new building, and the utility of their results in current rating tools remains limited (Thatcher & Milner, 2016). Thus, occupant comfort in sustainable buildings is strongly connected with the availability of ambient personal controls. Nevertheless, very few rating tools demonstrate how usable those controls are for people who have to use them (DGNB, 2014). Points about health are usually limited to the level of contaminants, rarely addressing Sick Building Syndrome questions (Joshi, 2008). Accessibility should directly incorporate HFE factors, but it is surprisingly under-evaluated in the rating systems, probably relying on the respective mandatory norms. Numerous potential HFE points may be associated with serviceability. This focuses on optimization and operating performance, where the high quality of human operation may be facilitated by several characteristics of buildings, and its compliance may exert indirect but significant effects on resource use and environmental quality (Alayami & Rezgui, 2012). It may include safety and security issues, generally not present in a specific category, as well as maintenance, and building efficiency aspects. Considering space (volumetric and systems adaptability), the functional quality may be assessed through measuring building potentiality to consider the most suitable layout.

POSSIBLE FURTHER CONTRIBUTIONS OF HFE TO SUSTAINABLE BUILDING RATING SYSTEMS

HFE is a design-oriented discipline. Through HFE one is able to understand interactions between people and the environments that surround them, in order to optimize human well-being and overall system performance, including comprehending behavior, abilities, limitations, and other human characteristics, and applying that information to the design of systems and environments (Dul et al., 2012; Karwowski, 2006). But looking at the current sustainable building systems, the core contributions of HFE are currently generally disregarded. The ergonomics credits are awarded to occupational issues, far from the current approach to the role of work systems for sustainable development (Lange-Morales, Thatcher, & García-Acosta, 2014). The credits are limited to programs about ergonomic awareness and education of occupants, and to interventions on working environments, concerning spaces, places, and furniture, that can only ergonomically "accommodate" into "already built" rigid plans, and the compliance can only be demonstrated in the post-design stage. Looking at these different concerns, and according to the sustainable system of systems approach (Thatcher & Yeow, 2015), the need for a broader consideration of the role of HFE for enhancing the green building process must be considered (Attaianese, 2012), and possibly further contributions to building sustainability rating tools can be outlined.

Building Construction Process

- improving effectiveness of integrated sustainable building design teams, by supporting communication problems due to the need to harmonize different expertise (Chiocchio, Forgues, Paradis, & Iordanova, 2011);
- improving safety issues of construction workers, through prevention through design methods (Dewlaney & Hallowell, 2012; Thatcher, 2013);
- improving safety of the construction process, and the efficient use of resources, by including human-related criteria in sustainability assessment of building materials (Attaianese & Duca, 2012);
- improving the building deconstruction process, by implementing ergonomic design for disassembly (Charytonowicz, 2007; MIL-HDBK, 1997).

Occupant Well-Being

- increasing building accessibility by including credits about building compliance with design characteristics that allow for the accommodation of different levels of physical, sensory, and mental ability and different ages. This could include slip-resistant and easy-to-walk walking surfaces; simple, clear, and logical workplace layouts; intuitive, obvious, and accessible fire evacuation routes; accessible information; and good lighting and good visual contrast of walls, floors, doors, and signage (Attaianese, 2016; Capolongo et al., 2016; ISO, 2011);
- supporting occupant identity, by improving building participative design at the early stages of the design process if possible;
- improving building functionality by including building usability evaluations in the building rating tools (Alho, Nenonen, & Nissinen, 2008; Alzaed & Boussabaine, 2012; Haron et al. 2013; Kim et al., 2013);
- improving subjective perceptions of safety through participative design, also increasing post-occupancy evaluations on safety;
- improving points related to health in the building rating tools by including more factors related to assessing (and reducing) Sick Building Syndrome symptoms (Delos, 2016);

– increasing the use of longitudinal studies for improving the under-standing about the relationships between people's responses about well-being and productivity into the built environment in green buildings (Thatcher & Milner, 2016).

Comfort and Efficiency

– improving design and assessment of systems for adaptive comfort performance, particularly in the remote control of large environ-ments (Altomonte et al., 2015);
– improving design for easy-to-use personal ambient command and control systems, considering different user's variability (Nadadur & Parkinson, 2013);
– increasing the effectiveness of displays of control systems (of natural resources such as air and lighting and the removal of pollutant such as carbon dioxide and volatile organic compounds), for improving occupants sustainable behaviors, by improving usability;
– improving basic knowledge about the dynamic dimension of com-fort, by increasing ethnographic studies on cultural and social behaviors.

Building Operation and Maintenance

– increasing serviceability and maintainability by exploring an ergo-nomic approach to building maintainability and serviceability at the design stage (Attaianese, 2011; Attaianese & Duca, 2010);
– increasing building effectiveness and operability by improving occu-pants' considerations in maintenance programs;
– improving easy-to-use building guides and system manuals, accord-ing to the occupants' and operators' variability (Nadadur & Parkinson, 2013).

Workplace Layout and Work Design

– increasing building "workability" by connecting green building resource efficiency considerations with occupants' specific work needs, in the initial building plan of building interiors in order to promote building performance profiling and to be more responsive to the environmental and socio-cultural context;

  – improving the role of occupant surveys and POE, to understand the effects of building green features on a wider range of occupant behaviors, including ways to increase physical activity via workstation design.

CONCLUSIONS

From an ergonomics viewpoint, the built environment may be considered as a facility that is able to support people acting in and around it during their everyday life. Thus, the main objective of ergonomic design for the built environment, as with any other tool, is to balance dual outcomes: system performance and human well-being. On the other hand, the sustainability perspective requires building design and construction to contribute to addressing the social, economic, and environmental goals of sustainable development by encouraging an integrated approach, including a complex systems and life-cycle perspective. Green ergonomics for building design, in a sustainability framework, needs to focus on integrating building and construction performance with sustainability goals as the basis for a human-nature approach by optimizing the use of natural resources (i.e. energy, materials, soil, air quality, water, etc.), complete functionality and serviceability of buildings, and the efficiency and effectiveness of facility systems, with the whole spectrum of well-being, for all involved people, in terms of health, safety, comfort, and productivity.

Unfortunately, although it is generally accepted that the scientific domain of HFE may represent an effective driver for achieving sustainability goals in green buildings, and indeed this is advocated by many, experiences with the actual involvement of the HFE approach in sustainable building design are limited. This is probably due to a shortage of mature references, within the scope of the HFE community, about the ergonomics principles in building design. Although it is believed that ergonomics may act as a driver to encourage green behavior, it seems, to the contrary, that the design of green buildings is acting to increase the role of HFE due to the need to meet the difficult demands of resource optimization and to harmonize the frequent conflicts that human involvement arouses in sustainable development.

It is argued in current HFE literature reviews that green ergonomics principles to sustainable buildings are today only partially applied. In fact, current literature demonstrates that discussions about ergonomics in green building are related to workplaces and office buildings, but other application contexts, such residential or living environments, are still lacking. Further, more has been written about the effects on occupants of sustainable solutions to green buildings and little about identifying how to shape these solutions in an integrated design process that ensures efficient resource use as well as human well-being. On the other hand, from outside the HFE community, an increasing number of studies on sociotechnical aspects of energy efficiency in buildings, including adaptive comfort technologies, have been reported. HFE can make a significant contribution to the sustainable buildings project, but a greater synergy between ergonomists and building design professionals is needed. HFE expertise should be incorporated into an integrated design team; but architects, engineers, and interior designers should also acquire basic HFE principles in their education, for an effective integration of human-related issues and environmental goals in building design for sustainability.

HFE needs to go beyond the limited approach currently existing in the green building domain, by embracing all living and working places, and offering its expertise to directly implement integrated green building design, green building systems, and the related components of that design, for efficient, effective, healthy, safe, and satisfying building and construction processes (Brown & Legg, 2011). Outside the HFE community this trend has been perceived, and the need for improving HFE in the design of the built environment has been clearly expressed (Altomonte et al., 2015; Clements-Croome, 2014). In contrast, inside the HFE community, this opportunity still seems to be underestimated.

References

Abbaszadeh, S., Zagreus, L., Lehrer, D., & Huizenga, C. (2006). *Occupant satisfaction with indoor air environmental quality in green buildings.* Proceedings of the healthy buildings 2006 conference, Lisbon, Portugal, pp. 365–370.

Aickin, C., & Pollard, B. (2013). *GreenStar interiors rating tool – Ergonomic credit. What should CPE's know?* Proceedings of the 49th annual human factors and ergonomics society of Australia conference 2013, pp. 198–202.

Akadiri, P. O., Chinyio, E., & A. & Olomolaiye P.O. (2012). Design of a sustainable building: A conceptual framework for implementing sustainability in the building sector. *Buildings, 2012*(2), 126–152.

Alayami, S., & Rezgui, R. (2012). Sustainable building assessment tool development approach. *Sustainable Cities and Society, 5,* 52–62.

Alho, J., Nenonen, S., & Nissinen, K. (2008). *Usability assessment of shopping centres: Components of usability rating tool.* CIB usability of workplaces phase 2. http://www.irbnet.de/daten/iconda/CIB8912.pdf

Altomonte, S., Rutherford, P., & Wilson, R. (2015). Human factors in the design of sustainable built environments. *Intelligent Buildings International, 7*(4), 224–241.

Alwaer, H., & Clements-Croome, D. J. (2010). Key performance indicators (KPIs) and priority setting in using the multi-attribute approach for assessing sustainable intelligent buildings. *Building and Environment, 45*, 799–807.

Alzaed, A., & Boussabaine, A. H. (2012, September 3–5). Towards a new methodology for integrating user expectations into passive building design. In S. D. Smith (Ed.), *Proceedings 28th annual ARCOM conference* (pp. 1467–1477). Edinburgh, UK: Association of Researchers in Construction Management.

Asdrubali, F., Baldinelli, G., Bianchi, F., & Sambuco, S. (2015). A comparison between environmental sustainability rating systems LEED and ITACA for residential buildings. *Building and Environment, 86*, 98–108.

Attaianese, E. (2011). Human factors in maintenance for a sustainable management of built environment. In J. Lindfors, M. Savolainen, & S. Väyrynen (Eds.), *Wellbeing and innovation through ergonomics. Proceedings of NES2011 (NordikErgonomic Society)*, Oulu, Finland, pp. 129–134.

Attaianese, E. (2012). A broader consideration of human factors to enhance sustainable building design. *Work, 41*(Supplement 1), 2155–2159.

Attaianese, E. (2014). Human factors in design of sustainable buildings. In M. Soares & F. Rebelo (Eds.), *Advances in ergonomics is design and usability and special population part III*, USA: AHFE, pp. 392–403

Attaianese, E. (2016). Increasing sustainability by improving full use of public space: Human centred design for easy-to-walk built environment. In F. Rebelo & M. Soares (Eds.), *Ergonomics in design: Proceedings of the AHFE 2016 international conference on ergonomics in design*. Walt Disney World, FL: Springer.

Attaianese, E., & Duca, G. (2010). Human factors and ergonomic principles in building design for life and work activities: An applied methodology. *Theoretical Issues in Ergonomics Science, 13*(2), 187–202.

Attaianese, E., & Duca, G. (2012). The human component of sustainability: A study for assessing human performance of energy efficient construction blocks. *Work, 41*(Supplement 1), 2141–2146.

Barlow, S., & Fiala, D. (2007). Occupant comfort in UK offices—How adaptive comfort theories might influence future low energy office refurbishment strategies. *Energy and Buildings, 39*, 837–846.

Berardi, U. (2011). Sustainability assessment in the construction sector: Rating systems and rated buildings. *Sustainable Development, 20*, 411–424. November/December 2012.

Berardi, U. (2013). Clarifying the new interpretations of the concept of sustainable building. *Sustainable Cities and Society, 8*, 72–78.

Berardi, U. (2015). Sustainability assessments of buildings, communities, and cities. In J. J. Kleme (Ed.), *Assessing and measuring environmental impact and sustainability* (pp. 497–545). Oxford, UK: Elsevier.

Boerstra, A., Beuker, T., Loomans, M., & Hensen, J. (2013). Impact of available and perceived control on comfort and health in European offices. *Architectural Science Review, 56*(1), 30–41.

Brown, C., & Legg, S. (2011). Human factors and ergonomics for business sustainability. In G. Eweje & M. Perry (Eds.), *Business and sustainability: Concepts, strategies and changes, critical studies on corporate responsibility, governance and sustainability* (Vol. 3, pp. 61–81). Bingley, UK: Emerald Group Publishing Limited.

Capolongo, S., Gola, M., di Noia, M., Nickolova, M., Nachiero, D., Rebecchi, A., et al. (2016). Social sustainability in the healthcare facilities: A rating tool for analysing and improving social aspects in environment of care. *Ann Ist Sup Sanità, 52*(1), 15–23.

Ceylan, C., Dul, J., & Aytac, S. (2008). Can the office environment stimulate a manager's creativity? *Human Factors and Ergonomics in Manufacturing, 18*(6), 589–602.

Chan, E. (2014). Building maintenance strategy: A sustainable refurbishment perspective. *Universal Journal of Management, 2*, 19–25.

Charitonowicz, J., & Leszeks, S. (2001). Toward sustainable housing – Ergonomics of housing stock management organization as dematerialization. In M. J. Smith & G. Salvendy (Eds.), *Systems, social, and internationalization design aspects of human-computer interaction* (pp. 392–396). London: Laurence Lerbaum.

Charytonowicz, J. (2007). *Reconsumption and recycling in the ergonomic design of architecture.* Universal access, Lecture notes in computer science (Human-computer interaction. Ambient Interaction), Vol. 4555.

Chen, H. (2010). *CPWR: Green and healthy jobs.* Berkeley, CA: University of California. http://www.cpwr.com/sites/default/files/publications/Green-Healthy%20Jobs%20fnl%20for%20posting.pdf

Chiocchio, F., Forgues, D., Paradis, D., & Iordanova, I. (2011, December). Teamwork in integrated design projects: Understanding the effects of trust, conflict, and collaboration non performance. *Project Management Journal, 42*(6), 78–91.

Clements-Croome, D. (2000). Productivity and indoor environment. *Proceedings of healthy buildings conference* University of Technology, Helsinki, Vol 1, 629–634.

Clements-Croome, D. (2006). *Creating productive workplaces* (2nd ed.). Oxford, UK: Routledge.

Clements-Croome, D. (2013). Can intelligent buildings provide alternative approaches to heating, ventilating and air conditioning of buildings? *Droesti Lecture, RACA Journal, 29*(6), 22–33.

Clements-Croome, D. (2014). *Sustainable intelligent buildings for better health, comfort and well-being.* Report for Denzero Project supported by the TÁMOP-4.2.2.A- 11/1/KONV-2012-0041 co-financed by the European Union and the European Social Fund. http://www.derekcroome.com/Document%20Files/DENZERO.pdf

Cole, R. J. (2014). Comprehensive assessment system for building environmental efficiency. In S. Murakami, K. Iwamura, & R. Cole (Eds.), *CASBEE. A decade of development and application of an environmental assessment system for the built environment Institute for Building Environment and Energy Conservation* (pp. 12–25).

Cole, R. J., Brown, Z., & McKay, S. (2010). Building human agency: A timely manifesto. *Building Research and Information, 38*, 339–350.

Delos. (2016). *The WELL building standard v1.* New York.

Dewlaney, K., Hallowell, M. R., & Fortunato, B. R. (2012). Safety risk quantification for high performance sustainable building construction. *Journal of Construction Engineering and Management, 138*, 964–971.

Dewlaney, K. S., & Hallowell, M. (2012). Prevention through design and construction safety management strategies for high performance sustainable building construction. *Construction Management and Economics, 30*, 165–177.

DGNB CORE 14. (2014). *Criteria overview.* Office version. http://www.dgnb-system.de/en/system/criteria/core14/

Dirlich, S. (2011). Comparison of assessment and certification schemes for sustainable building and suggestions for an international standard system. *The IMRE Journal, 5*, 1–12.

Dorsey, J. (2016). Sustainable design in the workplace. In A. Hedge (Ed.), *Ergonomic workplace design for health, wellness, and productivity.* Boca Raton, FL: CRC Press/Taylor & Francis Group.

Dul, J., Bruder, R., Buckle, P., Carayon, P., Falzon, P., Marras, W. S., et al. (2012). A strategy for human factors/ergonomics: Developing the discipline and profession. *Ergonomics, 55*, 377–395.

European Agency for Safety and Health at Work – EASHW. (2013). *E-fact 70 occupational safety and health issues associated with green building.* https://osha.europa.eu/en/publications/e-facts/e-fact-70-occupational-safety-and-health-issues-associated-with-green-building/view

Feifer, L. (2011, July 13–15). *Sustainability indicators in buildings. Identifying key performance indicators.* 27th international conference on passive and low energy architecture. Architecture and sustainable development.

Fischer, E. A. (2011). *Issues in green building and the federal response: An introduction.* Congressional Research Service, 7-5700, R40147, Washington, DC, 2010. Retrieved on line August 28, 2011.

Flemming, S. A. C., Hilliard, A., & Jamieson, G. A. (2007, February 6). *Considering human factors perspectives on sustainable energy systems.* Poster presented at the International Society for Industrial Ecology Conference, Toronto.

Flemming, S., Hilliard, A., & Jamieson, G. A. (2008). *The need of human factors in the sustainability domain.* Proceedings of the human factors and ergonomics society 52nd annual meeting.

Fowler, K. M., & Rauch, E. M. (2006). *Sustainable building systems summary.* Pacific Northwest National Laboratory operated for the U.S. Department of Energy by Battelle. http://www.usgbc.org/Docs/Archive/General/Docs1915.pdf

Gambatese, J., & Hinze, J. (1999). Addressing construction worker safety in the design phase designing for construction worker safety. *Automation in Construction, 8,* 643–649.

GhaffarianHoseini, A. H., GhaffarianHoseini, A., Makaremi, N., & Ghaffarian Hoseini, M. (2012). The concept of zero energy intelligent buildings (ZEIB): A review of sustainable development for future cities. *The British Journal of Environment & Climate Change, 2,* 339–367.

Gram-Hanssen, K. (2014). New needs for better understanding of household's energy consumption – Behaviour, lifestyle or practices? *Architectural Engineering and Design Management, 10*(1–2), 91–107.

Green Star-Performance v1.1. (2016). *List of credits.* https://www.gbca.org.au/uploads/194/36034/List%20of%20Credits_v1.1.pdf

Hanson, M. A. (2013). Green ergonomics: Challenges and opportunities. *Ergonomics, 56*(3), 399–408.

Haron, S. N., Hamid, M. Y., & Talib, A. (2013). Using "USEtool": Usability evaluation method for quality architecture in-use. *Journal of Sustainable Development, 6,* 100–110.

Haslam, R., & Waterson, P. (2013). Ergonomics and sustainability. *Ergonomics, 56*(3), 343–347.

Hauge, A. L., Thomsen, J., & Berker, T. (2011). User evaluations of energy efficient buildings: Literature review and further research. *Advance in Building Energy Research, 5*(1), 109–127.

Hedge, A. (2000). Where we are in understanding the effects of where we are? *Ergonomics, 43,* 1019–1029.

Hedge, A. (2008). The sprouting of "green" ergonomics. *HFES Bulletin, 51,* 1–3.

Hedge, A., & Dorsey, J. (2013). Green buildings need good ergonomics. *Ergonomics, 56,* 492–506.

Hedge, A., Rollings, K., & Robinson, J. (2010). *"Green" ergonomics: Advocating for the human element in buildings.* Proceedings of the human factors and ergonomics society 54th annual meeting. Santa Monica, CA: HFES.

Heerwagen, J. (1998, March 12–14). *Design, productivity and well being: What are the links?* Paper presented at The American Institute of Architects conference on highly effective facilities Cincinnati, Ohio.

Heerwagen, J. (2011). *Investing in people: The social benefits of sustainable design.* Proceedings of the rethinking sustainable, Construction 06, Sarasota, FL.

Heerwagen, J. H. (2000). Green buildings, organizational success, and occupant productivity. *Building Research and Information, 28,* 353–367.

Heerwagen, J. H., & Zagreus, L. (2005, April). *The human factors of sustainability: A post occupancy evaluation of the Philip Merrill Environmental Center.* Summary report for U.S. Department of Energy, Center for the Built Environment, University of California, Berkeley, CA.

Hilliard, A. (2008). *Can human factors methods help design sustainable buildings?* Proceedings of the conference on sustainable building SB08 Melbourne, Vol. 1.

ISO 21542:2011. (2011). *Building construction. Accessibility and usability of the built environment.* Geneva, Switzerland.

ISO/DIS 15392. (2006). *Sustainability in building construction- general principles.* Geneva, Switzerland.

Janda, K. B. (2011). Buildings don't use energy: People do. *Architectural Science Review, 54,* 15–22.

Joshi, S. M. (2008). The sick building syndrome. *Indian Journal of Occupational Environmental Medicine, 12*(2), 61–64.

Karwowski, W. (2006). The discipline of ergonomics and human factors. In G. Salvendy (Ed.), *Handbook of human factors and ergonomics* (3rd ed.). Hoboken, NJ: Wiley.

Kibert, C. J. (2008). *Sustainable construction: Green building design and delivery.* Hoboken, NJ: Wiley.

Kim, M. J., Oh, M. W., & Kim, J. T. (2013). A method for evaluating the performance of green buildings with a focus on user experience. *Energy and Buildings, 66,* 203–210.

Kordjamshid, M. (2011). *House rating schemes: From energy to comfort base.* Berlin, Germany: Springer.

Lange-Morales, K., Thatcher, A., & García-Acosta, G. (2014). *Synergies between ergoecology and green ergonomics: A contribution towards a sustainability agenda for HFE.* Human factors in organizational design and management – Xi Nordic Ergonomics Society annual conference, Vol. 46.

Larsson, N. (2015). *SBTool for 2015.* iiSBE. http://www.iisbe.org/system/files/SBTool%20Overview%2018Jul15.pdf

Leaman, A., & Bordass, B. (2007). Are users more tolerant of 'green' buildings? *Building Research and Information, 35*(6), 662–673.

Lee, S. Y., & Kang, M. (2013). Innovation characteristics and intention to adopt sustainable facilities management practices. *Ergonomics, 56*(3), 480–491.

Leech, J. A., Raizenne, M., & Gusdorf, J. (2011). Health in occupants of energy efficient new homes. *Indoor Air, 14*(3), 169–173.

LEED. (2015). *New LEED pilot credit: Prevention through design.* http://www.usgbc.org/articles/new-leed-pilot-credit-prevention-through-design

LEED. (2016a). *LEED v4 for building design and construction.* http://www.usgbc.org/sites/default/files/LEED%20v4%20BDC_01.27.17_current.pdf

LEED. (2016b). *Ergonomic approach for computer users.* LEED v4. http://www.usgbc.org/credits/new-construction-schools-new-construction-retail-new-construction-healthcare-commercial-inte

Lutzenhiser, L. (1993). Social and behavioral aspects of energy use. *Annual Review of Energy and the Environment, 18*, 1–664.

Martin, K., Legg, S., & Brown, C. (2013). Designing for sustainability: Ergonomics – Carpe diem. *Ergonomics, 56*(3), 365–388.

MIL-HDBK-470. (1997). *Department of defense handbook: Designing and developing maintainable products and systems (Volume I)*. (04 DEC 1997).

Muehleisen, R. T. (2010). Acoustics of green buildings. *InformeDesign, 8*, 1–7.

Nadadur, G., & Parkinson, M. B. (2013). The role of anthropometry in designing for sustainability. *Ergonomics, 56*(3), 422–439.

Newsham, G. R., Mancini, S., & Birt, B. J. (2013). Do LEED-certified buildings save energy? Yes, but.... *Energy and Buildings, 41*, 897–905.

Nicol, F., & Roaf, S. (2005). Post occupancy evaluation and field studies of thermal comfort. *Building Research and Information, 33*(4), 338–346.

Obiozo, R., & Smallwood, J. (2013, September 2–4). The role of "greening" and an ecosystem approach to enhancing construction ergonomics. In S. D. Smith & D. D. Ahiaga-Dagbui (Eds.), *Proceedings 29th annual ARCOM conference*. Reading, UK: Association of Researchers in Construction Management.

Peffer, T., Perry, D., Pritoni, M., Aragon, C., & Meier, A. (2013). Facilitating energy savings with programmable thermostats: Evaluation and guidelines for the thermostat user interface. *Ergonomics, 56*(3), 463–479.

Persons, K. C. (2000). Environmental ergonomics: A review of principles, methods and models. *Applied Ergonomics, 31*(6), 581–594.

Rajendran, S., Gambatese, J., & Behm, M. (2009). Impact of green building design and construction on worker safety and health. *Journal of Construction Engineering Management, 135*(10), 1058–1066.

Reed, R., Wilkinson, S., Bilos, A., & Shulte, K. (2011, January 16–19). *A comparison of international sustainable building tools – An update*. Proceedings of the 17th annual Pacific Rim Real Estate Society conference, Gold Coast. https://www.academia.edu/897809/A_Comparison_of_International_Sustainable_Building_Tools_An_Update

Sass, C., & Smallwood, J. (2015, August 9–14). *The role of ergonomics in green building*. Proceedings of the 19th triennial congress of the IEA, Melbourne.

Siew, R. Y. J., Balabat, M. C. A., & Carmichael, D. G. (2013). A review of building/infrastructure sustainability reporting tools (SRTs). *Smart and Sustainable Built Environment, 2*, 106–139.

Sorrento, L. (2012). A natural balance: Interior design, humans and sustainability. *Journal of Interior Design, 37*(2), ix–xxiii.

Stedmon, A. W., Winslow, R., & Langley, A. (2013). Micro-generation schemes: User behaviours and attitudes towards energy consumption. *Ergonomics, 56*(3), 440–450.

Sutcliffe, M., Hooper, P., & Howell, R. (2008). Can eco-foot printing analysis be used successfully to encourage more sustainable behaviour at the household level? *Sustainable Development, 16*, 1–16.

Thatcher, A. (2013). Green ergonomics: Definition and scope. *Ergonomics, 56*(3), 389–398.

Thatcher, A. & Chunilal, H. (2015, November 1–4). *Development and validation of a workspace layout scale for use in green building certifications.* Creating sustainable work-environments Nordic Ergonomics Society 47th annual conference, Lillehammer, Norway, pp. A4–A9.

Thatcher, A., & Milner, K. (2012). The impact of a 'green' building on employees' physical and psychological wellbeing. *Work, 41*, 381–893.

Thatcher, A., & Milner, K. (2014). Changes in productivity, psychological wellbeing and physical wellbeing from working in a 'green' building. *Work, 49*, 381–393.

Thatcher, A., & Milner, K. (2016). Is a green building really better for building occupants? A longitudinal evaluation. *Building and Environment, 108*, 194–206.

Thatcher, A., & Yeow, P. H. P. (2015). A sustainable system of systems approach: A new HFE paradigm. *Ergonomics, 59*(2), 167–178.

Tidwell, J., & Murphy, J. J. (2010). *Bridging the gap- fire safety and green buildings.* National Association for Fire Marshals. http://www.firemarshalsarchives. org/pdf/FireSafetyGreenBuildingHiResFINALv3sec.pdf

Toole, T. M., & Gambatese, J. (2008). The trajectories of prevention through design in construction. *Journal of Safety Research, 39*, 225–230.

U.S. Environmental Protection Agency EPA. (2009). *Green building.* http://www.epa.gov/greenbuilding/pubs/about.htm.

UN. (1987). *Report of the World Commission on Environment and Development.* United Nations.

UNI/PdR 13.0:2015. (2015). *Environmental sustainability of construction works – Operational tools for sustainability assessment.* General framework and methodological principles.

UNI/PdR 13.1:2015. (2015). *Environmental sustainability of construction works – Operational tools for sustainability assessment.* Residential buildings.

US EPA. (2009) *Buildings and their impact on the environment: A statistical summary.* US EPA Archive document. https://archive.epa.gov/greenbuilding/web/pdf/gbstats.pdf

USGBC. (2008a). *Administrative credit interpretation ruling innovation in design and innovation in operations credit for an ergonomics strategy.* http://www.usgbc.org/Docs/Archive/General/Docs5408.pdf

USGBC. (2008b). *Ergonomics requirements for innovation and design point: Example of a user survey for ergonomics issues.* http://ergo.human.cornell.edu/USGBC/USGBC_Ergonomic_Survey.pdf

Vierra, S. (2014). Green building standards and certification systems. In *Whole building design guide.* https://www.wbdg.org/resources/gbs.php

Vink, P., & Hallbeck, S. (2012). Editorial: Comfort and discomfort studies demonstrate the need for a new model. *Applied Ergonomics, 43*(2), 271–276.

Waris, M., Liew, M. S., Kamidi, M. F., & Idrus, A. (2014). Criteria for the selection of sustainable onsite construction equipment. *International Journal of Sustainable Built Environment, 3*, 96–110.

Zuo, J., & Zhao, Z. (2014). Green building research-current status and future agenda: A review. *Renewable and Sustainable Energy Reviews, 30*, 271–281.

CHAPTER 10

Human Factors and Ergonomics: Contribution to Sustainability and Decent Work in Global Supply Chains

Klaus J. Zink and Klaus Fischer

Introduction: Globalised Value Creation in the Context of Sustainable Development

Adopting the new 2030 Agenda for Sustainable Development in September 2015 (United Nations, 2015), the international community of states re-confirmed sustainable development as a global paradigm. Much more so than the preceding policy schemes of the Agenda 21 (UNCED, 1992a) or the Millennium Development Goals (United Nations, 2000), this agenda emphasises the need for a balanced set of Sustainable Development Goals. It also targets addressing industrialised and industrially developing countries as well as social, ecological, and economic aspects in an integrated manner (United Nations, 2015, p. 13ff).

One of these 17 goals directly addresses the way of value creation in our globalised economy: Goal 8 postulates the promotion of "sustained, inclusive and sustainable economic growth, full and productive employment

K. J. Zink (✉) • K. Fischer
Institut für Technologie und Arbeit e.V., Center for Human Factors, University of Kaiserslautern, Kaiserslautern, Germany

© The Author(s) 2018   243
A. Thatcher, P. H. P. Yeow (eds.), *Ergonomics and Human Factors for a Sustainable Future*, https://doi.org/10.1007/978-981-10-8072-2_10

and decent work for all" with its associated 12 targets (UN, 2015). The question remains how this goal is to be deployed to the diverse settings of work systems worldwide.

Looking at current challenges with regard to decent work and sustainability of global value creation, this chapter discusses the contributions of Human Factors/Ergonomics (HFE) to finding adequate solutions. In the past, these questions have mainly focused on "blue collar work" and physical production (Zink, 2013). However, as work systems become increasingly digitised, knowledge work driven by an internet based crowdsourcing (as crowd work) will also be included in this discussion.

In the first section, we will introduce the relevant trends of global (out-) sourcing and value creation on the one hand and the increasing digitisation of work systems on the other hand, both posing specific challenges to sustainable development.

Second, we will discuss how HFE is currently positioned with regard to the need for research and development arising from the requirements of sustainable development: Is the discipline's mindset already prepared for a holistic view on work systems in a "systems of systems" perspective that would be needed to deal with questions of sustainability? What approaches and experiences do already exist in the fields of blue collar work and digitisation and what tasks and opportunities will arise from those for HFE?

In the third part, three areas necessary for the further development of HFE as a discipline will be pointed out which seem necessary to better prepare it for dealing with the challenges described.

Global (Out-)Sourcing and Supply Chain Management

Over the last few decades, global (out-)sourcing and the transfer of labour-intensive production steps to industrially developing countries (IDCs) developed into an important factor of cost competition: Due to lower wage levels and lower costs for meeting local social and environmental standards, purchasing from IDCs is often considered to be much cheaper than the production in industrialised countries. Consequently, more and more supply chains spread globally, and the manufacturing depth of "final producers" or original equipment manufacturers declined in a lot of industries (Jentsch & Zink, 2016).

With regard to sustainable development, global (out-)sourcing is of particular importance as the allocation of value and damage between IDCs and the industrialised world often does not comply with the principles of intra- and intergenerational equity. Large parts of the value created within the international division of labour remain in the industrialised world. Thus, final branding and marketing of products as well as the know-how for product design and innovation are mainly located in industrialised countries. At the same time, the IDCs' resources are depleted, including long-term impacts for their human and social capital (e.g. through inhuman and underpaid working conditions, child labour or corruption) as well as for their ecosystems (e.g. through the uprooting of forests, monocultures and the use or disposal of toxics) (cp. Fischer, Hobelsberger, & Zink, 2009).

However, countries seeking foreign investments even provide special incentives for multinationally acting companies (e.g. through export processing zones) on a "global market" of low social and environmental standards (Hiß, 2006). They cannot always achieve a positive balance through their efforts (ICFTU, 2004) and rather often attract short-term investments in underpaid low-tech workplaces (Zink, 2009).

In the last years, multinationally acting and purchasing companies are increasingly forced to take a higher degree of responsibility for the entire supply chain of their products. International standards and guiding principles with regard to corporate social responsibility[1] (CSR) call, amongst others, for decent working conditions in globalised supply chains (ILO, 1998; ISO, 2010; United Nations Global Compact Office, 2014). Besides non-binding soft law, the number of legal regulations is also growing, for example, concerning the use of "conflict minerals"[2] from African mines that are coupled with violations of human rights ("Dodd-Frank Act"[3]; European Council, 2016) or the increase of requirements for sustainability reporting and public procurement through EU regulation (European Parliament, 2014a, 2014b, 2014c). Additionally, issues such as fatal accidents due to fires and collapses of buildings in the Asian garment industry (cp. Foxvog et al., 2013) led to an increasing public interest in working conditions in global supply chains.

Digitisation of Work

Whereas "supply chain ergonomics" in the past focused on blue collar work (e.g. Zink, 2013), a new dimension of globalisation is introduced by the digitisation of work—and here especially by the concept of crowdsourcing. Anyone with access to the internet can perform either micro-tasks or multi-hour tasks offered by respective internet platforms (Kittur et al., 2013).

Micro-tasks in particular are based on "hyperspecialization" as Malone and others formulated in their Harvard Business Review (HBR) article titled "The Big Idea: The Age of Hyperspecialization" (Malone, Laubacher, & Johns, 2011). Referring to Adam Smith's *The Wealth of Nations* published in 1776, who described the division of labour as central driver of economic progress, they see huge "productivity gains of dividing work into ever smaller tasks performed by ever more specialised workers [...] thanks to the rise of knowledge work and communication technology". They are thus concluding that "[w]e are entering an era of hyperspecialization – a very different, and not yet widely understood, world of work" (Malone et al., 2011, p. 58).

Crowdsourcing mobilises a growing number of people to accomplish tasks on a global scale. A study of MBO Partners from 2015 expects that the growth rate of the freelance workforce will be more than four times higher than the expected growth rate of the overall workforce (MBO Partners, 2015).

As international competition today focuses more and more on knowledge work, it is not sufficient to discuss about blue collar work in the context of sustainability, but to emphasise on crowd work, too. There mainly exist two types: crowd work based on collaboration and crowd work based on competition. As implied by the term collaboration, crowd workers work together/in a team and deliver a joint solution or product—though the individual crowd workers do not have to know each other personally. Regarding remuneration, this is considered a type of piecework.

Independent crowd work based on competition takes either a result-oriented form, which means that only the best solution will be paid, or a time-oriented approach. Here, remuneration is based on a first-come-first-serve approach, with the fee being paid for all solutions fulfilling the quality standards defined in advance. However, the basic problem of crowd work is that no legal framework is (yet) established (Leimeister, Zogaj, & Blohm, 2015, p. 28).

For companies, the advantage of crowd work lies in the possibility to get faster solutions of a higher quality based on a broader range of ideas as compared to traditional organisational structures. However, crowd workers can benefit as well: In principle, there is the possibility for the crowd worker to select different types of tasks which leads to a higher self-determination regarding these tasks but also with regard to their working times and working places. In addition, the flexibility regarding the decision to accept a task or not is increased, thereby contributing to a better work-life balance. Another advantage lies in new employment possibilities for target groups not being able to leave their homes (e.g. people with disabilities). Also, there is in general the possibility to exchange experiences concerning the principals with other crowd workers, for example, regarding the quality of calls for bids and payment questions. Lastly, crowd workers will be able to charge higher rates for their services by becoming highly specialised in their respective fields (Leimeister et al., 2015, p. 34).

However, from a human factors perspective, there are also several risks to be considered: Especially micro-tasks may lead to a poor remuneration ("digital exploitation"). For all types of (self-employed) tasks, there is no social security. The tasks could be (very) monotonous based on a high standardisation or on decomposition into very small pieces (digital "Taylorism"), as, for example, simple and repetitive steps of manual data entry and maintenance or inspection tasks. As mentioned above, the competition for crowd work is global which means that crowd workers may have to compete with (lower paid) offers from IDCs, which could lead to "self-exploitation". Legislation regarding continuity of employment, participation in decisions, and demands for pension or holidays does not exist (Leimeister et al., 2015, p. 34). This ties in with the question, whether this type of employment can really be called independent or if it should not rather be treated as a dependent one, which would result in exactly these employee rights and social security benefits.

We will return to the issues of globalised blue collar and knowledge work in the following section where we discuss the possible contributions of HFE with regard to sustainability and decent work.

Human Factors/Ergonomics: State of the Art with Regard to Sustainable Global Value Creation

In the previous section, we identified two major trends leading to specific challenges in the context of sustainability and decent work: the increasing globalisation of supply chains and the digitisation of work.

In this section, we will discuss the current role and state of the art of HFE in this context. Firstly, we focus on the mindset of this discipline, manifested through its normative basis and the "lens" we are looking through while modelling and designing work systems. Secondly, we present some already existing approaches dealing with sustainability and decent work in global supply chains and discuss how HFE could contribute to their further development.

The Mindset of Human Factors/Ergonomics

As dealing with the challenges of globalisation, sustainability, and decent work requires the adherence to the context of global human development, we first of all need to look at the mindset of HFE as a discipline by asking on the one hand whether its goals and purposes are (already) formulated broadly enough. On the other hand, we need to ask whether the way we are modelling and designing work systems complies with a holistic "systems in systems" perspective necessary for dealing with sustainability aspects.

What We Want to Achieve: Our Normative Basis

Ergonomics has been defined by the International Ergonomics Association (IEA) as the scientific discipline concerned with the interaction amongst humans and other elements of a system in order to optimise human well-being and overall systems performance (IEA, 2000). The term "optimising" in itself contains a certain normative orientation, but we need to concretise our underlying target system as well as our understanding of "human well-being" and "overall systems performance". In this context, Hancock and Drury (2011) are asking whether HFE only contributes to the quality of life for a selected segment. This leads to questions regarding the target groups of HFE and the goal of our improvements. Are we really improving working conditions and quality of life in industrially developing regions as well? Is our main goal to increase productivity by downsizing or by increased intensity of work? Who is paying for our research and consultancy work?

Of course, with regard to globalisation, ergonomists can no longer solely focus on the needs of their stakeholders in the industrialised world. "Overall systems performance" cannot be solely defined through an economic perspective, but needs to comprise criteria of social and ecologic performance as well, leading to an "overall" perspective in the sense of a "systems of systems" view.

Therefore, HFE needs to incorporate adequate instruments and concepts, for example, a life-cycle perspective (see below) or more specifically and normative in the sense of the International Labour Organization's (ILO) concept of decent work: The term decent work sums up the aspirations of people in their working lives including the opportunity for work that is productive and delivers a fair income, providing security at the workplace and social protection for families. Furthermore, decent work strives to achieve better prospects for personal development and social integration as well as freedom for people to express their concerns and organise and participate in the decisions that affect their lives based on equality of opportunity and treatment for all (ILO, 2016). This concept is far from being realised worldwide, and it is also newly developing into an increasing problem in so-called developed countries (with a growing precariat (Standing, 2011)). Thus, the ILO placed decent work on the global Agenda 2030 for sustainable development (see above goal 8 "Promote inclusive and sustainable economic growth, employment and decent work for all" of the UN Sustainability Development Goals) (ILO, 2016).

What We See and What We Don't See: How We Are Modelling Work Systems
Work system models are the basic element for macro- and microergonomic analysis. However, our models are often too reductionistic:

From a microergonomic point of view, work systems are classically modelled as linear processes transforming a particular input into a well-defined output of products and residuals. Main system elements are one (or more) worker(s) interacting with different means of work, such as machines or tools, used in a work process (Alter, 2009; Schlick, Bruder, & Luczak, 2010). In this way, however, it is not considered where the input (as knowledge, working capacity, energy or resources) for the work process is coming from, how it is (re-) generated and what impact the emerging residuals and matter have in the surroundings of the work system, in particular when they are harmful and exceed their load carrying capacity (Fischer & Zink, 2012).

Macroergonomic system design already leads to a more comprehensive understanding of work systems, insofar as on the one hand it considers interactions between people and technology as important objects of system design (Hendrick & Kleiner, 2002; Kleiner, 2006) and on the other highlights the relevance of human and social capital for the quality of the work process, as shown in the famous coal-mining study from Trist and Bamforth (1951). But as mentioned above, here also relevant sustainability parameters remain unconsidered.

At the beginning of the new millennium, the concept of "sustainable work system design" was introduced by different authors (Docherty, Forslin, Shani, & Kira, 2002; Docherty, Kira, & Shani, 2009; Eijnatten, 2000). According to this concept, sustainable work systems are characterised by reproducing at a minimum the resources which were used, prohibiting the generation of one kind of capital (e.g. economic capital) at the expense of another (e.g. social or ecological capital stocks) and investing in overall system viability (Docherty et al., 2009, p. 3; Eijnatten, 2000, p. 9). In addition to these characteristics, Eijnatten (2000) distinguishes different "sustainability purposes" of work systems at individual, organisational, and societal/ecological system levels. He thus illustrates the relations between different system levels and considers work systems as being embedded in super-ordinated social, ecological, and economic surroundings, a concept which is highly compatible with the concept of sustainable development (Fischer & Zink, 2012).

Recently, Thatcher and Yeow (2016) further called for a "systems of systems approach" (see also Zink, 2014), referring to Costanza and Patten's succinct understanding of a "nested hierarchy of systems" in the early sustainability debate (Costanza & Patten, 1995, p. 196).

Summing up, when looking at the role of HFE in globalised value creation, we need to broaden our perspective from a rather classical focus on a single work system located in the industrialised world to a "systems of systems"-thinking, realising that work systems are often embedded in globally spread value creation networks. The same is true regarding the normative basis for optimising overall systems performance: Sustainability and decent work strive for a more comprehensive target set than economic efficiency gains or ergonomic improvements on a micro-scale level.

Ergonomic Tasks and Approaches in the Fields of "Blue Collar" and "Knowledge Work"

Having discussed the normative basis of HFE and the way we are modelling work systems, we now explicitly refer to the above-mentioned fields of blue collar and knowledge work and show several links to ergonomic tasks and approaches accruing from the call for sustainability and decent work.

Decent Blue Collar Work in (Global) Value Creation Chains
There are some international experiences regarding the improvement of working conditions in globalised value creation, which are related to ergonomic interventions (cp. Fischer et al., 2009). These include, amongst others, approaches of microergonomic work system design helping to overcome the negative spiral of poor working and living conditions (see Scott, 2008a, 2008b) as well as macroergonomic change management leading to

participatory approaches that enable self-help and intrinsically motivated changes of behaviour (see Kawakami, Kogi, Toyama, & Yoshikawa, 2004; Kogi, 2008; Imada, 2008).

With regard to decent work in globalised supply chains, the "oldest" approach is to define so-called codes of conduct. They lead to the voluntary self-obligation of a multinational enterprise to ensure minimum social and ecological standards at their production and supplier sites worldwide. Their implementation is mostly driven by non-governmental organisations (NGOs) and critical customers, leading to respective requirements in B2B procurement. In the meantime, a multitude of different codes of conduct have emerged and their multiplicity might pose a problem in itself at times.

Of course, a successful implementation of codes of conduct first of all assumes knowledge about their existence by managers and employees at all global production and supplier sites likewise as well as knowledge on how to transform these codes into effective measures for improved working conditions. However, respective capacity development has to be seen as a neglected field. Therefore, Locke, Amagual, and Mangla (2009) came to the conclusion that "voluntary compliance programs, promoted by global corporations and non-governmental organisations alike, have produced only modest improvements in working conditions and labour rights in global supply chains" (Locke et al., 2009).

As an alternative, Worldwide Enhancement of Social Quality (WE) together with Deutsche Gesellschaft für Internationale Zusammenarbeit (GIZ)[4] and the Non-Food division of a German coffee retailer (Tchibo) realised pilot projects in Bangladesh, China, and Thailand to improve working conditions[5] (Knolle, 2012). The improvement of the dialogue between management and employees was seen as a core element and precondition of this approach. To realise this, first a training infrastructure had to be established in order to implement a concept for better social standards, efficiency, and cooperation at the workplace. The dialogue between retailers and suppliers also was a goal for improvement.

Although the impacts vary significantly between the countries involved, an impact study shows positive changes regarding participation and communication, as well as improvements of social benefits, of occupational health and safety (OHS) and the introduction of minimum wages but also regarding economic results like productivity and quality of products. Amongst others, management commitment and recognition of the business case of employee participation and dialogue were identified as key factors for success (Ramboll, 2010).

A similar approach, which is focused even more strongly on improving economic results or competitiveness as a precondition to improve working conditions, is realised by the ILO Better Work Programme. Better Work is based on an agreement between the International Labour Organization (ILO) and the International Finance Corporation (IFC) (a member of the World Bank Group) to develop a global programme for better labour standards in global supply chains (Better Work, 2016). It includes various industries, for example, garments and footwear, plantations, electronic equipment, and light manufacturing. Pilot projects are set up in countries of the Middle East, Southern Africa, and East Asia. The programme offers different services: assessment, training, and advisory services. The assessment tool creates a "framework for assessing compliance with core international labour standards and national labour laws but also the impact of improvements activities on quality and productivity" (Better Work, 2016). The tool is used by trained enterprise advisors who also support improvement activities. Better Work also offers targeted training courses to managers, supervisors, and workers. Topics include ILO core labour standards and workers' rights and responsibilities to human resource management, supervisory skills, and occupational health and safety (Better Work, 2016). Impact research has been put in place, the results of which are regularly published (Better Work, 2016).

The idea to improve working conditions by augmenting the competitiveness of a company is not new. In 2007 Locke and Romis published a comparison of working conditions and labour rights at two Mexican factories and showed that not codes of conduct but the implementation of a new management system was successful in improving the conditions. Their analyses showed that the better performing factory applied a TQM-based management approach that allowed to effectively address the root causes of poor working conditions in global supply chains. It thereby combined micro- and macroergonomic interventions like work content design (job rotation, job enrichment, and enlargement) as well as employee participation and work organisation, leading to multi-skilled work groups. Thus, new forms of work organisation and human resource management systems that promoted not only healthier and more equitable workplaces but also new sources of competitive advantage have been identified as better solutions than codes of conduct (Locke & Romis, 2007).

Though many companies (like Nike, ABC, Hewlett-Packard) started multiple private initiatives across different countries and economic sectors to improve working conditions and labour standards, Locke (2013) is concluding that there are limits of private (non-state) power concerning the governance of global supply chains with regard to sustainability aspects (for a closer look on different forms of sustainability governance in global supply chains see Jentsch & Fischer, 2017). Accordingly, a stronger private-public partnership is needed as "each of the strategies [...] – private compliance efforts, capability building efforts, and even innovative state enforcement strategies – are necessary and important components to this strategy but none alone is sufficient to tackle this complex set of issues" (Locke, 2013, p. 177).

Returning to the question of how the knowledge and approaches of HFE are able to contribute to more sustainability and decent work in global value creation chains, we can thus summarise that micro- and macroergonomic interventions:

- need to be embedded in holistic (management) approaches accompanied by measures on the political and systemic level (as inclusion of local governments, organisations of civil society, and addressing cultural aspects) (cp. Locke, 2013),
- can build the basis for achieving profound and long-term transformations towards sustainability and decent work through participatory, efficiency-gaining measures relying on the principles of ownership and responsibility, not on enforcing standards through "external" instructions (cp. Kawakami et al., 2004; Kogi, 2008; Scott, 2008a, 2008b), and
- are already in use in different countries, industries, and steps of value creation but not yet systematically integrated in a consistent approach of (sustainable) supply chain management and governance (cp. Better Work, 2016; Jentsch & Fischer, 2017).

Decent Knowledge Work in (Global) Supply Chains
As described above, there is a growing market for (global) supply chains regarding knowledge work (especially as crowd work) which is not in itself fulfilling the demands of decent work (e.g. regarding remuneration and social security). At the moment, there are different approaches to handling this problem.

Taking the example of Germany, we can see a first code of conduct formulated by some leading crowdsourcing providers (amongst others, Testbirds, Streetspotr, and clickworker) which is supported by the German Crowdsourcing Association (Testbirds GmbH, 2015).[6]

It contains the following principles[7]:

1. Tasks in conformance with the law (no tasks with illegal, discriminating, fraudulent, demagogic, violent, or anti-constitutional content; considering age limitations, etc.)
2. Clarification on legal situations (information of the crowd workers about legal and tax regulations connected to crowd working)
3. Fair payment (all subscribers pay a fair and appropriate wage which has to be clarified in advance)
4. Motivating and good work (not only financial reimbursement but also intrinsic motivational factors play an important part, e.g. implemented through prices and awards, training possibilities, and user-friendly, intuitive platforms)
5. Respectful interaction (providers are aware of their responsibility to respect and consider the interests of both parties)
6. Clear tasks and reasonable timing (detailed description of all the criteria regarding timing and content that need to be met in order to successfully complete a crowdsourcing project)
7. Freedom and flexibility (crowd working takes place on a voluntary basis, no negative consequences for the crowd worker through the refusal of an offered task)
8. Constructive feedback and open communication (crowdsourcing companies are available for questions regarding the task, give best possible assistance and technical support, and prompt feedback)
9. Regulated approval process and rework (transparent approval process which are justified and based on the project description; fair and neutral complaint process for crowd workers)
10. Data protection and privacy (providers are obligated to act under confidentiality and can only be relieved from this responsibility by the client)

Although this code of conduct was designed by crowdsourcing providers and thus includes company's interests, some of the above cited statements directly address core elements of decent work (ILO, 2016) as, for example, fair income or aspects of social and workplace security as well as of social integration.

More employee (or union) orientation can be found when looking at demands of (German) unions (Brandl, 2015; Wedde & Spoo, 2015).

According to them, fair basic standards should be based on:

1. Adequate possibilities of co-determination of the works council combined with a new definition of "firm" and "employment" to exclude pseudo-self-employment
2. Securing a minimum of holistic work contents
3. No violation of personality rights (e.g. protection of privacy)
4. Definition of a minimum remuneration
5. Regulations concerning social security

Internationally, some scientists (but still only a few ergonomists) are discussing the problem of decent crowd work (e.g. Kittur et al., 2013; Silberman, Irani, & Ross, 2010). Lilly Irani, Assistant Professor at UC San Diego and Six Silberman, PhD student at the University of California, developed a platform for crowd workers[8] where they are able to evaluate crowdsourcers (so-called requesters) of Amazon's Mechanical Turk (AMT) with regard to the following questions: How was the payment made and was it adequate? Have results been refused without explanation? Have questions been answered to the crowd workers? and so on (Nagrale, 2012). This platform helps the crowd workers of AMT to stay in contact with each other and improve their dealings with requesters. The Turkopticon Toolbar shows the requester ratings by other Turk workers and thus reduces the information asymmetry between workers and requesters which is a precondition for improving working conditions (Silberman et al., 2010, p. 40).

With regard to the question of how HFE can contribute to decent work and sustainability in (global) supply chains of knowledge work, we can refer to Kittur et al. (2013) who formulated some ideas for the future of crowd work:

- Develop tools to support not only the work itself but also those performing the work
- Job design with "traditional" criteria: providing skill variety, task identity, and task significance; timely and task specific feedback, as well as the opportunity to self-assessments to help workers to learn, preserve, and produce better work
- Create a broad set of motivations including fair payment but also reputation and credentials (like certifications)
- Create career ladders
- Improve task design through better communication
- Facilitate learning

These ideas can be understood as "to-do-items" from a human factors perspective. They show that micro- and macroergonomic interventions could help to ensure that also the "remote" workplaces of crowd workers are (at least partly) designed due to criteria of decent work and sustainable work systems. Referring to the above-mentioned aspects of sustainable work systems, we can state that here again ethics of work (or especially ethics of professional crowd work) is the topic to be handled as fundamental principles and achievements of our modern analogue working world (remuneration, legal protection, feedback about working results) seem not yet to have been fully transferred to the digitalised one (Silberman et al., 2010; see also LaPlante & Silberman, 2015). In the end there should be better systems, better requests, and better work—requirements which highly fit with the claim of HFE to improve overall systems performance (IEA, 2000) provided that our normative target system also covers the often anonymous work systems of crowd workers.

In order to complete the above-mentioned examples for already existing as well as potential HFE contributions to decent work and sustainability in global supply chains, Fig. 10.1 illustrates several possible applications for HFE interventions in this context along several value creation phases:

- During the phase of product design and conception, the preconditions and characteristics of a product are defined for the whole life cycle along its value chain—starting from the extraction of raw materials to production phases, sales, and use as well as disposal and recycling. In the context of the different work systems that are associated with value creation, the phase of design and conception thus is essential with regard to sustainability and decent work. Here it is determined, for example, which materials are used and how they need to be processed (e.g. socially and environmentally sound or burdened by resource conflicts and human rights abuse) and if the product will be ergonomic for both the user (e.g. concerning consumer safety) and the workers involved in production, maintenance, refurbishment, and recycling (e.g. safe and ergonomic processes or dangerous and hazardous for health). Life-cycle ergonomics and approaches of a "design for sustainability" are mainly relevant here.

- Concerning the different phases of production in global supply chains, the approaches of sustainable work system design can help to improve the often non-sustainable conditions. These could be coupled with awareness raising and capacity building by microergonomic

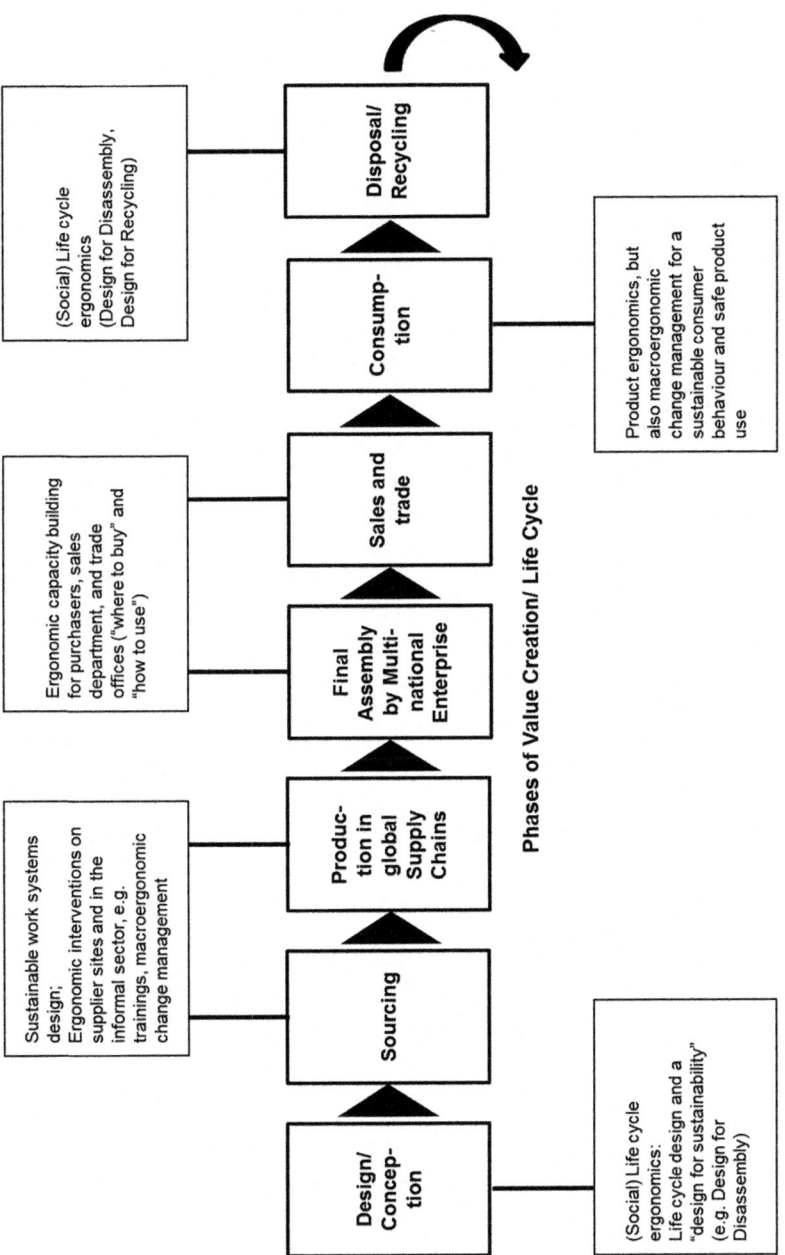

Fig. 10.1 Ergonomic interventions in different phases of global value creation (Adapted from Kubek, Fischer, & Zink, 2015, p. 74)

trainings for improving working conditions on site as well as with macroergonomic change management considering cultural differences. Fostering intrinsic motivation and supplier ownership are also relevant. Scott (2008a, 2008b) and Kawakami et al. (2004) as well as Imada (2008) show that low-cost and low-threshold participatory ergonomic approaches can help to improve working and living conditions in IDCs, thus contributing to decent work and a more sustainable development in global supply chains.

- Looking at production steps and final assembly at the original equipment manufacturers' sites, for example, wider knowledge about ergonomic criteria and work system design would help to increase the level of competence of purchasers and decision makers with regard to the questions where and under which conditions suppliers or production sites should be chosen. And of course, this knowledge would also be useful for retailers and trade offices as costumers become increasingly aware of quality criteria beyond technological features, comprising the eco- and socio-balance of products.

- For the consumer, ergonomic criteria of course play an important role concerning product ergonomics and usability. But even more importantly in the context of sustainability, ergonomic approaches could also help to ensure a safe and environmentally friendly use of the product as well as to support a more sustainable consumer behaviour, for example, through modular and easy-to-upgrade or easy-to-repair products and bundles of services.

- Concerning disposal and recycling, again the above-mentioned aspects of life-cycle ergonomics during the conception phase come into effect.

Although Fig. 10.1 shows a classical supply chain model and the above-mentioned examples for ergonomic approaches mainly address blue collar work which is classically organised in physical supply chains, most of these applications can also be transferred to knowledge work taking place in the crowd. Of course, some phases (as disposal and recycling) are not relevant here, but awareness raising and capacity building for ergonomic principles and know-how on sustainable work system design would improve the situation of crowd workers worldwide. However, the target group of an agile "community" of crowd workers and their costumers probably needs to be addressed in a specific way. But it is undeniable that ergonomists could and should contribute to the future development of this field, too.

Identifying Fields for Further Development of HFE

Assuming that sustainability is more than a "buzzword" for ergonomics and also that globalisation is accepted as a challenge for ergonomics, the previous sections already have shown some need for further developments of our discipline. In this section, we identify three areas for the further development of HFE which can be seen as essential for the future role of our discipline in this context: (1) attaining thinking in whole life cycles, (2) adapting the curricula of (macro-)ergonomics education, and (3) building (further) partnerships with relevant key actors.

Attaining Thinking in Whole Life Cycles

The preceding paragraphs have shown that when talking about "globalised value creation", not only the conditions of production at the original equipment manufacturer or first-tier supplier sites are of interest but also those of all formal and informal workplaces along the entire life cycle of a product or service (cp. Fig. 10.1):

- in the case of physical products starting with the extraction of raw materials over production and maintenance up to the recycling or re-use of the finished product
- in the case of knowledge work or services comprising all forms of brainwork, including that of crowd work organised in anonymous networks

Thus, while calling for decent work and sustainability, it is not sufficient for HFE to focus on a single phase of value creation (e.g. design or production). In fact, we need an integrated view on design, production, use, and (for physical products) recycling/disposal in the sense of life-cycle ergonomics.

Life-cycle-oriented design (e.g. design for manufacturing or assembly) and evaluation concepts (as technology assessment or product life-cycle management) are traditionally based on technical and economic aspects, including also ecological aspects.

A similar approach can be found looking at newer publications referring to "Total Life-Cycle Management" where sustainability and life-cycle orientation is discussed from a company's point of view (Herrmann, Bergmann, Thiede, & Halubek, 2007). In most of these publications the economic and ecological dimension is broadly described, whereas the assessment of the social dimension is kept to only a few pages. An explanation for this could be that—in contrast to economic and environmental aspects where quantitative data and indicators for sustainability are already quite sophisticated—social sustainability is more difficult to measure, including the problem of dealing with qualitative data and "soft" aspects as, for example, cultural differences. However, these aspects have to be discussed intensively as they are relevant influencing factors for the (un-)successful implementation of sustainability standards, for example, in the field of Occupational Health and Safety.

Coming from a sustainability perspective, the discussion of new technologies like nuclear power plants or genetic engineering has always been accompanied by a life-cycle assessment focused on the impacts on society and citizens. As a consequence of the Rio Declaration in 1992 (UNCED, 1992b) and the following Agenda 21, the development of "criteria and methodologies for the assessment of environmental impacts and resource requirements throughout the full life cycle of products and processes" have become a topic of interest (UNCED, 1992a).

But as sustainability is based on a three-pillar approach, dealing with ecological aspects in a narrow sense would not be enough. Social and economic assessments have to be included, too. Whereas the discussion of life-cycle costs is not new, the assessment of social impacts emerged only within the last years (Zamagni, Amerighi, & Buttol, 2011).

There already exist concepts of a Social Life-Cycle Analysis based on sustainability, but they are not yet satisfying in regard of detailed ergonomics aspects. One example is the "Guidelines for Social Life Cycle Assessment of Products" which have been published in 2009 by the United Nations Environment Programme (UNEP, 2009). UNEP (2009, p. 37) defines Social (and socio-economic) Life-Cycle Assessment as "a social impact (real and potential impacts) assessment technique that aims to assess the social and socio-economic aspects of products and their positive and negative impacts along their life cycle encompassing extraction and processing of raw materials; manufacturing; distribution; use; re-use; maintenance; recycling; and final disposal".

The stakeholders taken into regard are workers/employees, local community, society (national and global), consumers (covering end-consumers as well as the consumers who are part of each step of the supply chain), and value chain actors (cp. UNEP, 2009, p. 46).

Taking first of all a more "traditional" ergonomic perspective, workers/employees and consumers are of primary interest. But as shown in the preceding sections, defining ergonomics in and for a globalised world has to include several value chain actors as well as community and society, referring to the above-mentioned "systems of systems" approach linked with the concept for sustainable work systems. Thus, an ergonomic life-cycle assessment has to start with the development process itself, which should include the analysis of working conditions regarding raw materials, for example, in the electronic industry, followed by the analysis of working conditions in manufacturing and assembly, but also including maintenance, repair, disassembly, and recycling or re-use (Zink & Eberhard, 2006). In addition, the analysis should include impacts on the user of the product.

In this context, the "ergonomic quality in design" approach of the International Ergonomics Association is of interest.[9] In the further development of this concept, it should include the Social Life-Cycle Assessment criteria not discussed until now. Besides the ergonomic quality in design products, the ergonomic quality in design workplaces or systems could be a helpful approach. However, as already discussed when looking at the current mindset of our discipline in the section above: Again a broader view is needed, accompanied by a systems perspective, as well as the readiness to deal with more comprehensive aspects related to work system design as, for example, the measurement of decent work according to the proposition of ILO (2008).

Adapting HFE Curricula

The topics discussed in this chapter—supply chain ergonomics for blue collar and knowledge work as well as sustainable work systems coupled with a life-cycle perspective—would also imply that new contents for a curriculum in micro- and/or macroergonomics have to be developed. As sustainability is built on a three-pillar understanding, some basic knowledge in economic and ecological assessments of work systems should exist. Furthermore, bringing sustainability to practice is not free of contradictions or paradoxes (Ehnert, 2014). Therefore, the process of introducing such concepts requires the balance between different approaches and goals. Traditional ergonomics curricula cover these fields only in part. If we discuss supply chains, we cannot focus on a single organisation or the single work systems within it any longer. In consequence the understanding of macroergonomics and systems ergonomics has to be increased (Zink, 2014). And last but not least we have to scrutinise the moral basis of HFE as described by Hancock and Drury (2011) again and again and have to continue developing the whole discipline (Dul et al., 2012).

Developing Cooperation with Key Actors
Reflecting the role of HFE in the above shown cases, ergonomists might have been involved as part of ILO, but mostly they were not involved at all. If we are interested in designing sustainable work systems in a globalised world, ergonomics has to act as a "global player", too. Thus, ergonomists should strengthen their cooperation with available partners in IDCs or with the networks of ILO or World Health Organization (WHO) and use international CSR platforms such as UN Global Compact (United Nations Global Compact Office, 2014) or Business for Social Responsibility (www.bsr.org). Other partners could be unions—where established in respective countries—or non-governmental organisations (NGOs) dealing with working conditions. As shown in the preceding section, our know-how, instruments, and approaches are highly relevant for achieving long-lasting and self-sustaining interventions beyond and complementary to the enforcement of standards through codes of conduct or other governance instruments.

Also in Western companies, ergonomists need new partners and contact persons. Purchasers play a crucial role as they decide where and at which costs suppliers will be engaged. They thus need sufficient know-how about sustainability in global supply chains, a field where ergonomists could give support. But (non-)decent work at their own suppliers could also be of interest for the risk management department of a company and CSR managers in general.

To be an adequate partner for these target groups, ergonomists have to know international standards of work (as, e.g. defined by OECD or ILO), specific labour laws in respective countries, and the concepts of CSR (Zink, 2003) (as comprehensively defined in ISO 26000) and sustainable supply chain management (Jentsch & Zink, 2016). As the above shown examples of micro- and macroergonomic interventions at workplaces in global supply chains for blue collar and knowledge work show, HFE approaches have a lot experience to offer in this context and could complement and augment existing instruments, for example, from human resource management. Thereby, the role of ergonomics does not end with OHS but needs more discussion of comprehensive approaches in macroergonomics combined with sustainability (Zink, 2015).

CONCLUSIONS

As we have shown in this chapter, the mega trends of globalisation and digitisation are directly affecting work systems in various settings worldwide and are thus immediately concerning the fields of research and activity of our discipline. This is in particular true when these trends are put in relation to the requirements of sustainability and decent work: HFE in its strive for optimising human well-being and overall systems performance needs to answer which demands and contributions it could make in this context.

Discussing our normative basis, our modelling approaches, and a couple of different examples for possible and already applied ergonomic interventions in global value creation, we have described the current status of HFE and which fields for further development still exist.

Summarising our discussion, we can state that HFE provides a lot of know-how, methods, and instruments directly applicable for contributing to adequate working conditions and a more sustainable works system design for blue collar and knowledge work in global supply chains. However, we need to further develop and profile our approaches to become more visible as a competent partner in the global discussion about sustainable development and decent work. The newly installed Sustainable Development Goals (United Nations, 2015) which are currently deployed to a countless number of sustainability strategies, policy programmes, and indicator sets worldwide could provide the right impetus for that.

NOTES

1. Corporate social responsibility is defined as the responsibility of an enterprise for its impacts on society (European Commission, 2011).
2. According to the OECD Due Diligence Guidance for Responsible Supply Chains of Minerals, conflict minerals are defined as minerals mined in conflict-affected and high-risk areas, "identified by the presence of armed conflict, widespread violence or other risks of harm to people" (OECD, 2013, p. 13). The list of conflict minerals includes the ores of tin, tantalum and tungsten, as well as gold (OECD, 2013).
3. In the so-called Dodd-Frank Act, conflict minerals are defined due to their origin from the Democratic Republic of the Congo or an adjoining country (U.S. Government Publishing Office, 2010, p. 2218).
4. German Society for International Cooperation.
5. https://www.we-socialquality.com.

6. Deutscher Crowdsourcing Verband e.V.
7. See http://www.crowdsourcing-code.com.
8. https://turkopticon.ucsd.edu/.
9. http://www.iea.cc/project/project_equid.html.

REFERENCES

Alter, S. (2009). Service system fundamentals: Work system, value chain, and life cycle. *IBM Systems Journal, 47*, 71–85.

Better Work. (2016). Retrieved June 30, 2016, from https://www.betterwork. org/global/

Brandl, K.-H. (2015). Gefordert: Faire Mindeststandards [Demanded: Fair minimum standards]. In *Arbeitsrecht Im Betrieb, Sonderausgabe, September 2015: Crowdworking – Gute Arbeit für die Crowd* (pp. 40–42). Frankfurt a.M.: Bund Verlag.

Costanza, R., & Patten, B. C. (1995). Defining and predicting sustainability. *Ecological Economics, 15*, 193–196.

Docherty, P., Forslin, J., Shani, A. B., & Kira, M. (2002). Emerging work systems. In P. Docherty, J. Forslin, & A. B. Shani (Eds.), *Creating sustainable work systems* (pp. 3–14). London/New York: Routledge.

Docherty, P., Kira, M., & Shani, A. B. (2009). What the world needs now is sustainable work systems. In P. Docherty, M. Kira, & A. B. Shani (Eds.), *Creating sustainable work systems* (2nd ed., pp. 1–21). London/New York: Routledge.

Dul, J., Bruder, R., Buckle, P., Carayon, P., Falzon, P., Marras, W. S., et al. (2012). A strategy for human factors/ergonomics: Developing the discipline and profession. *Ergonomics, 55*(4), 377–395.

Ehnert, I. (2014). Paradox as a lense for theorizing sustainable HRM: Mapping and coping with paradoxes and tensions. In I. Ehnert, W. Harry, & K. J. Zink (Eds.), *Sustainability and human resource management: Developing sustainable business organizations* (pp. 247–271). Heidelberg, Germany/New York/ Dordrecht, The Netherlands/London: Springer.

Eijnatten, F. M. V. (2000). From intensive to sustainable work systems: The quest for a new paradigm of work. In M. Sapir (Ed.), *Working without limits? Re-organizing work and reconsidering workers' health* (pp. 47–66). Brussels, Belgium: TUTB-SALTSA.

European Commission. (2011). Communication from the Commission to the European Parliament, the Council, the European Economic and Social Committee and the Committee of the Regions: A renewed EU strategy 2011–2014 for Corporate Social Responsibility. COM(2011) 681 final. Brussels, Belgium: Author.

European Council. (2016). Trade in conflict minerals: Presidency agreement with the European Parliament. Press release 677/16 (22/11/2016) of the Council of the European Union. Retrieved February 2, 2017, from http://www.consilium. europa.eu/press-releases-pdf/2016/11/47244650625_en.pdf

European Parliament. (2014a). Directive 2014/24/EU of the European Parliament and of the Council of 26 February 2014 on public procurement and repealing Directive 2004/18/EC. Retrieved February 2, 2017, from http:// eur-lex.europa.eu/legal-content/EN/TXT/PDF/?uri=CELEX:32014L0024 &from=de

European Parliament. (2014b). Directive 2014/95/EU of the European Parliament and of the Council of 22 October 2014 amending Directive 2013/34/EU as regards disclosure of non-financial and diversity information by certain large undertakings and groups. Retrieved February 2, 2017, from http://eur-lex.europa.eu/legal-content/EN/TXT/PDF/?uri=CELEX:3201 4L0095&from=EN

European Parliament. (2014c). Directive 2014/24/EU of the European Parliament and of the Council of 26 February 2014 on public procurement and repealing Directive 2004/18/EC. Retrieved February 2, 2017, http://eur-lex.europa.eu/legal-content/EN/TXT/PDF/?uri=CELEX:32014L0024&from=de

Fischer, K., Hobelsberger, C., & Zink, K. J. (2009). Human factors and sustainable development in global value creation. In International Ergonomics Association (Ed.), *17th world congress on ergonomics of the IEA 2009*. Beijing, China: IEA Press. CD-ROM.

Fischer, K., & Zink, K. J. (2012). Defining elements of sustainable work systems— A system-oriented approach. *A Journal of Prevention, Assessment and Rehabilitation, 41*(1), 3900–3905.

Foxvog, L., Gearhart, J., Maher, S., Parker, L., Vanpeperstraete, B., & Zeldenrust, I. (2013). Still waiting: Six months after history's deadliest apparel industry disaster, workers continue to fight for compensation. Retrieved February 1, 2017, from http://www.cleanclothes.org/resources/ publications/still-waiting

Hancock, P. A., & Drury, C. G. (2011). Does human factors/ergonomics contribute to the quality of life? *Theoretical Issues in Ergonomics Science, 12*(5), 416–426.

Hendrick, H. W., & Kleiner, B. M. (2002). *Macroergonomics: Theory, methods, and application*. London: Lawrence Erlbaum.

Herrmann, C., Bergmann, L., Thiede, S., & Halubek, P. (2007, August 27–29). *Total life cycle management – An integrated approach towards sustainability*. Paper presented at the "Third International Conference on Life Cycle Management", University of Zurich, Irchel, Switzerland.

Hiß, S. (2006). *Warum übernehmen Unternehmen gesellschaftliche Verantwortung?* [Why do companies take on social responsibility?]. Frankfurt, Germany: Campus Verlag.

IEA (International Ergonomics Association). (2000). *Ergonomics international news and information – August 2000*. London: Marshall Associates.

ILO (International Labour Office). (1998). Declaration on fundamental principles and rights at work. Retrieved August 11, 2016, from http://www.ilo.org/wcmsp5/groups/public/---ed_norm/---declaration/documents/publication/wcms_467653.pdf

ILO (International Labour Organization). (Ed.) (2008). *Measurement of decent work*. Discussion paper for the Tripartite Meeting of Experts on the Measurement of Decent Work, ILO, Geneva, Switzerland.

ILO (International Labour Organization). (Ed.) (2016). Decent work and the 2030 Agenda for sustainable development. Retrieved July 5, 2016, from http://www.ilo.org/global/topics/sdg-2030/lang--en/index.htm

Imada. (2008). Achieving sustainability through macroergonomic change management and participation. In K. J. Zink (Ed.), *Corporate sustainability as a challenge for comprehensive management* (pp. 129–138). Heidelberg, Germany: PhysicaVerlag.

International Confederation of Free Trade Unions (ICFTU). (2004). Behind the brand names. *Working conditions and labour rights in export processing zones*. Retrieved August 11, 2016, from http://www.newunionism.net/library/internationalism/ICFTU%20-%20Working%20Conditions%20amd%20Labour%20Rights%20in%20Export%20Processiong%20Zones%20-%202004.pdf

ISO (International Organization for Standardization). (2010). *Guidance on social responsibility*. Geneva, Switzerland: ISO.

Jentsch, M., & Fischer, K. (2017). Sustainability governance of global supply chains. In P. Schukat, M. Schmidt, D. Giovannucci, B. Hansmann, & D. Palekhov (Eds.), *Towards sustainable global value chains: Concepts, instruments and approaches, Natural resource management in transition*. Heidelberg, Germany: Springer. (in print).

Jentsch, M., & Zink, K. (2016). Strategische Bedeutung eines nachhaltigen Lieferkettenmanagements. In T. Wunder (Ed.), *CSR und strategisches Management* [CSR and strategic management] (pp. 199–215). Berlin, Germany/Heidelberg, Germany: Springer.

Kawakami, T., Kogi, K., Toyama, N., & Yoshikawa, T. (2004). Participatory approaches to improving safety and health under trade union initiative – Experiences of POSITIVE training program Asia. *Industrial Health, 42*, 196–206.

Kittur, A., Nickerson, J. V., Bernstein, M. S., Gerber, E. M., Shaw, A., Zimmerman, J., et al. (2013, February 23–27). *The future of crowd work*. In CSCW Conference Proceedings. San Antonio, Texas. Retrieved July 6, 2016, from https://www.Iri.fr/~mbl/ENS/CSCW/2012/papers/Kittur-CSCW13.pdf

Kleiner, B. M. (2006). Macro-ergonomics: Analysis and design of work systems. *Ergonomics, 37*, 81–89.

Knolle, M. (2012). *Influence of participatory organisation structures on the implementation of social standards: An empirical study of Chinese garment factories.* Luneburg: Leuphana University.

Kogi, K. (2008). Participation as precondition for sustainable success: Effective workplace improvement procedures in small-scale sectors in developing countries. In K. J. Zink (Ed.), *Corporate sustainability as a challenge for comprehensive management* (pp. 183–198). Heidelberg, Germany: Physica-Verlag.

Kubek, V., Fischer, K., & Zink, K. J. (2015). Sustainable work systems: A challenge for macroergonomics? *IIE Transactions on Occupational Ergonomics and Human Factors, 3*(1), 72–80.

LaPlante, R. & Silberman, S. (2015). Design notes for a future crowd work market. Retrieved June 15, 2016, from https://medium.com/@silberman/design-notes-for-a-future-crowd-work-market-2d7557105805#.2uon97cy8

Leimeister, J. M., Zogaj, S., & Blohm, I. (2015). Crowdwork – digitale Wertschöpfung in der Wolke [Crowdwork – Digital value creation in the cloud]. In C. Benner (Ed.), *Crowdwork – zurück in die Zukunft? Perspektiven digitaler Arbeit* (pp. 9–41). Frankfurt am Main, Germany: Bund Verlag.

Locke, R. (2013). *The promise and limits of private power: Promoting labor standards in a global economy.* Cambridge, UK: Cambridge University Press.

Locke, R., Amagual, M., & Mangla, A. (2009). Virtue out of necessity? Compliance, commitment and the improvement of labor conditions in global supply chains. *Politics & Society, 37*(3), 319–351.

Locke, R., & Romis, M. (2007). Improving working conditions in a global supply chain. *MIT Sloan Management Review, 48*(2), 54–62.

Malone, T. W., Laubacher, R., & Johns, T. (2011). The big idea: The age of hyperspecialization. *Harvard Business Review, 89*(7/8), 56–65.

MBO Partners. (2015). MBO Partners State of Independence in America 2015. Retrieved August 11, 2016, from https://www.mbopartners.com/uploads/files/state-of-independence-reports/MBO-SOI-REPORT-FINAL-9-28-2015.pdf

Nagrale, P. (2012). What is Turkopticon Toolbar? How to Use This in Turk. Retrieved July 6, 2016, from http://moneyconnexion.com/what-is-turkopticon-toolbar-how-to-use-this-in-mturk.htm

OECD. (2013). *OECD due diligence guidance for responsible supply chains of minerals from conflict-affected and high-risk areas.* Paris: OECD Publishing.

Ramboll. (2010). *Impact assessment of the public private partnership of GTZ and Tchibo – WE Project.* Berlin: Ramboll. Retrieved February 22, 2017, from https://www.we-socialquality.com/DownloadDocument.aspx?id=70

Schlick, C. M., Bruder, R., & Luczak, H. (2010). *Arbeitswissenschaften* [Ergonomics]. Munich, Germany: Springer.

Scott, P. A. (2008a). Global inequality, and the challenge for ergonomics to take a more dynamic role to redress the situation. *Applied Ergonomics, 39*, 495–499.

Scott, P. A. (2008b). The role of ergonomics in securing sustainability in developing countries. In K. J. Zink (Ed.), *Corporate sustainability as a challenge for comprehensive management* (pp. 171–181). Heidelberg, Germany: Physica-Verlag.

Silberman, S., Irani, L., & Ross, J. (2010). Ethics and tactics of professional crowdwork. *XRDS, 17*(2), 39–43.

Standing, G. (2011). *The precariat: The new dangerous class.* London/New York: Bloomsbury Academic.

Testbirds GmbH. (Ed.) (2015). Code of conduct: Paid crowdsourcing for the better, guideline for a prosperous and fair cooperation between companies, clients & crowd workers. Retrieved July 6, 2016, from www.crowdsourcing-code.com

Thatcher, A., & Yeow, P. H. P. (2016). A sustainable system of systems approach: A new HFE paradigm. *Ergonomics, 59*(2), 167–178.

Trist, E. L., & Bamforth, K. W. (1951). Some social and psychological consequences of the long wall method of coal getting. *Human Relations, 4*, 3–38.

U.S. Government Publishing Office. (2010). Dodd-Frank Wall Street Reform and Consumer Protection Act. Public Law 111-203-July 21, 2010. Retrieved February 2, 2017, from http://www.gpo.gov/fdsys/pkg/PLAW-111publ203/pdf/PLAW-111publ203.pdf

UNCED. (1992a). *Agenda 21 – The United Nations program of action from Rio.* United Nations Conference on Environment and Development, New York.

UNCED. (1992b). Report of the United Nations Conference on Environment and Development, Annex I: Rio Declaration on Environment and Development. A/CONF.151/26 (Vol. I). Retrieved August 5, 2016, from http://www.un.org/documents/ga/confl51/aconf15126-1annex1.htm

UNEP (United Nations Environment Program). (2009). *Guidelines for social life cycle assessment of products.* Nairobi, Kenya: UNEP.

United Nations. (2000). United Nations millennium declaration. Retrieved February 2, 2017, from http://www.un.org/millennium/declaration/ares552e.htm

United Nations. (2015). Transforming our world: The 2030 Agenda for Sustainable Development. Retrieved February 2, 2017, from http://www.un.org/ga/search/view_doc.asp?symbol=A/RES/70/1

United Nations Global Compact Office. (2014). Corporate sustainability in the world economy. Retrieved August 11, 2016, from https://www.unglobalcompact.org/docs/news_events/8.1/GC_brochure_FINAL.pdf

Wedde, P., & Spoo, S. (2015). Mitbestimmung in der digitalen Arbeitswelt [Co-determination in a digitized world of work]. In ver.di-Bereich Innovation und gute Arbeit (Ed.), *Gute Arbeit und Digitalisierung: Prozessanalysen und Gestaltungsperspektiven für eine humane digitale Arbeitswelt* (pp. 35–38). Berlin, Germany: ver.di.

Zamagni, A., Amerighi, O., & Buttol, P. (2011). Strengths or bias in social LCA? *International Journal of Life Cycle Assessment, 16,* 596–598.

Zink, K. J. (2003). Corporate social responsibility promoting ergonomics. In H. Luczak & K. J. Zink (Eds.), *Human factors in organizational design and management – VII: Re-designing work and macroergonomics – Future perspectives and challenges* (pp. 63–72). Santa Monica, CA: IEA Press.

Zink, K. J. (2009). Human factors and ergonomics in industrially developing countries: Necessity and contribution. In P. A. Scott (Ed.), *Ergonomics in developing regions: Needs and applications* (pp. 15–27). Boca Raton, FL/London/ New York: CRC Press.

Zink, K. J. (2013). Designing sustainable work systems in a globalized world: A new challenge for ergonomics? In Human Factors and Ergonomics Society (Ed.), *Proceedings of the human factors and ergonomics society 57th annual meeting* (pp. 1075–1079). Santa Monica, CA: HFES.

Zink, K. J. (2014). Designing sustainable work systems: The need for a systems approach. *Applied Ergonomics, 45*(1), 126–132.

Zink, K. J. (2015). Digitalisierung der Arbeit als arbeitswissenschaftliche Herausforderung: ein Zwischenruf [Digitisation of labour as a challenge to work science: A break-in]. *Zeitschrift für Arbeitswissenschaft, 69*(4), 227–232.

Zink, K. J., & Eberhard, D. (2006). Product and production ergonomics as part of a newly defined product management. In R. N. Pikaar, E. A. P. Koningsveld, & P. J. M. Settels (Eds.), *Proceedings of the IEA 2006, 16th world congress on ergonomics.* Maastricht, The Netherlands: Elsevier Science B.V. (CD Rom).

CHAPTER 11

Natural Resource Use, Institutions, and Green Ergonomics

Ashutosh Sarker, Wai-Ching Poon, and Gamini Herath

INTRODUCTION

Human interactions with natural resources for economic activities generate pollutants (such as heavy metals and greenhouse gases) that are discharged into the natural environment, causing irreversible damage. These interactions also contribute to green ergonomic issues involving human health and sustainable natural resource use (Thatcher, 2013; Thatcher, Garcia-Acosta, & Lange-Morales, 2013). Green ergonomics, which is a new subfield of human factors and ergonomics (HFE), emphasises bi-directional relationships between humans and nature (Thatcher, 2013; Thatcher et al., 2013). The central goal of green ergonomics is to focus on systems designed for human use that would minimise adverse health effects from natural resource use.

The world's population has reached 7.5 billion in 2017 and is estimated to be 9.8 billion by mid-2050 (Population Reference Bureau, 2017). The

A. Sarker (✉) • W.-C. Poon • G. Herath
Department of Economics, Monash University Malaysia,
Bandar Sunway, Malaysia

increasing population growth and human activities in production will exacerbate green ergonomic problems with consequences for community health and sustainable natural resource use. A delicate balance exists between the productive potential and long-run deterioration of the natural resource base, determined by the combined effects of government's natural resource policies, technology, and institutions (Couttenier, 2008; Massa, 2015). Government policies in many developing countries have encouraged increased agricultural production by subsidising the widespread adoption of new crop varieties and use of industrial inputs, such as fertiliser and pesticides, which have caused water pollution.

Human factors and ergonomics (HFE) was originally developed to examine the relationships between humans and other elements (such as system machines and equipment) that they interact with to optimise human well-being and system performance (International Ergonomics Association, 2015). Ergonomists have recently endeavoured to broaden the conventional scope of ergonomics by accommodating the relationships between humans and the natural environment into ergonomics and have created a new subfield of "green ergonomics" (García-Acosta, Pinilla, Larrahondo, & Morales, 2014; Hanson, 2010, 2013; Radjiyev, Qiu, Xiong, & Nam, 2015; Thatcher, 2013; Zink, 2014).

They have argued that complex social arrangements that accommodate interactive relationships between natural resources, social institutions, economic institutions, governance, and humans are essential to understanding the issues that affect aquatic systems and public health (Romiszowski, 2016; Tapiola & Paloviita, 2015). Sustainable development of a natural resource system is based on complex interconnections between public policies, technology, and institutions (Couttenier, 2008; Massa, 2015). The formulation of environmental policy, institutions, and technology had been active areas of scholarly interest for decades; however, policy failure occurred in sustainable development because neoclassical economics ignored the important role of natural capital and institutions in natural resource use (Ghosh, Mukhopadhyay, Shah, & Panda, 2015; Kapp, 2012). Nevertheless, more inclusive institutional forms, which accommodate green ergonomics discipline to improve the human-nature relationships, allow humans to optimise human well-being in agriculture. Therefore, the specific objectives of this chapter are as follows:

- provide an overview of the link between natural resource use and green ergonomics;

- develop a framework to portray the complex interconnections between institutions for natural resource use, water pollution, and community health;
- highlight the value of institutional approach and propose a new institutional approach;
- evaluate the role of institutions and ergonomics in the Malaysian context; and
- identify policy implications and challenges for ergonomics with a focus on natural resource use.

ERGONOMICS AND RIVER WATER MANAGEMENT

Ergonomic impacts in the agricultural and manufacturing industries have been recently highlighted (Fathallah, 2010; Twomlow, O'Neill, Sims, Ellis-Jones, & Jafry, 2002). Agrochemicals and heavy metals in water and food continue to affect community health (Volety, 2008; Zheng et al., 2007). Growing water pollution and increasingly degraded water quality threaten the public health and aquatic ecosystems with a new, complex global water quality challenge (Davidson, Myers, & Chakraborty, 1992; UNESCO, 2015).

Water demand in Malaysia has increased over the years and is expected to grow from 14,069 million litres per day (MLD) in 2010 to 25,884 MLD in 2050 (Ministry of Natural Resources and Environment, 2011). The Pahang-Selangor Inter-State Raw Water Transfer Project (Pahang-Selangor ISRWT) from Pahang state (Pahang River Basin) to Selangor, Kuala Lumpur, and Putrajaya (Langat River Basin) was initiated in the Ninth Malaysia Plan 2006–2010 to address the anticipated increase in water demand. The Pahang-Selangor ISRWT project has caused some problems for the indigenous (Orang Asli) families whose traditional land was used in the development of the Kelau Dam. In Malaysia, the Water Supply Enactment (1955) and Environmental Quality Act (1974) are two important acts to prevent water contamination. Moreover, the National Monitoring Network was established in 1978 to monitor river water quality.

Nevertheless, many rivers in Malaysia have become polluted due to manufacturing and agro-based industries, domestic sewerage, effluents from mining, logging activities, clearing of forest, and heavy metals from factories. Every month, 2200 tons of garbage is dumped into Malaysian rivers, drains, and waterways. Furthermore, 700 kg of rubbish is dumped daily into the Sungai Klang that flows through downtown Kuala Lumpur and Greater Kuala Lumpur (The Straits Times Online, July 26, 2016).

The turbidity level of a major river in Malaysia may even reach 6000 nephelometric unit (a measure of turbidity) (The Free Malaysia Today Online, May 25, 2016). The 2013 Malaysia Environmental Quality Report indicated that 5.3% of 473 Malaysian rivers were polluted and 36.6% were slightly contaminated (The Rakyat Post, September 24, 2015).

The Sungai Semenyih, which is the main waterway from the Semenyih Dam to the treatment plant, provides more than 630 million litres of clean water daily for consumption (The Star Online, September 29, 2016b). The Sungai Semenyih Water Treatment Plant was closed several times because of contamination concerns, interrupting water supply to more than 330,000 premises. Furthermore, the pollution from a factory adjacent to a building material company in Jalan Sungai Lalang, Semenyih, resulted in the emission of a strong odour (The Star Online, October 5, 2016a).

Water from the Sungai Buah, Negeri Sembilan, flowing into the Sungai Semenyih, was affected by smell pollution that, at the source of contamination, was more than five times worse than the contaminated water of the Selangor River. The contaminated water, which contained 4-bromodiphenyl ether (a flame retardant compound), was extremely toxic (The Malay Mail Online, October 27, 2016). The water samples gave off a foul odour with a threshold odour number of four, indicating that the water required treatment; dead fish were spotted in the water from where the samples were collected (New Straits Times Online, March 4, 2017).

Malaysia has introduced a Water Quality Index (WQI) that records biological oxygen demand (BOD), total suspended solids (TSS), chemical oxygen demand (COD), ammoniacal nitrogen (NH3-N), pH value, and dissolved oxygen (DO) (Economic Planning Unit, 2002). According to the index, the percentage of clean rivers decreased significantly from 53.3% in 1990 to 28.3% in 2000. The number of contaminated river basins increased in the 1990s primarily due to an increase in pollution and a decline in rainfall. The major sources of pollution include sewerage from livestock farms, effluents from the agro-based industry and manufacturing sector, as well as soil erosion.

Untreated industrial toxic and hazardous waste, wastewater, and sewage account for 90% of the total industrial pollution load in local rivers (Abdullah, 1995). Economic development activities, including land use activities (such as forestland conversion), deforestation, agro-based industries (such as rubber plantations during the rubber boom of the 1990s), logging activities in response to the increasing demand for timber export during the 1970s and 1980s, and pollution by palm oil mills effluent, have impacted waterways significantly. For instance, poor mining regulations during 1909–1939 resulted in 16.26 million tons of sediment being dumped into the river drainage systems (Balamurugan, 1991).

Pollutants, such as heavy metals, and pesticides, in river water cause health threats to human beings and aquatic life. Consumption of aquatic food (such as fish, prawn, or cockles) that has accumulated heavy metal pollutants affects human reproduction rates and life spans. Lead poisoning, particularly from polluted water, can cause memory problems, tingling of hands and feet, muscle pain, malaise, fatigue, decreased libido, and sleeping problems. Nitrate contamination is common in most polluted waters, causing methaemoglobinaemia, or blue-baby syndrome, in bottle-fed infants under three months of age. The World Health Organization (WHO) has proposed a guideline value of 50 micrograms per litre (mg/l) nitrate based on studies in which the condition was rarely seen below that concentration but was increasingly seen above 50–100 mg/l (Ahamed, 2014).

During the 1980s, the presence of endemic arsenicosis was recognised in mainland China, with the arsenic concentration in groundwater in the 220–2000 µg/l range (Mandal & Suzuki, 2002). Heavy metals have been developed in agricultural soils in South China, especially in mining areas contaminated by manganese (Li, Luo, & Su, 2007), lead (Wong, Li, Zhang, Qi, & Min, 2002), and cadmium (Li et al., 2007; Wong et al., 2002).

Chinese villagers near long-closed lead mines suffer from painful swellings all over their bodies, caused by cadmium poisoning; these swellings prevent some of the villagers from working (Daily Mail Online, December 3, 2014). The Guangxi Environment and Geology Research Centre reported that water in the affected area had 17.4 times higher cadmium levels than the national standard.

The use of contaminated drinking water containing toxic chemicals, such as fluoride or arsenic, may cause various disorders, including headaches, poor performance of immune system, and lower back pain. In Indonesia, tests have indicated the presence of dissolved mercury, the highest in a mining area being 2.78 mg/l (The Water Environment

Partnership in Asia, WEPA, http://www.wepa-db.net/policies/state/indonesia/indonesia.htm); moreover, significant differences have been observed in the concentrations of metals (particularly lead and zinc) in aquatic biota (Widianarko, Verweij, Van Gestel, & Van Straalen, 2000).

China has experienced an increase in industrial discharges and excessive application of fertilisers, insecticides, and pesticides in agricultural areas, over the past several years. Measures to reduce the annual discharge loads of arsenic and mercury (60–70%) into water and construction of more than 60,000 industrial waste water treatment plants have reduced industrial water contamination in the country (Zhang et al., 2010). Nevertheless, improvements in water quality due to these measures are not well recorded because of ineffective monitoring.

In Bangladesh, 35% of deep wells contain arsenic (Smith, Lingas, & Rahman, 2000), and drinking from these has been prohibited (Millennium Ecosystem Assessment [MEA], 2005). More than 20,000 individuals (mainly rice farmers) from the dry zone of Sri Lanka have died over the last ten years due to chronic kidney disease (CKD) caused by toxic metals from deep well water (Bandarage, 2013). High arsenic content in water has also had dermatological manifestations and caused arsenicosis in India (Mazumder, 2008).

In the United States, contamination of drinking water causes lower back pain (APEC Water, n.d.). Contaminants in drinking water that contribute to back pain or lower back pain include chlorite (affects the nervous system and consequently produces back pain), chlorine dioxide, cadmium (causes kidney damage and acute back pain), fluoride (causes bone damage that results in back pain), mercury (causes kidney damage), and lead (causes kidney damage) (APEC Water, n.d.). Moreover, heavy metal concentrations of cadmium and nickel in freshwater fish in the region between Norway and Russia have increased with an increased proximity to smelting plants (Amundsen et al., 1997).

NATURAL RESOURCES AND ERGONOMICS: AN INTEGRATED APPROACH

Human activities, such as water management, pollution control, cultural operations and manual farming practices, and biocide management in agriculture, involve mostly manual labour and poor technology. Plants and animals, water, soils, and rivers provide ecosystem and waste disposal functions. The environmental groups are not organised within a systems framework, and therefore may not yield optimal outcomes. We propose a model that integrates technology, institutions, and agricultural policy to create an overarching management system of natural capital assets within agriculture because the market alone cannot properly coordinate such a natural system. We, therefore, adopt an agricultural system approach in which natural capital is used within the extant policy, technology, and institutional system because we believe that a better integration of agricultural practices, organisational forms, and technologies with natural capital embedded in agriculture will provide a useful framework for ergonomic interaction.

Figure 11.1 presents the elements of the integrated agricultural systems model, which includes three broad dimensions: (1) the policy environment, (2) technological innovation, and (3) the institutional environment. The policy, institutional, and technology environments are firmly interconnected. The arrows in the model indicate relations among the dimensions and the health and ergonomic effects of agriculture in terms of these dimensions. Technologies can save water resources, maintaining water quality and ecological integrity for future generations. The excessive use of water, land, forests, groundwater, and biocides reflect the type of technology in use.

The declining quality of water in many water basins is due to heavy abstraction supported by wrong technologies (Seckler, 1996). The major health and ergonomic effects in agriculture are the result of the collective impact of the use of natural resources, such as water, forestry, land, soil, and air, to produce agricultural output. Figure 11.1 identifies the ergonomic impacts, compiled from published studies of different countries. It is developed here to highlight the interconnectedness of human action, technology, institutions, and the environment. Complexities in natural resources arise due to the common property nature of natural resources. Hardin's (1968) "tragedy of the commons" argues that overexploitation of common property resources could occur in the long term. The tragedy of the commons applies to contexts in which the actions of independent and self-interested people are collectively interdependent; in such circumstances, these actors overuse the commons or their common-pool resources (CPRs), eventually destroying the resources (e.g. irrigation systems, fisheries, and forestry).

Nevertheless, commons have not inevitably led to catastrophes (cf. Ostrom, 1990). A CPR is defined as a large, natural resource system having two specific characteristics: excludability, the difficulty of excluding

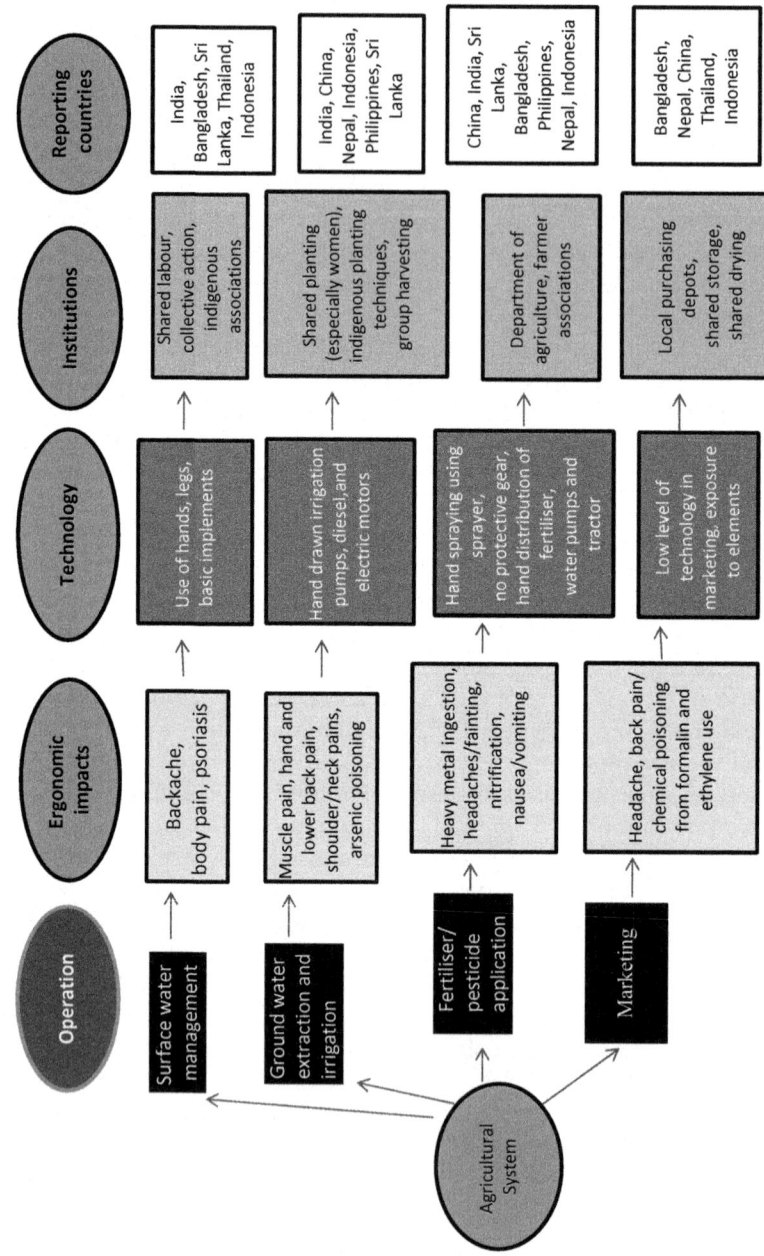

Fig. 11.1 Schematic representation of the integrated model

beneficiaries from appropriating a benefit from the resource, and subtractability, when the benefit is appropriated by a beneficiary, it is no longer available to other beneficiaries (Ostrom, 1990, 2005). The role of institutions, which constrain the activities of individuals to achieve coordinated action, can help avoid the tragedy associated with a CPR (North, 1990, p. 3).

Institutions represent formal and informal rules that humans collectively devise to promote behavioural changes that improve and preserve shared natural systems (North, 1990; Ostrom, 1990). Institutions integrate the environmental, social, and economic values of natural resources. Many developing countries have adopted cooperative institutions, locally focused collective action, local knowledge, and self-monitored norms of cooperation (Ostrom, 1990)[1] to integrate diverse ecological and socioeconomic constraints and improve natural resource use (Coase, 1960; North, 1990; Ostrom, 1990). Watershed committees and associations, water user groups, and community forestry facilitate collective action by mobilising the community to overcome the common property and the tragedy of the commons.

Ergonomics has an important institutional role in agriculture; alleviation of the ergonomic and health effects requires the cooperation of stakeholders to integrate the opinions and concerns of the farmers, managers, and ergonomists. Workers in developing countries are at risk, and therefore focusing on their ergonomic issues is essential. Ergonomic interventions must be developed in consultation with society. Poor consultation produces inappropriate technology and/or even worse technology, as evinced by some interventions. The following example exemplifies the role of institutions in ergonomics: small tank communities in Sri Lanka and India have used collective action and shared labour for water management; subsequently, other inputs have disappeared, resulting in the loss of work sharing, heavier burden on individual farmers, water pollution, and ergonomic issues. However, interventions can be more scientifically implemented using the most appropriate institutional arrangements. Many innovative institutions have been tested and found wanting (e.g. the Irrigation Management Transfer, widely deployed in Asia to create collective action, failed because the concept was not properly conceptualised [Herath, 2012]). The synergies and the nuances of institutions must be correctly understood and duly incorporated for better ergonomic results. In the following section, we present such a model, referred to as the polycentric governance approach.

THE POLYCENTRIC GOVERNANCE APPROACH

A polycentric governance system comprises multiple centres of decision-making authority at various levels; these centres are interrelated but formally independent of each other (McGinnis & Ostrom, 2012; Ostrom, 2010, 2014; Ostrom, Tiebout, & Warren, 1961). The polycentric model can address water quality and pollution issues more effectively (Sarker, Ross, & Shrestha, 2008). When water quality is defined as both a resource unit and an attribute of water, and when river water is defined as a resource system in the context of the CPR literature, we identify diverse individuals and groups from multiple levels who are involved with and responsible for degradation in water quality. Thus, the polycentric governance approach is appropriate for natural resources, including water quality, to address green ergonomic issues in agriculture. User self-governance is a better policy alternative for green ergonomics. Ostrom (1990) made substantial contributions to the establishment of user self-governance as a viable alternative for self-governance (Ostrom, 2007; Toonen, 2010).

A polycentric governance system (Ostrom et al., 1961) combines multiple levels and various types of associations from the public, private, nonprofit, and community sectors, which in turn have overlapping jurisdictional and functional areas (McGinnis & Ostrom, 2012). This governance system may involve three distinguishable entities with overlapping levels of authority that affect actions by users and outcomes achieved by managing a CPR (Kiser & Ostrom, 1982). Typically, a set of rules is developed by a group of actors and used to conduct recursive activities that affect those actors and, at times, certain non-actors (Ostrom, 1992).

Multiple policy alternatives are accommodated and configured in polycentric governance systems, and independent authorities operate interdependently to address certain problems of natural resources at the local community level (Arrow, Keohane, & Levin, 2012; Bish, 2014; McGinnis & Ostrom, 2012; Ostrom et al., 1961).

Table 11.1 presents the link among constitutional-, collective-, and operational-choice rules (or institutions) and their scales of analysis in a polycentric governance system. The process of use of resources, maintenance of resources, monitoring of rule compliance, and implementation occur at the operational scale. The processes of rule-based actions, administration, and formal dispute resolution occur at the collective-choice scale. Creation of rules, rule-based regulation, formal dispute resolution, and modification of rules occur at the constitutional sphere. Constitutional-choice rules affect the collective-choice rules that then affect the operational rules (Ostrom, 1990). Because CPR users have the ability to self-govern, they alternate

Table 11.1   Connections among rules and levels of analysis

Types of rules	Constitutional rules directly affecting activities at operational sphere	Collective rules indirectly affecting activities at operational sphere	Operational rules directly affecting daily decisions at operational sphere
Scales of analysis	Constitutional sphere	Collective sphere	Operational sphere
Courses of actions	Creation of rules Rule-based regulation Formal dispute resolution Modification of rules	Rule-based actions Administration Formal dispute resolution	Use of resources Maintenance of resources Monitoring of rule compliance implementation

Sources: Adapted from Ostrom (1990, 2005)

between operational-, collective-, and constitutional-choice arenas when resolving CPR problems (Ostrom, 1990). A polycentric governance system accommodates government regulation, market-based solutions, co-management, and user self-governance, and addresses the limitations of resolving specific CPR problems at the local level (Andersson & Ostrom, 2008; McGinnis & Ostrom, 2012).

Figure 11.2 presents a conceptual model of polycentric governance approach for addressing the water quality issue in a river. The model comprises national, state, local, and river basin levels and involves state and non-state stakeholders from these levels.

Stakeholders, such as villagers, industries, and farmers, have their separate self-governing organisations. Although each organisation interacts and is active at an individual level, they can also extend beyond their respective levels to change operational rules in the collective-choice arena. Operational rules include informal and formal rules that are developed and amended through cooperation and networking among the water quality management stakeholders while maintaining discrete, yet well-connected, authoritative boundaries.

Abundant literature on environmental pollution has supported a polycentric governance system (e.g. Ballet, Koffi, & Pelenc, 2013; Bartelmus, 2010; Baumgärtner & Quaas, 2010a, 2010b; Bazin, Ballet, & Touahri, 2004; Binder & Witt, 2012; Birkin & Polesie, 2013; Bithas, 2011; Lejano & Stokols, 2013; Söderbaum, 2011; Van den Bergh, 2010; White, 2013). Baumgärtner and Quaas (2010b, p. 449) noted that the worldview on the human-nature relationship covers multiple and interacting spatial scales, from local to global, and includes the analysis of feedbacks and interactions as well as the emergence of systematic properties.

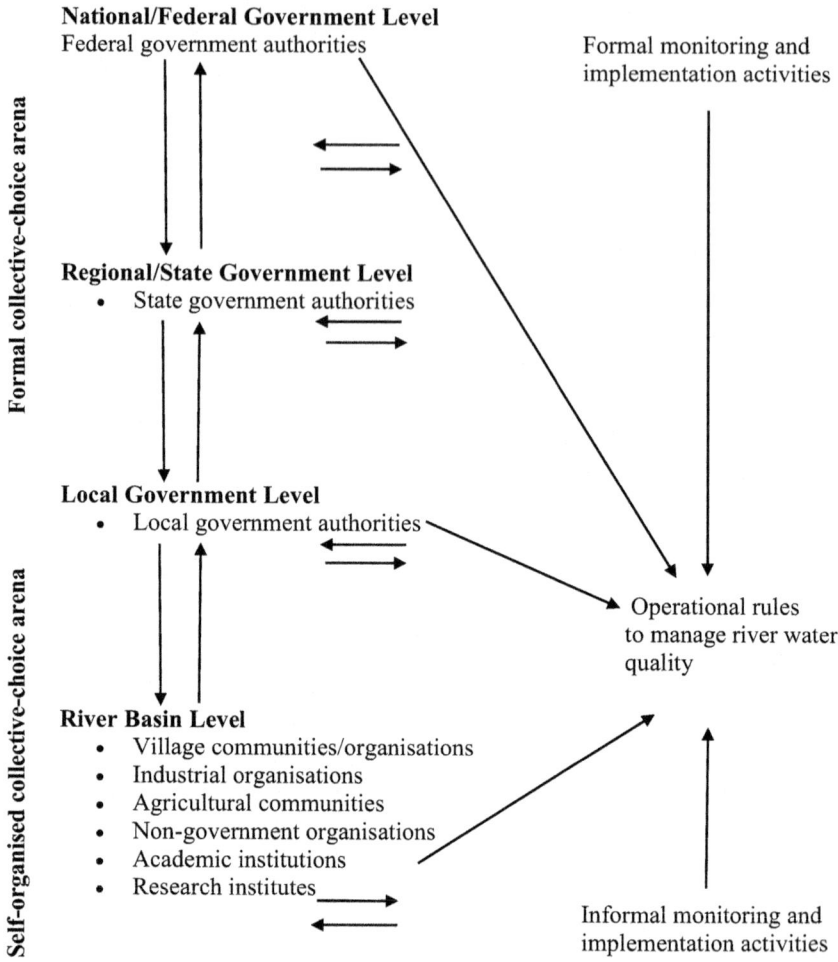

Fig. 11.2 Polycentric governance to manage river water quality (Sources: Adapted from Ostrom (1990, 2005))

Referencing Van den Bergh (2010), Bithas's (2011) internalisation of externalities addresses the environmental welfare of the present generation and ensures that environmental rights of future generations are preserved. This view reflects the statement by Baumgärtner and Quaas (2010b) regarding intra- and inter-generational environmental justice, particularly concerning present and future human-nature relationships in a polycentric governance system. It is important that agricultural households have independent, yet interrelated, organisations, to monitor pollution at the local level. Furthermore, Bithas (2011) added that when considering externality issues regarding sustainability economics, both intra- and inter-generational interests must be accommodated.

Relevance of the Polycentric Governance Approach for Malaysia

Natural resources typically have two components: a resource unit and a resource system (Gardner, Ostrom, & Walker, 1990; Ostrom, Gardner, & Walker, 1994; Ostrom & Ostrom, 1977). One user's appropriation of a certain amount of water from the resource system reduces the water amount for all users. Subsequently, a rational, self-interested irrigator is likely to withdraw as much water as he or she wants, until the average amount of water becomes zero, resulting in Hardin's (1968) tragedy of the commons, which is relevant to water pollution in the Malaysian context.

Upstream users of a river may pollute river water by releasing pesticides or eroding soil into the river basin, affecting midstream and downstream users. In the absence of proper institutions, this upstream river pollution reduces the welfare of all people because the average level of river water decreases when a user withdraws clean water upstream, which eventually decreases the amount of water for all users; thus, all parties are affected.

These externalities can be best addressed by the informal institutional elements in the polycentric model. Since free riding occurs in water pollution, local rules can be established. The need for institutional collaborations with research organisations and universities is highlighted by the WQI use in Malaysia. For example, the WQI in Malaysia does not address the presence of heavy metals in polluted water, although it is an issue in Malaysia. Many research institutes and universities have conducted research on heavy metal pollution, but the lack of communication has resulted in the exclusion of this research from policy development (Poon & Herath, 2012).

Therefore, as shown in Fig. 11.2, cooperation and communication among village communities/organisations, industrial organisations, agricultural communities, non-government organisations, academic institutions, and research institutes at the river basin level are of vital importance.

In Malaysia, the disposal of industrial, agricultural, and household wastes into rivers has degraded the river water quality. Human factors, such as the absence of communication among state, regional, local authorities and users of the rivers, lack of institutional arrangements among the stakeholders, and poorly defined boundaries of government authorities, have caused a weak human-nature relationship, thereby reducing the water quality. Thus, the polycentric model can be adopted in Malaysia to refine and strengthen the relations and directions among the different decision-making levels.

For example, the pollution of the Selangor water treatment plant would not have occurred if there was greater coordination between the Semenyih River management authorities at the local/national levels and the water utility industry in Selangor. This proved to be very costly and could have been avoided by more clearly defining the conditions of collaboration and rules of compliance. In Japan, the rules of compliance are strictly adhered to, implying strong institutions. Malaysia is weak in this regard and strengthening the institutional standards and regulation compliance is imperative to establish a robust polycentric system.

Moreover, Japanese society is very law abiding and polite, and political interference in people's affairs is minimal. The success of the polycentric system in Japan is due to this non-interference. Malaysia must take note of this requirement because governance in Malaysia is still replete with instances of political interference in all affairs that can reduce the effectiveness of the proposed model.

WATER AND FISH POLLUTION IN MALAYSIAN RIVERS

Water Pollution in Malaysia

We argue that green ergonomics still lacks details and that most health impacts of agriculture, such as poisoning, food contamination, and water quality, must be included in this new paradigm. While existing water institutions in Malaysia are weak, human activities cause considerable river pollution in Malaysia. The Klang River is polluted by sewage, sullage, and discharges from septic tanks and industrial areas. The following are few examples of water pollution in Malaysia: first, the Sungai Semenyih Water Treatment Plant was shut down in September 2010 and again in September 2016 due to ammonia contamination; the closure of the Semenyih Dam affected 1.6 million people in the Selangor. Second, the Selangor River water contamination, beginning from Negeri Sembilan to Selangor, remains at a critical level (The Malay Mail Online,

October 27, 2016). Third, the Sepang District Kampung Ginching Water Treatment Plant[2] was closed in July 2012 following high ammonia levels detection; the contamination was evident from the foaming, bubbles, and a thin oil film over the water as well as a foul smell. Fourth, more than one million consumers were affected by water disruptions due to the closure of the Selangor River following a diesel spillage in 2013. Subsequently, four treatment plants that supplied 57% of the total water demand in Selangor, Kuala Lumpur, and Putrajaya were closed (The Star Online, August 31, 2013). Thus, weak and inappropriate institutions, unregulated human activities, weak rules, and poor compliance have contributed to water pollution in Malaysia (Adnan, Zakaria, Juahir, & Ali, 2012; Mokhtar, Toriman, Hossain, Abraham, & Tan, 2011; Sany, Salleh, Rezayi, et al., 2013; Sany, Salleh, Sulaiman, et al., 2013); particularly, river water was contaminated due to heavy metal concentrations (Ahmad, Mushrifah, & Shuhaimi-Othman, 2009; Poon, Herath, Sarker, Masuda, & Kada, 2016).

The consumption of food contaminated with heavy metals, such as fish from polluted river water, affects human health (Khan et al., 2008). According to the Department of Environment (2010), the Malaysian industry produced more than 1.8 million metric tons of scheduled waste in 2010. Poon et al. (2016) examined river water and fish samples from the Klang River Basin (KRB) and Langat River Basin (LRB). They investigated the impacts of human activities on water and fish resources from an ergonomics perspective and found that heavy metals in river water contaminate fish, causing health risks and health costs for residents. In parts of Malaysia, intensive upland agriculture relies heavily on chemical fertilisers, causing the water of downstream rivers to become so severely contaminated that farmers refrain from using the river water for irrigation (Othman, 2008). According to Othman, farm products are sometimes so contaminated due to polluted river water that it is illegal to sell them in the public markets.

Findings of the Socioeconomic Survey

Some villagers consumed fish from these rivers, and the potential for health risk due to fish contamination remains a threat. The study conducted a socioeconomic survey to improve the understanding of the link between river pollution, sustainability, and ergonomics. The survey identified rural villagers' understanding of health issues, their health and food security concerns, and their vulnerability to change as well as perceive threats from river water pollution.

This study reports on the research on water pollution, which is embedded within green ergonomics. It discusses water pollution through heavy metals in the Langat and Klang rivers in Malaysia. It examines water pollution through heavy metals that endangers the water, fish, and the health of the stakeholders. To determine the heavy metal contamination, we conducted field visits along KRB and LRB and interviewed villagers living in the river basin using a semi-structured questionnaire. The sample was selected randomly. Along the LRB,[3] we interviewed a total of 41 households (14 from downstream, 15 midstream, and 12 from upstream). Along the KRB,[4] we interviewed a total of 45 households (15 each from downstream, midstream, and upstream). We conducted the interview surveys in October 2012 and May 2013 along the LRB and in October 2013 along the KRB.

Prior to the survey, the project team consulted officers from the National Water Services Commission and the Malaysian Water Association, local community groups, local rural villagers, and village leaders. We used a four-part pretested survey questionnaire. Part one of the questionnaire collected information on gender, age, educational level, and occupation, the length of stay in the village, monthly income, expenditure, and some basic health check of the household condition. The second part focused on the respondents' understanding of environmental quality exposure, knowledge about environmental issues, and the impact of environmental degradation on human health and the community. The third part of the questionnaire obtained information on sanitation and hygiene practices, trash disposal methods, and available sources of water supply. Finally, the fourth part of the questionnaire examined respondents' perceptions of water and fish from the river. To better represent the project outcome, we developed a package of visual aids, such as pictures of different species of fish, vegetables, and maps that was translated into Bahasa Malaysia verbally for the illiterate villagers.

In addition to the perceptions, we collected data on the villagers' fish consumption, awareness of heavy metal pollution and the food-health risk security nexus, and understanding and knowledge that can have significant implications for environmental policy.

The results of the survey were as follows: regarding the demographic features, of the total respondents, 24% were female; the majority were aged between 40 and 59 years; nearly 63% had a family size of more than five members in each household; only 41% had received tertiary education; 68% were lower-income earners, with a salary of less than RM 2000 per month; and only 24.4% had been living at the river basin for less than ten years. According to Table 11.2, approximately 44% of the respondents had septic tanks at their home, and 70% had flush toilet facilities; some of them suffered from heart disease, dengue, and cancer.

In the downstream areas of the LRB, we observed significant pollution of water through household garbage. The upstream water was cleaner, and a waterfall located upstream in the Hulu Langat District is channelled to the villages for fruit farming. However, households'

Table 11.2   Understanding of environmental awareness and health issues

Items	LRB (n = 41) (%)	KRB (n = 45) (%)
Use water supply from SYABAS (the distributor of treated water)	100	96
Sanitation and hygiene practices		
Septic tank at home	44	62
Flush toilet at home	71	69
Segregate garbage	71	27
Perception on toxicant exposure		
Exposed to substances in the surrounding that affect health	90	29
What types of substance? (N = 13)		
Chemical	41	23
Heavy metals	5	0
Smoke/dust	73	62
Which emerging health issues are addressed by current health programs? [*multiple responses]		
Dengue	61	71
Cancer	34	0
Pulmonary disease	37	0
Drugs	27	0
How do you dispose waste garbage? [*multiple responses]		
Burning	66	13
Collected by garbage collector	37	87
Dump in a vacant lot or open space	46	27

Source: Partially extracted from Poon et al. (2016). Date of interviews: 2013/10/01–2013/10/03. Authors' Socioeconomic Survey

disposal of liquid and wet-trash waste at the riverbank had caused the river water contamination.

The villagers in the midstream and downstream areas did not use the river water because it was polluted by emissions from palm oil, paper mill, electronics, rubber, plastics, tires, aluminium, steel, and wood factories. The drainage system was poor; drains were an ideal breeding ground for the *Aedes* mosquitoes, causing the dengue haemorrhage disease. In addition, the downstream water quality was poor due to the various land use activities and waste. Across all villages, the majority burned their garbage in the open air (66%), while that of the others was collected by trash workers (37%) or dumped into the river (46%) (Table 11.2). Although 90% of the respondents were aware that the exposure to chemicals could affect their health, only 5% corresponded that heavy metals had contaminated the river water; this substantiates the critical need for educational programmes.

Most villagers used pipe-borne water facilities for their daily activities (Table 11.3). Despite regular upgrades in the overall water quality in the LRB by Syarikat Bekalan Air Selangor (SYABAS, the distributor of treated water), the respondents asserted that this treated water had a light yellowish colour. The majority of the villagers boiled water before drinking, and 49% of them also installed a filter at home to evade water-borne problems.

Some villagers caught and consumed the Langat River fish. Tilapia and catfish (*ikan keli*) were the most eaten by the local communities. However, a minority of the villagers did not eat the fish from the river because they perceived that industrial effluents had contaminated the fish. However, we did not find any reported cases of sickness, ailment, or serious skin disease that were directly related to the consumption of the river water fish.

The analysis of the river water samples and data from the existing studies and interviews with the residents revealed that water pollution primarily resulted from industrial, agricultural, and domestic/sewage activities. Moreover, because river water pollution entails varied and disparate sources, the resolution requires multidisciplinary knowledge, implying that a single method, level of authority, or disciplinary approach is likely to be insufficient.

This analysis suggests that the polycentric governance approach, originally developed by Ostrom et al. (1961), has enormous potential for tackling this type of challenge.

Table 11.3 Daily water usage routine

Items	LRB (n = 41) (%)	KRB (n = 45) (%)
Water source and activities		
Drinking		
Ground/well water	2	4
Rain water	0	0
Pipe water	85	80
River water	0	0
Processed water	12	16
Cooking		
Ground/well water	2	4
Rain water	0	0
Pipe water	85	84
River water	0	0
Processed water	12	11
Bathing		
Ground/well water	7	4
Rain water	0	0
Pipe water	80	93
River water	0	0
Processed water	12	2
Washing clothes		
Ground/well water	7	4
Rain water	0	0
Pipe water	80	96
River water	0	0
Processed water	12	0
How is the drinking water transported from the water source to your home?		
Conveyed through water pipes	95	89
Fetched from well/water pump	5	2
Delivered from the water-refilling station	0	9
How is the drinking water treated?		
Boiled	90	84
Filtered/strained	49	53
Processed	5	4
Perception of fish from the river nearby		
I regularly eat fish from the river	56	11
I am aware of the warning that fish from the river nearby is contaminated	7	27
I heard someone getting sick after eating fish from the river nearby	0	4

Note: Partially extracted from Poon et al. (2016). Date of interviews: 2013/10/01–2013/10/03.
Authors' Socioeconomic Survey

LIMITATIONS AND POLICY IMPLICATIONS

This study has some limitations. It was not originally cast within the ergonomic approach. A broader approach could have incorporated several other important issues related to green ergonomics. Although we focused on heavy metals in water, we could have addressed other water contaminants. The samples were essentially limited because the in-depth analytical approach adopted to heavy metal pollution precluded large samples due to cost and time.

Nevertheless, this study provides several important policy implications. Unfettered discharge of waste into river systems may imply the absence of appropriate waste treatment methods, treatment plants, and weak enforcement of environmental policies. Malaysian policy makers have failed to incorporate heavy metal pollution in river water quality assessment and should address this concern. The adoption of safe minimum standards of heavy metal concentration in the WQI may be a useful policy. The government could provide taxes and other incentives for industries to develop environmentally benign waste treatment plants.

Second, households' disposal of wet trash and garbage into the river should be controlled by the state authorities by allocating sufficient funds to improve garbage collection; moreover, the authorities must raise awareness among the resident population to avoid drinking water and eating fish from contaminated rivers. Although the literacy rate among the residents is reasonable, they do not have pollution-specific knowledge; this should be improved. The local and regional authorities must use local leadership and social capital at the village level, which is the final segment of the polycentric model. Opinion leaders, policy makers, scientists, and resident populations must increase focus and discussions on river pollution. Furthermore, monitoring and research by multidisciplinary teams and non-governmental organisations can improve policy effectiveness. Interventions should be well-formulated to avoid any long-term health problems due to river water pollution. In the long term, however, addressing the green ergonomic issues that link humans to natural systems is imperative.

Malaysian bureaucracy and governance systems are weak and political interference in local organisations, such as water user associations, is detrimental to providing fair service and level-playing field. This aspect must be particularly addressed to ensure that the polycentric model becomes a robust pollution control mechanism in Malaysia.

CONCLUSION

This study demonstrated that human factors, such as the absence of communication among state, regional, local authorities and users of rivers, lack of institutional arrangements among the stakeholders, and poorly defined boundaries of government authorities, have caused weak human-nature relationships, river water pollution, and associated green ergonomic issues in Malaysia. The lack of proper institutional mechanisms can be improved using a polycentric governance model, which is primarily used in the common-pool resource literature. The proposed model helps analyse how relevant state, regional, and local authorities and users can work together in reducing the pollution within the scope of green ergonomics to enhance human well-being. A polycentric governance system is an institutional approach that comprises multiple centres of decision-making authorities at multiple levels; these centres are interrelated but formally independent of each other. When this system is accommodated within the green ergonomics discipline, it can help improve the human-nature relationship, allowing the optimisation of human well-being and river systems' performance.

With the identification of diverse pollution sources, individuals, and groups linked to water quality reduction, the water commons presents itself as a complex CPR that necessitates a sophisticated management approach to contain pollution. This approach must include various policy alternatives to facilitate cooperation and communication among multiple centres of state and non-state actors. Thus, a polycentric governance system should be implemented as a specific CPR approach to address river water quality.

It is crucial for the relevant authorities to develop policies that will successfully control the input of hazardous substances, and other water pollutants, into the aquatic ecosystems. We should stress the importance of this responsibility that is shared by those in local and national governments, scientists, and civil society. We suggest that stakeholders from diverse horizons, with interconnected activities, share responsibilities and accountabilities to address green ergonomic problems and strengthen the sustainable relationship between human well-being and natural resource systems.

NOTES

1. The second-order collective action is an important issue.
2. The Salak Tinggi water treatment plant draws its supply from Sungai Labu and has a design capacity of 10.8 MLD.

3. Along the LRB, we began upstream (Hulu Langat District), followed by midstream (Sepang District), and finally downstream (Kuala Langat District).
4. Along the KRB, we began upstream [from Kg. Sungai Machang, Tmn Melawati to Ampang Waterfront], followed by midstream [from Kg. Seri Andalas, Jln Sultan Hishamudin, Medan Pasar Besar, Leboh Pasar Besar, Jln Tun Sambanthan to Puchong, and Subang Jaya], and finally downstream [from Kg. Kebun Bunga to Lebuh Sultan Abdul Samad].

REFERENCES

Abdullah, A. R. (1995). Environmental pollution in Malaysia: Trends and prospects. *Trends in Analytical Chemistry, 14*, 191–198.

Adnan, N. H., Zakaria, M. P., Juahir, H., & Ali, M. M. (2012). Faecal sterols as sewage markers in the Langat River, Malaysia: Integration of biomarker and multivariate statistical approaches. *Journal of Environmental Sciences, 24*, 1600–1608.

Ahamed, M. I. N. (2014). A review on quality of drinking water and associated health risks. *Octa Journal of Environmental Research, 2*, 255–261.

Ahmad, A. K., Mushrifah, I., & Shuhaimi-Othman, M. (2009). Water quality and heavy metal concentrations in sediment of Sungai Kelantan, Kelantan, Malaysia: A baseline study. *Sains Malaysiana, 38*, 435–442.

Amundsen, P. A., Staldvik, F. J., Lukin, A. A., Kashulin, N. A., Popova, O. A., & Reshetnikov, Y. S. (1997). Heavy metal contamination in freshwater fish from the border region between Norway and Russia. *Science of the Total Environment, 201*, 211–224.

Andersson, K. P., & Ostrom, E. (2008). Analyzing decentralized resource regimes from a polycentric perspective. *Policy Sciences, 41*, 71–93.

Apec Water. (n.d.). *Water health: Lower back pain & water contamination.* Available from http://www.freedrinkingwater.com/water_health/lower-back-pain-drinking-water-pollution.htm

Arrow, K. J., Keohane, R. O., & Levin, S. A. (2012). Elinor Ostrom: An uncommon woman for the commons. *Proceedings of the National Academy of Sciences, 109*, 13135–13136.

Balamurugan, G. (1991). The mining and sediment supply in Malaysia with special reference to the Kelang River Basin. *The Environmentalist, 11*, 281–291.

Ballet, J., Koffi, J. M., & Pelenc, J. (2013). Environment, justice and the capability approach. *Ecological Economics, 85*, 28–34.

Bandarage, A. (2013, October–December). Political economy of epidemic kidney disease in Sri Lanka. *Sage Open*, 1–13.

Bartelmus, P. (2010). Use and usefulness of sustainability economics. *Ecological Economics, 69*, 2053–2055.

Baumgärtner, S., & Quaas, M. (2010a). Sustainability economics—General versus specific, and conceptual versus practical. *Ecological Economics, 69*(11), 2056–2059.

Baumgärtner, S., & Quaas, M. (2010b). What is sustainability economics? *Ecological Economics, 69*, 445–450.

Bazin, D., Ballet, J., & Touahri, D. (2004). Environmental responsibility versus taxation. *Ecological Economics, 49*, 129–134.

Binder, M., & Witt, U. (2012). A critical note on the role of the capability approach for sustainability economics. *The Journal of Socio-Economics, 41*(5), 721–725.

Birkin, F., & Polesie, T. (2013). The relevance of epistemic analysis to sustainability economics and the capability approach. *Ecological Economics, 89*, 144–152.

Bish, R. L. (2014). Vincent Ostrom's contributions to political economy. *Publius: The Journal of Federalism, 44*, 227–248.

Bithas, K. (2011). Sustainability and externalities: Is the internalization of externalities a sufficient condition for sustainability? *Ecological Economics, 70*(10), 1703–1706.

Coase, R. (1960). The problem of social cost. *Journal of Law and Economics, 3*, 1–44.

Couttenier, M. (2008). *Relationship between Natural Resources and Institutions.* Documents.

Daily Mail Online. (2014, December 3). *Water polluted with heavy metal causes Chinese villagers to develop horrific, painful swellings.* http://www.dailymail. co.uk/health/article-2859324/Water-polluted-heavy-metal-causes-Chinese-villagers-develop-horrific-painful-swellings.html

Davidson, J., Myers, D., & Chakraborty, M. (1992). *No time to waste: Poverty and the global environment* (p. 4). Oxford, UK: Oxfam.

Department of Environment. (2010). *Ministry of natural resources and environment Malaysia.* Malaysia quality environmental report 2010. http://www. malaysia.ahk.de/fileadmin/ahk_malaysia/Market_reports_2012/Market_ Watch_2012_-_Environmental.pdf. Accessed 20 Feb 2013.

Economic Planning Unit, Malaysia. (2002). *Malaysian quality of life 2002.* Kuala Lumpur.

Fathallah, F. A. (2010). Musculoskeletal disorders in labour-intensive agriculture. *Applied Ergonomics, 41*(6), 738–743.

García-Acosta, G., Pinilla, M. H. S., Larrahondo, P. A. R., & Morales, K. L. (2014). Ergoecology: Fundamentals of a new multidisciplinary field. *Theoretical Issues in Ergonomics Science, 15*(2), 111–133.

Gardner, R., Ostrom, E., & Walker, J. M. (1990). The nature of common-pool resource problems. *Rationality and Society, 2*, 335–358.

Ghosh, N., Mukhopadhyay, P., Shah, A., & Panda, M. (2015). *Nature, economy and society.* New Delhi, India: Springer.

Hanson, M. A. (2010). Green ergonomics: Embracing the challenges of climate change. *The Ergonomist, 480*, 12–13.

Hanson, M. A. (2013). Green ergonomics: Challenges and opportunities. *Ergonomics, 56*, 399–408.

Hardin, G. (1968). The tragedy of the commons. *Science, 162*(3859), 1243–1248.

Herath, G. (2012). *Institutional aspects of water resources management.* New York: Nova Publishers.

International Ergonomics Association (IEA). (2015). *Definitions and domain ergonomics.* http://www.iea.cc/whats/

Kapp, K. W. (2012). In S. Berger & R. Steppacher (Eds.), *The foundations of institutional economics.* Abingdon, UK: Routledge.

Khan, S., Cao, Q., Zheng, Y. M., Huang, Y. Z., & Zhu, Y. G. (2008). Health risks of heavy metals in contaminated soils and food crops irrigated with wastewater in Beijing, China. *Environmental Pollution, 152*, 686–692.

Kiser, L. L., & Ostrom, E. (1982). The three worlds of action. In E. Ostrom (Ed.), *Strategies of political inquiry.* Beverly Hills, CA: Sage Publications.

Lejano, R. P., & Stokols, D. (2013). Social ecology, sustainability, and economics. *Ecological Economics, 89*, 1–6.

Li, M. S., Luo, Y. P., & Su, Z. Y. (2007). Heavy metal concentrations in soils and plant accumulation in a restored manganese mineland in Guangxi, South China. *Environmental Pollution, 147*(1), 168–175.

Mandal, B. K., & Suzuki, K. T. (2002). Arsenic round the world: A review. *Talanta, 58*(1), 201–235.

Massa, I. (2015). *Technological change in developing countries: Trade-offs between economic, social and environmental sustainability (No. 051).* United Nations University-Maastricht Economic and Social Research Institute on Innovation and Technology (MERIT).

Mazumder, D. N. G. (2008). Chronic arsenic toxicity & human health. *Indian Journal of Medical Research, 128*, 436–447.

McGinnis, M. D., & Ostrom, E. (2012). Reflections on Vincent Ostrom, public administration, and polycentricity. *Public Administration Review, 72*(1), 15–25.

MEA (Millennium Ecosystem Assessment). (2005). *Ecosystems and human well-being.* Washington, DC: Island Press.

Ministry of Natural Resources and Environment. (2011, August). *Review of the national water resources study (2000–2050) and formulation of national water resources policy.* Final report, Department of Irrigation and Drainage Malaysia.

Mokhtar, M. B., Toriman, M. E. H., Hossain, M., Abraham, A., & Tan, K. W. (2011). Institutional challenges for integrated river basin management in Langat River Basin, Malaysia. *Water and Environment Journal, 25*(4), 495–503.

New Straits Times Online. (2017, March 4). *Sg Semenyih pollution crisis: Tests reveal presence of poison.* http://www.nst.com.my/news/2016/10/183918/sg-semenyih-pollution-crisis-tests-reveal-presence-poison

North, D. C. (1990). *Institutions, institutional change and economic performance.* New York: Cambridge University Press.

Ostrom, E. (1990). *Governing the commons: The evolution of institutions for collective action.* Cambridge, UK: Cambridge University Press.

Ostrom, E. (1992). *Crafting institutions for self-governing irrigation systems.* Office of Water Conservation, Panoche Water and Drainage District, Calif. (USA), Water Management Research Laboratory (USA), California, USA.

Ostrom, E. (2005). *Understanding institutional diversity.* Princeton University Press. http://press.princeton.edu/chapters/s8085.pdf

Ostrom, E. (2007). A diagnostic approach for going beyond panaceas. *Proceedings of the National Academy of Sciences, 104,* 15181–15187.

Ostrom, E. (2010). Beyond markets and states: Polycentric governance of complex economic systems. *American Economic Review, 100,* 641–672.

Ostrom, E. (2014). A polycentric approach for coping with climate change. *Annals of Economics and Finance, 15,* 97–134.

Ostrom, E., Gardner, R., & Walker, J. (1994). *Rules, games and common-pool resources.* Ann Arbor, MI: University of Michigan Press.

Ostrom, V., & Ostrom, E. (1977). Public goods and public choices. In E. S. Savas (Ed.), *Alternatives for delivering public services: Towards improved performances* (pp. 7–49). Boulder, CO: Westview Press.

Ostrom, V., Tiebout, C. M., & Warren, R. (1961). The organisation of government in metropolitan areas: A theoretical inquiry. *American Political Science Review, 55,* 831–842.

Othman, J. (2008, February 23). *Agricultural development, food security and sustainability issues in Malaysia.* Proceedings of the symposium on the emerging ecological risks and food security in Asia. Japan, Yokohama National University.

Poon, W. C., & Herath, G. (2012). Institutional issues in the provision of water and sanitation services in Malaysia. In G. Herath (Ed.), *Institutional aspects of water management: Evaluating the experience* (pp. 193–214). New York: Nova Science Publishers.

Poon, W. C., Herath, G., Sarker, A., Masuda, T., & Kada, R. (2016). River and fish pollution in Malaysia: An ergonomics perspective. *Applied Ergonomics, 57,* 80–93.

Population Reference Bureau. (2017). *World population data sheet: With a special focus on youth.* http://www.prb.org/pdf17/2017_World_Population.pdf

Radjiyev, A., Qiu, H., Xiong, S., & Nam, K. (2015). Ergonomics and sustainable development in the past two decades (1992–2011): Research trends and how ergonomics can contribute to sustainable development. *Applied Ergonomics, 46,* 67–75.

Romiszowski, A. J. (2016). *Designing instructional systems: Decision making in course planning and curriculum design.* New York: Routledge.

Sany, S. B. T., Salleh, A., Rezayi, M., Saadati, N., Narimany, L., & Tehrani, G. M. (2013). Distribution and contamination of heavy metal in the coastal sediments of Port Klang, Selangor, Malaysia. *Water, Air, & Soil Pollution, 224*, 1–18.

Sany, S. B. T., Salleh, A., Sulaiman, A. H., Sasekumar, A., Rezayi, M., & Tehrani, G. M. (2013). Heavy metal contamination in water and sediment of the Port Klang coastal area, Selangor, Malaysia. *Environmental Earth Sciences, 69*, 2013–2025.

Sarker, A., Ross, H., & Shrestha, K. K. (2008). A common-pool resource approach for water quality management: An Australian case study. *Ecological Economics, 68*(1), 461–471.

Seckler, D. (1996). *The new era of water resources management: From dry to wet water savings*. International Water Management Institute (IWMI), Research report no. 1, International Irrigation Management Institute, Colombo, Sri Lanka. https://core.ac.uk/download/pdf/6405320.pdf

Smith, A. H., Lingas, E. O., & Rahman, M. (2000). Contamination of drinking-water by arsenic in Bangladesh: A public health emergency. *Bulletin of the World Health Organisation, 78*(9), 1093–1103.

Söderbaum, P. (2011). Sustainability economics as a contested concept. *Ecological Economics, 70*, 1019–1020.

Tapiola, T., & Paloviita, A. (2015). *Building resilient food supply chains for the future. Climate change adaptation and food supply chain management*. London: Routledge.

Thatcher, A. (2013). Green ergonomics: Definition and scope. *Ergonomics, 56*(3), 389–398.

Thatcher, A., Garcia-Acosta, G., & Lange-Morales, K. (2013). Design principles for green ergonomics. In M. Anderson (Ed.), *Contemporary ergonomics and human factors* (pp. 319–326). Boca Raton, FL: CRC Press.

The Free Malaysia Today Online. (2016, May 25). *43 rivers in Malaysia polluted, says Minister*.http://www.freemalaysiatoday.com/category/nation/2016/05/25/43-rivers-in-malaysia-polluted-says-minister/

The Malay Mail Online. (2016, October 27). *Contamination of water from NS into Selangor remains "critical", exco says*. http://www.themalaymailonline.com/malaysia/article/selangor-water-contamination-still-critical-says-exco-by-ram-anand

The Rakyat Post. (2015, September 24). *Problem of water pollution in Malaysia becoming serious, says WWF*. http://www.therakyatpost.com/news/2015/09/24/problem-of-water-pollution-in-malaysia-becoming-serious-says-wwf/

The Star Online. (2013, August 31). *A million in Klang Valley to go without water – Maybe for days*. http://www.thestar.com.my/news/nation/2013/08/31/code-red-after-diesel-spill-a-million-in-klang-valley-to-go-without-water-maybe-for-days/

The Star Online. (2016a, October 5). *Contamination forces Semenyih plant to be closed again*. http://www.thestar.com.my/news/nation/2016/10/05/

another-water-disruption-contamination-forces-semenyih-plant-to-be-closed-again/

The Star Online. (2016b, September 29). *Semenyih factory main culprit of river pollution.* http://www.thestar.com.my/news/nation/2016/09/29/semenyih-factory-main-culprit-of-river-pollution-initial-probe-confirms-discharge-from-industrial-pa/#YdH4yjJkJhAaHVol.99

The Straits Times Online. (2016, July 26). *Loads of rubbish clogging up Malaysian rivers.* http://www.straitstimes.com/asia/se-asia/loads-of-rubbish-polluting-choking-malaysian-rivers

The Water Environment Partnership in Asia, WEPA. *State of water environmental issues- Indonesia.* http://www.wepa-db.net/policies/state/indonesia/indonesia.htm

Toonen, T. (2010). Resilience in public administration: The work of Elinor and Vincent Ostrom from a public administration perspective. *Public Administration Review, 70*(2), 193–202.

Twomlow, S., O'Neill, D., Sims, B., Ellis-Jones, J., & Jafry, T. (2002). An engineering perspective on sustainable smallholder farming in developing countries. *Biosystems Engineering, 81*(3), 355–362.

UNESCO (United Nations Educational, Scientific and Cultural Organisation). (2015). *International initiative on water quality.* http://unesdoc.unesco.org/images/0024/002436/243651e.pdf

Van den Bergh, J. C. (2010). Externality or sustainability economics? *Ecological Economics, 69*(11), 2047–2052.

Volety, A. K. (2008). Effects of salinity, heavy metals and pesticides on health and physiology of oysters in the Caloosahatchee Estuary, Florida. *Ecotoxicology, 17,* 579–590.

White, M. A. (2013). Sustainability: I know it when I see it. *Ecological Economics, 86,* 213–217.

Widianarko, B., Verweij, R. A., Van Gestel, C. A. M., & Van Straalen, N. M. (2000). Spatial distribution of trace metals in sediments from urban streams of Semarang, Central Java, Indonesia. *Ecotoxicology and Environmental Safety, 46,* 95–100.

Wong, S. C., Li, X. D., Zhang, G., Qi, S. H., & Min, Y. S. (2002). Heavy metals in agricultural soils of the Pearl River Delta, South China. *Environmental Pollution, 119*(1), 33–44.

Zhang, J., Mauzerall, D. L., Zhu, T., Liang, S., Ezzati, M., & Remais, J. V. (2010). Environmental health in China: Progress towards clean air and safe water. *The Lancet, 375*(9720), 1110–1119.

Zheng, N., Wang, Q., Zhang, X., Zheng, D., Zhang, Z., & Zhang, S. (2007). Population health risk due to dietary intake of heavy metals in the industrial area of Huludao city, China. *Science of the Total Environment, 387,* 96–104.

Zink, K. J. (2014). Designing sustainable work systems: The need for a systems approach. *Applied Ergonomics, 45,* 126–132.

CHAPTER 12

Examining the Challenges of Responsible Consumption in an Emerging Market

Fandy Tjiptono

INTRODUCTION

Sustainability has been suggested as an emerging business megatrend that will profoundly affect firm survival and competitiveness (Lubin & Esty, 2010; Mittelstaedt, Shultz, Kilbourne, & Peterson, 2014). It has been defined as 'meeting the needs of the present without compromising the ability of the future generations to meet their own needs' (United Nations World Commission on Environment and Development, 1987, no page number). In essence, it focuses on the balance of people, planet, and profits (triple bottom line) or the 'Three Es': environment (ecological), equity (social), and economic (financial) dimensions (Savitz & Weber, 2006).

Companies, governments, non-governmental organizations, and consumers are increasingly interested in sustainability-related issues. Companies, for instance, have initiated, managed, and communicated their sustainable marketing activities through many programs, including corporate social responsibility (CSR) initiatives. The *Financial Times* (2014) reported that Fortune 500 companies spent more than US$15 billion on CSR initiatives. However, most CSR programs have not met

F. Tjiptono (✉)
School of Business, Monash University Malaysia, Bandar Sunway, Malaysia

their objectives (Dans, 2015). In order for sustainable marketing practices to succeed, they need to be aligned with consumer interests, because consumers are the ultimate determinant of CSR success (Morrison & Bridwell, 2011). Olander and Thogersen (1995) highlighted that understanding consumer behaviour is a prerequisite for successful sustainability efforts. For example, if a company offers 'green energy' products, there should be enough environmentally conscious consumers to fuel the demand for such products (Vitell, 2015). Unfortunately, to date few studies have examined consumer social responsibility (CnSR) that reflects a broad range of consumer-oriented responsibilities towards society (Caruana & Chatzidakis, 2014; Quazi, Amran, & Nejati, 2016; Vitell, 2015).

Furthermore, an extensive literature review reveals that there are three main challenges with regard to responsible or sustainable consumption: (1) consumer segments are not only either green or non-green groups (McDonald, Oates, Alevizou, Young, & Hwang, 2012); (2) the attitude-behaviour gap phenomenon, where positive attitudes towards environmental issues do not necessarily translate into actual green purchase behaviour, has been found consistently in many sustainable consumption studies (Carrington, Neville, & Whitwell, 2010; Grimmer & Miles, 2017; Prothero et al., 2011); and (3) consumers tend to perceive certain barriers to green behaviour, which in turn affect their readiness to be green (Arli, Tan, Tjiptono, & Yang, 2018; Johnstone & Tan, 2015). These three challenges prevent many consumers from engaging in responsible consumption (Arli, Tan, Tjiptono, & Yang, 2015; Grimmer & Miles, 2017; McDonald et al., 2012). For instance, highly environmentally oriented consumers may not show consistent green product purchase due to a lack of perceived readiness to be green from the organizations that provide products (Arli et al., 2015). The mismatch between companies' sustainable consumption initiatives and consumer interests as well as targeting the wrong segments may lead to ineffective sustainable consumption programs (McDonagh & Prothero, 2014; Morrison & Bridwell, 2011). While most of the literature on responsible/sustainable consumption tends to focus on the developed country context, research in the emerging market context has been very limited (Arli et al., 2018; Newholm & Shaw, 2007).

Therefore, this chapter aims to examine these three specific challenges (i.e. responsible consumption segmentation, the attitude-behaviour gap phenomenon, and perceived readiness to be green) for the development of responsible consumption in an emerging market context. The scope of the study is sustainability issues at the micro or individual level (Thatcher & Yeow, 2016), because consumers' responsible consumption tends to be neglected in consumer research (Oberseder, Schlegelmilch, & Murphy, 2013; Quazi et al., 2016).

Indonesia was selected as the main focus as it is the world's fourth largest population with around 256 million people (CIA, 2016) and is the largest economy in Southeast Asia with a GDP of US$873 billion in 2015 (CIA, 2016) and a gross national income (GNI) of US$9788 per capita in 2011 (UNDP, 2016). Like many other developing countries, Indonesia has a young population: around 42% of its people are 24 years old or younger. Furthermore, a survey conducted by the Pew Research Center (2010) reported that concerns for environmental issues were diverse across countries. About 61% of Indonesians, for instance, believe that protecting the environment should be given priority, but less than half (47%) perceive global climate change as a very serious problem, and only 32% were willing to pay higher prices to address global climate change. The same survey also showed diverse concerns for environmental issues across countries.

The structure of this chapter is organized as follows. It will briefly discuss the sustainability marketing practices and then present the arguments for the importance of responsible consumption. The three main challenges of responsible consumption (i.e. green segmentation, attitude-behaviour gap, and readiness to be green) will be examined by using two new studies (i.e. a typology of responsible consumption segments and consumer social responsibility) and a review of previous research on perceived readiness to be green as illustrations. Finally, several other challenges of responsible consumption in emerging markets and future research directions are identified.

SUSTAINABLE MARKETING PRACTICES

At the individual consumer level of sustainability issues, it is important to model the system on how the interactions of consumers and companies (marketers) work (see Fig. 12.1). On the one hand, marketers decide to produce and market a set of market offerings (sustainable products and services) as a means to achieve their objectives (i.e. profitability, growth, competitive strength, innovativeness, contribution to owners and society) within their competence and limited capacity. On the other hand, consumers have many specific needs and wants that have to be satisfied within

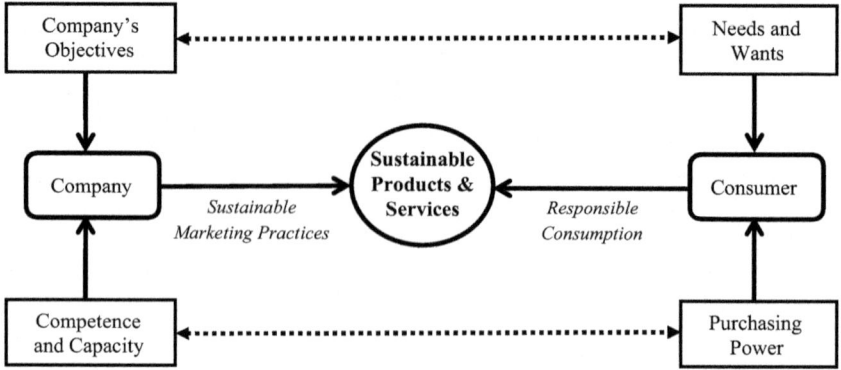

Fig. 12.1   A basic marketing system perspective

their limited purchasing power. Therefore, successful, responsible market-ing and consumption practices can be realized when the market offerings serve those needs and wants effectively.

From the marketing perspective, companies are increasingly aware of the importance to adopt a sustainable perspective in their strategies (McDonagh & Prothero, 2014). Peter Drucker was the first expert to integrate sustainability issues into the marketing domain (Connelly, Ketchen, & Slater, 2011). He highlighted the need to create value for customers through socially, environmentally, and ethically responsible actions. A number of marketing practices have been developed to incorpo-rate the triple bottom line (Cronin et al., 2011; Peattie, 2001). Kotler and Armstrong (2014), for instance, used two dimensions (needs of business and needs of customers) to identify four sustainable marketing practices: the marketing concept, the strategic planning concept, the societal mar-keting concept, and the sustainable marketing concept. Suggested as the ideal practice, the sustainable marketing concept was defined as 'socially and environmentally responsible actions that meet both the immediate and future needs of customers and the company' (Kotler & Armstrong, 2014, p. 583).

Furthermore, Peattie (2001) suggested that the development of sustain-able marketing practices can be classified into three inter-related stages: ecological marketing, environmental marketing, and sustainable marketing. Ecological marketing focuses on particular environmental problems, includ-ing water and air pollution, depletion of oil reserves, and the impact of pesticide usage on the environment. Environmental marketing emphasizes the adoption of clean technology, understanding and targeting the green consumer segments, and implementation of socio-environmental performance as a competitive advantage, whereas sustainable marketing strives to create sustainable development and the economy.

Empirical studies have indicated that sustainable marketing practices may lead to greater financial gains, higher market share, high levels of employee commitment, increased firm performance, increased capabilities, increased customer satisfaction and loyalty, improved brand image, greater firm value, lower firm-idiosyncratic risk, and cost-saving advantages (Baker & Sinkula, 2005; Cronin et al., 2011; Ganesan, George, Jap, Palmatier, & Weitz, 2009; Lash & Wellington, 2007; Luo & Bhattacharya, 2006; Maignan & Ferrell, 2001; Maignan, Ferrell, & Hult, 1999; Menguc & Ozanne, 2005; Porter & van der Linde, 1995; Pujari, Wright, & Peattie, 2003).

THE NEED FOR CONSUMER RESPONSIBLE CONSUMPTION

Responsible consumption has received significant attention in the literature in recent years (Newholm & Shaw, 2007; Phipps et al., 2013; Valor & Carrero, 2014; Webb, Mohr, & Harris, 2008). However, to date there is no single universally accepted definition of responsible consumption (Valor & Carrero, 2014). Just like many other marketing and consumer behaviour constructs (e.g. social responsibility, consumer satisfaction, and customer loyalty), the term 'responsible consumption' has been defined differently for different contexts. Narrow definitions include a variety of concepts: ethical consumption, consumer activism, green consumption, environmental consumption, sustainable consumption, and political consumption (McDonald et al., 2012; Valor & Carrero, 2014). A broader definition was proposed by Barnett, Cloke, Clarke, and Malpass (2005, p. 29) who defined it as 'any practice of consumption in which explicitly registering commitment or obligation toward distant or absent others is an important dimension of the meaning of the activity to the actors involved'. Similarly, Ulusoy (2016, p. 285) formulated it as 'the consumption that has less negative impact or more positive impact on the environment, society, the self, and the other-beings'. She argues that the definition covers various types of consumption, such as sustainable consumption, ethical consumption, consumer citizenship, socially responsible consumption, and green consumption. The absence of a common definition of responsible consumption suggests three important aspects: (1) it is a complex phenomenon with multiple dimensions (Peattie & Collins, 2009;

Phipps et al., 2013; Ulusoy, 2016); (2) responsible consumption reflects a growing awareness of the impacts of consumption practices on consumer health, society well-being, and the environment (Giesler & Veresiu, 2014); and (3) responsible consumption remains a 'work in progress' (Szmigin, Carrigan, & McEachern, 2009).

Why do we need to focus on responsible consumption? First, empirical studies have suggested that responsible consumption is relevant to all areas of consumption (Peattie & Collins, 2009) and consumption practices have social, ethical, and environmental consequences (Kotler & Armstrong, 2014; Mohr, Webb, & Harris, 2001). For instance, in a comprehensive analysis of the environmental impacts of 255 product types, Tukker et al. (2005) found that about 70–80% of total impacts relate to food and drink consumption, housing, and transportation services. In other words, what we buy, use, and dispose of now may affect both current and future generations (Luchs et al., 2011).

Second, it is argued that without the approval and support of consumers, sustainable marketing programs (including corporate social responsibility or CSR) cannot work effectively (Vitell, 2015). One of the main issues is that existing sustainability strategies do not directly focus on consumers (Sheth, Sethia, & Srinivas, 2011). The second issue is that consumers are responsible for creating positive social impacts by using their power in the marketplace (Dickinson & Carsky, 2005). In other words, consumers have a responsibility towards society as a whole, where they must minimize or eliminate societal harm and act proactively based on moral principles and standards for social benefit as they obtain, use, and dispose of goods and services (Mohr et al., 2001; Muncy & Vitell, 1992; Vitell, 2015). Such responsibility is called CnSR (Devinney, Auger, Eckhardt, & Birtchnell, 2006; Quazi et al., 2016; Vitell, 2015). Another issue is that there must be an alignment between sustainable marketing practices and responsible consumer consumption (see Fig. 12.1). What is important for marketers needs to be perceived similarly by consumers; otherwise the sustainable initiatives from marketers will not be effective.

Third, government policies to encourage responsible consumption behaviour have produced mixed results across different consumer segments in different countries. For example, plastic bag bans and taxes were reported to be effective in cutting the usage of plastic bags by at least 70% in several developed countries, such as the UK, the USA, Wales, Scotland, Northern Ireland, Hong Kong, Italy, and Australia (Barkham, 2016; Chow, 2016; Morley, 2016). Despite inconsistent compliances across the country, the plastic bag ban policy in China has been considered as considerably effective in reducing plastic bag use (Block, 2016). However, no significant behavioural changes were found since plastic bag bans and taxes started nationwide in 2011 in Malaysia (Bavani & Wong, 2016). Similarly, plastic bag ban remains a dream in Indonesia, the world's second largest plastic waste producer after China (Handayani, 2016). Due to public objections, the 'pay-for-plastic bag' campaign in Indonesia has been stopped (Ribka, 2016). Building awareness of the importance of reducing waste to landfill and reducing pollution is one thing; however, behavioural change is a different issue. It seems that implementing such policies in emerging markets has its own challenges.

Government regulation and control as well as company and industry associations' support are necessary but insufficient, because consumer acceptance and active support are an equally (if not more) important key success factor (Bavani & Wong, 2016; Block, 2016; Ribka, 2016).

Consumption behaviour, sustainable marketing practices, and government policies are interconnected as sustainability is related to what consumers consume, while sustainable marketing practices and government policies need approval or support from consumers to be effective. Therefore, sustainability is the overarching factor that determines the success of sustainability initiatives.

Several theories have been used as the framework to examine the antecedents of responsible consumption. These theories include the theory of planned behaviour (Ajzen, 1991), the norm-activation-theory (Schwartz & Howard, 1981), the value-belief-norm-theory (Stern, 2000; Stern, Dietz, Abel, Guagnano, & Kalof, 1999), and the motivation-opportunity-abilities (MAO) model (Olander & Thogersen, 1995). The phenomenon has been studied under several different terms, such as ethical consumption, green consumption, environmental consumption, sustainable consumption, and mindful consumption (McDonald et al., 2012; Phipps et al., 2013; Sheth et al., 2011; Valor & Carrero, 2014).

Existing literature provides at least three important insights. First, consumers are not either green or non-green (McDonald et al., 2012). Purchase decisions depend on the context in which they are made (e.g. individual purchase, household purchase, buying for self vs. buying for others) and on specific product category considered. Second, the attitude-behaviour gap phenomenon (i.e. expressed attitudes, behavioural intentions, and behaviour discrepancies; Belk, 1985) has been found consistently in many studies about green/sustainable consumption. Consumers' positive attitudes about environmental issues do not necessarily translate into actual green purchase behaviour (Carrington et al., 2010; Chatzidakis, Hibbert, Mittusis, & Smith, 2004; Devinney et al., 2006; Eckhardt, Belk, & Devinney, 2010; Pickett-Baker & Ozaki, 2008). A study by Pew Research Center (2010), for instance, reveals that despite most respondents in 22 surveyed countries agreeing that the environment should be protected, only one-third of the consumers were willing to pay higher prices to address global climate change. Third, some of the reasons why consumers decided not to buy greener products include price, economical rationalization, brand, green product availability, perceived performance, cynicism, confusion, trust, situational factors (e.g. economic constraints, lack of choice), and consumers' internal obstacles (e.g. ethical standards, sense of responsibility, etc.) (e.g. Bray, Johns, & Kilburn, 2011; Chan, Wong, & Leung, 2008; Eckhardt et al., 2010; Gleim, Smith, Andrews, & Cronin, 2013; Gupta & Ogden, 2009; McDonald et al., 2012; Pickett-Baker & Ozaki, 2008; Tanner & Kast, 2003). Johnstone and Tan (2015) classified the obstacles to green behaviour into three types: 'it is too hard to be green', 'the green stigma' (a mark of disgrace towards green consumers), and 'green reservations' (consumers' uncertainty that greener consumption practices will make a difference to the environment).

However, it is important to note that most of the existing research focused on the developed country contexts. Newholm and Shaw (2007, p. 259) suggest that responsible consumption might be seen as 'a cultural phenomenon within affluent consumer cultures'. Responsible consumption in an emerging market context remains under-researched. It is expected that different socio-cultural, political, economic, and natural environment factors may contribute to different responsible behaviours between developed and emerging markets.

McCarty and Shrum (2001) suggested that the development of responsible consumer behaviour is difficult to predict. However, the three insights discussed earlier (i.e. green segmentation, the attitude-behaviour gap, and readiness to be green) are worth investigating to better understand the responsible consumption phenomenon in an emerging country context. These insights or challenges are addressed by investigating three inter-related topics: consumer social responsibility (CnSR), typology of responsible consumption segments, and perceived readiness to be green.

First, a study on CnSR was conducted to investigate consumer perceptions of social responsibility dimensions. It shows that consumers assess different responsibility domains with varying degrees of importance. What is important for marketers/companies may not be perceived in a similar way. Therefore, if a CSR program is not aligned with consumer interests, the support from consumers will be low. This may explain why the impact of many programs, including CSR, on consumer purchasing decisions has been minimal (Mohr et al., 2001; Oberseder, Schlegelmilch, & Gruber, 2011). The second study focuses on the attitude-behaviour gap. This widely acknowledged gap found in many studies may be due to the fact that consumers are not ready to consume responsibly (Arli et al., 2015, 2018) and/or because of ineffective segmentation and targeting of consumers. Most of the extant literature focuses on grouping consumers into either green or non-green consumers (McDonald et al., 2012) or using traditional segmentation variables, predominantly demographic characteristics, such as age, education, ethnicity, and socio-economic status (e.g. Bhate & Lawler, 1997; Laroche, Bergeron, & Barbaro-Forleo, 2001; Roberts, 1996; Sener & Hazer, 2008; Zelezny, Chua, & Aldrich, 2000). The second study proposes and examines a different typology of responsible consumption segments using two dimensions: attitude towards responsible consumption and responsible consumption behaviour. This typology directly addresses the issue of attitude and behaviour discrepancies. The third study examines the role of perceived readiness to be green as one of the predictors of green product purchase intention. Only when consumers think that they are ready to be green, then their positive attitudes towards green product purchases may translate into intentions to purchase a green product.

STUDY 1: CONSUMER SOCIAL RESPONSIBILITY (CNSR)

This study aims to examine how consumers assess the importance of seven social responsibility domains: community, employee, shareholder, environmental, societal, customer, and supplier (Oberseder, Schlegelmilch, Murphy, & Gruber, 2014). Each domain encompasses different issues with regard to various stakeholder groups. Vitell (2015, p. 767) argued that while businesses try to 'proactively offer social benefits or public service, and voluntarily minimize practices that harm society', such initiatives will not be successful without approval and support from consumers. In other words, corporate social responsibility needs to be accompanied by consumer social responsibility (CnSR) (Devinney et al. 2006; Quazi et al., 2016; Vitell 2015). In this context, CnSR can be defined as 'the conscious and deliberate choice to make certain consumption choices based on personal and moral beliefs' (Devinney et al., 2006, p. 32). An understanding of CnSR may provide insights into specific social responsibility domains or sustainability issues perceived to be important by consumers. When consumers perceive a domain as important, it is more likely that they will have a more positive attitude towards relevant initiatives/practices dealing with the domain. Such positive attitude may translate into a more consistent behaviour. Moreover, a better understanding of CnSR may help companies and governments design and implement more effective sustainability programs.

Using a convenience sampling approach, 550 self-administered questionnaires were distributed to undergraduate students at a large private university in Semarang, Central Java, Indonesia. Semarang is the fifth most populous city in Indonesia (± 1.8 million people) and the fifth largest Indonesian city (Wikipedia, 2016). Incomplete questionnaires were excluded, resulting in 461 usable questionnaires (a response rate of 83.8%). The majority of the respondents were female (64%), Muslims (95.9%), aged between 19 and 20 years old (63.8%).

The CnSR measure was adopted from Oberseder et al. (2014). The questionnaire items were translated from English to Bahasa Indonesia and then back-translated to ensure consistency. Respondents were asked to rate the importance of each item using a 5-point Likert (1 = Not at all important; 5 = Extremely important). The reliabilities of the seven dimensions of social responsibility were as follows: community (3 items; α = 0.66), employee (6 items; α = 0.76), shareholder (3 items; α = 0.71), environmental (5 items; α = 0.80), societal (6 items; α = 0.79), customer (5 items; α = 0.79), and supplier (5 items; α = 0.83). Table 12.1 presents the scale items used in Study 1.

An ANOVA analysis was conducted to examine the mean differences between consumer perceptions of each social responsibility domain (see Table 12.2). Higher mean scores suggest higher importance of the domains, while lower mean scores indicate the opposite. The results show that consumers did not put equal importance on each social responsibility domain. The top three most important domains were community (M = 4.38), customer (M = 4.38), and employee (M = 4.28), while societal (M = 3.82) was perceived as the least important domain. The top three domains were related directly to consumer needs and wants, where they can assess the actual benefits for themselves in the short term. In contrast, shareholder, environmental, supplier, and societal domains represent indirect benefits for the consumers and may take a longer time to be effective.

Table 12.1 Consumer social responsibility measures

Variable	Item	Mean	SD	Alpha
Community domain	Create jobs for people in the region	4.14	0.731	0.665
	Source products and raw materials locally	4.62	0.564	
	Respect regional values, customs, and culture	4.40	0.679	
Employee domain	Respect human rights of employees	4.40	0.598	0.762
	Set working conditions which are safe and not hazardous to health	4.62	0.572	
	Set decent working conditions	4.02	0.744	
	Treat employees equally	4.46	0.568	
	Offer adequate remuneration	4.44	0.636	
	Develop, support, and train employees	3.80	0.817	
Shareholder domain	Ensure economic success of the company by doing successful business	4.16	0.674	0.714
	Invest capital of shareholders correctly	4.28	0.685	
	Communicate openly and honestly with shareholders	4.25	0.681	
Environmental domain	Reduce energy consumption	3.88	0.840	0.809
	Reduce emissions like CO_2	4.11	0.842	
	Prevent waste	4.35	0.680	
	Recycle	4.32	0.723	
	Dispose of waste correctly	3.95	0.855	
Societal domain	Employ people with disabilities	3.67	0.889	0.791
	Employ long-term unemployed	3.78	0.813	
	Make donations to social facilities	3.95	0.728	
	Support employees who are involved in social projects during working hours	3.64	0.795	
	Invest in the education of young people	3.94	0.678	
	Contribute to solving societal problems	3.95	0.715	
Customer domain	Implement fair sales practices	4.40	0.644	0.797
	Label products clearly and in a comprehensible way	4.35	0.638	
	Meet quality standards	4.55	0.579	
	Set fair prices for products	4.36	0.612	
	Offer the possibility to file complaints	4.25	0.632	
Supplier domain	Provide fair terms and conditions for suppliers	4.07	0.639	0.834
	Communicate openly and honestly with suppliers	4.16	0.687	
	Negotiate fairly with suppliers	4.12	0.669	
	Select suppliers thoroughly with regard to respecting decent employment conditions	4.13	0.689	
	Control working conditions at suppliers	3.76	0.783	

Notes: The scale was adopted from Oberseder et al. (2014); 1 = Not at all important; 5 = Extremely important

SD standard deviation

Table 12.2 Mean differences between social responsibility domains

No.	Social responsibility domain	Mean	SD
1	Community	4.387	0.512
2	Customer	4.383	0.462
3	Employee	4.289	0.448
4	Shareholder	4.229	0.542
5	Environment	4.122	0.596
6	Supplier	4.048	0.539
7	Societal	3.821	0.540
Overall		**4.183**	**0.554**

Notes: $F = 69.923$ ($\rho = 0.000$)

Based on Tukey HSD, no significant differences were found between 1 and 2, 1 and 3, 2 and 3, 2 and 4, 3 and 4, and 5 and 6

Interestingly, the environmental domain ($M = 4.12$) did not receive a top priority among university students who have a higher education level than average Indonesian consumers. Previous empirical studies suggest that education level has a positive relationship with environmental attitudes (Roberts, 1996; Zimmer, Stafford, & Stafford, 1994) and environmental consciousness (Manieri, Barnett, Valdero, Unipan, & Oskamp, 1997). Therefore, it is both interesting and worth investigating for future studies to explore how the general public in Indonesia and other emerging markets perceive the importance of the environmental domain as part of social responsibility dimensions.

The finding of Study 1 is slightly different from Oberseder et al.'s (2014) research in Austria that found the customer, the employee, and the environment as the most important domains. It may suggest that the importance of environmental concerns in developed and developing countries is different.

Regarding the CSR and CnSR relationship, the results of Study 1 suggest that consumers evaluate different domains of responsibility with varying importance levels (Oberseder et al., 2014). It is different from the managerial perspective as suggested by most CSR literatures that managers tend to perceive social responsibility domains as integrated elements of their CSR programs (Oberseder et al., 2014). As a consequence, CEOs and CSR managers need to focus on the top priority domains to gain consumer approval and support. This, in turn, will lead to a more alignment between sustainable marketing initiatives and responsible consumption behaviour.

Study 2: Responsible Consumption Segments

While it is well established that attitude is a positive determinant of behaviour or behavioural intention (Ajzen, 1991, 2005; Bredahl, 2001), many studies on socially responsible or green consumption have found that those who claimed to have a positive attitude towards environmental or social issues do not 'walk their talk' (Carrington et al., 2010; Fraj & Martinez, 2007; Moisander, 2007; Szmigin et al., 2009). Although consumers describe themselves as 'caring' individuals, when it comes to purchase decisions, they simply ignore social/environmental issues and repeat their usual product preferences and purchases (Devinney et al., 2006; Eckhardt et al., 2010). This discrepancy is known as the attitude-behaviour gap or green gap.

Study 2 aimed to propose an alternative typology of responsible consumption segments and provide empirical evidence for it using the purchase of environmentally friendly household products as the product context. While most of responsible consumption segmentation uses demographic variables as key dimensions, the proposed typology employs two dimensions of the attitude-behaviour gap: attitudes towards responsible consumption and responsible consumption behaviour. Drawing on Dick and Basu's (1994) customer loyalty framework, responsible consumption is viewed as the strength of relationship between an individual's attitude towards responsible consumption and responsible consumption behaviour (see Fig. 12.2). Attitudes towards responsible consumption refer to the degree to which an individual consumer has a favourable or unfavourable evaluation of responsible consumption (Ajzen, 1991). Responsible consumption behaviour refers to the purchase intention or the actual purchase of environmentally friendly or green products.

As depicted in Fig. 12.2, there are four responsible consumption segments. The ideal one is the 'truly responsible segment', where both attitudes towards responsible consumption and actual responsible behaviour are favourable or high. This segment represents consumers who 'walk their talk'. The opposite of this segment is the 'irresponsible segment' that has a combination of unfavourable attitudes and low actual responsible behaviour. This segment includes skeptics or non-believers who simply do not support the sustainability or responsible consumption ideas (McDonagh & Prothero, 2014).

A favourable attitude accompanied by low responsible behaviour is the 'latent (potential) responsible segment', which is a serious concern for mar-

Attitude towards Responsible Consumption

		Unfavourable	Favourable
Responsible Consumption Behaviour	**Low**	Irresponsible Segment	Latent (Potential) Responsible Segment
	High	Spurious Responsible Segment	Truly Responsible Segment

Fig. 12.2 Responsible consumption segments

keters. This segment represents the attitude-behaviour gap identified in many previous empirical studies. These types of consumers claim to care for sustainable-related issues, but it is not well translated into responsible consumption behaviours (Carrington et al., 2010; Devinney et al., 2006; Prothero et al., 2011). Furthermore, an unfavourable attitude combined with high responsible behaviour signifies a 'spurious responsible segment'. In some cases, it can also represent the 'enforced responsible segment', where consumers consume responsibly in compliance with the legal requirements. For instance, some consumers do not shop or refuse to use plastic bags on Saturday to avoid paying for the plastic bag charge on the day.

A survey was carried out to examine the typology outlined in Fig. 12.2. Data were collected in Daerah Istimewa Yogyakarta (DIY), a region that is commonly conceived as 'miniature Indonesia', due to its diverse origins and cultures of citizens (Zudianto, 2010). A total of 600 self-administered questionnaires were distributed in two big shopping malls and several residential areas in the region; 523 returned, but only 510 were usable, thereby offering an overall response rate of 85%. The demographic profiles of respondents were as follows: 56.9% of the respondents were female, 53.3% aged 26 years old or older, 48.2% were married, about 37.8% had undergraduate degrees, and 42.4% were Muslims.

Attitudes towards purchasing environmentally friendly products were used as a proxy measure for attitudes towards responsible consumption. Two separate proxies of responsible consumption behaviour were adopted (i.e. the intention to purchase environmentally friendly products and past purchase experience). All measures were adapted from Fishbein and Ajzen

(1975). Attitudes towards purchasing environmentally friendly products were measured using one item, that is, 'In general, my attitude towards purchasing an environmentally-friendly product is...' (1 = Very unfavourable; 5 = Very favourable). Purchase intentions were measured using two items: 'In the next six weeks, how likely are you to purchase environmentally-friendly household products?' (1 = No chance; 5 = Most definitely), and 'I intend to buy environmentally-friendly household products during the next six weeks' (1 = Strongly disagree; 5 = Strongly agree). 'Past purchase experiences' was used as a proxy measure for actual behaviour. It was measured using one item: *In the last six months, have you purchased household products that have been promoted as environmentally friendly?* In the questionnaire, this question was followed up with another question: *If Yes, please tick the products you have purchased, you can tick more than one.* The options were laundry detergent, dishwashing liquids, toilet paper rolls, soaps, and others. Soaps and laundry detergents were mentioned as the most purchased green household products during the last six months.

The measures used in the typology of responsible consumption segments needed a procedure to convert the scales into two categories (cf. Garland & Gendall, 2004). Samples were grouped into favourable and unfavourable attitudes as well as high and low purchase intention using medians as the cut-off points. Any scores equal to or higher than the medians were considered as favourable attitude or high purchase intention. The medians for attitude and purchase intention were 4 and 3, respectively. Moreover, past purchase experience was classified as experienced and inexperienced (never purchased before).

Chi-square (X^2) test was conducted to determine whether there was a significant association between attitude towards purchasing green products and intention to buy green products (see Fig. 12.3). The result indicates that the association was significant ($X^2 = 48.84$, $\rho = 0.000$). Similarly, a significant result was also found for the association between attitude towards purchasing green products and past purchase of green products ($X^2 = 10.39$, $\rho = 0.001$) (see Fig. 12.4). The findings suggest that the four segments were distinct groups, which provide empirical evidence for the proposed typology (Fig. 12.2). Since different segments reflect different combinations of attitude and behaviour, the typology can be used for market targeting and integrated marketing communication purposes. A CSR program or responsible consumption initiative can be most effective if it is directed to the 'truly responsible' segment, while 'potentially responsible' and perhaps 'spurious responsible' segments may be used as secondary targets.

Interestingly, as shown in Figs. 12.3 and 12.4, using purchase intention and past purchase experience as proxy measures for responsible consumption behaviour produced consistent findings of the significance of the four identified segments. For the green household product context in the

Attitude towards Purchasing Green Products

		Unfavourable	Favourable
Intention to Buy Green Products	**Low**	*Irresponsible* **62 people** (12.2%)	*Potential Responsible* **60 people** (11.8%)
	High	*Spurious Responsible* **73 people** (14.3%)	*Truly Responsible* **315 people** (61.8%)

Fig. 12.3 Responsible consumption segments in Indonesia (*Attitude * Intention to Buy Green Products*) (**Notes:** $X^2 = 48.847$, $\rho = 0.000$)

Attitude towards Purchasing Green Products

		Unfavourable	Favourable
Past Purchase of Green Products	**No (Never)**	*Irresponsible* **46 people** (9%)	*Potential Responsible* **76 people** (14.9%)
	Yes	*Spurious Responsible* **89 people** (17.5%)	*Truly Responsible* **299 people** (58.6%)

Fig. 12.4 Responsible consumption segments in Indonesia (*Attitude * Past Purchase of Green Products*) (**Notes:** $X^2 = 10.398$, $\rho = 0.001$)

Indonesian market, the majority of the consumers can be considered as falling into the 'truly responsible' segment (58.6% and 61.8%). The 'latent responsible' (or attitude-behaviour gap) segment was found to be only between 11.8% and 14.9%. One possible explanation is that the product category is something familiar for the respondents. It is commonly available to them and easy to understand. Another explanation may be attributed to the slightly higher number of female samples (56.9%). Previous studies revealed that females tend to have stronger environmental attitudes and behaviour than their male counterparts (Zelezny et al., 2000).

Despite these interesting findings, the typology of responsible consumption segments needs further examination with different products and different country contexts. In addition, different proxy measures for responsible consumption behaviour may be explored (e.g. consumption or purchase frequency or actual purchase measured in a longitudinal study (cf. Ajzen, 2002, 2011; Ajzen & Driver, 1992; Hrubes, Ajzen, & Daigle, 2001; Madden, Ellen, & Ajzen, 1992)).

The Role of Perceived Readiness to Be Green

As explained earlier, the relationship between green attitudes and actual behaviour has been debatable. The literature suggests that several theoretical frameworks have been proposed to explain the attitude-behaviour gap but no definitive explanation has yet been found (Kollmuss & Agyeman, 2002). Johnstone and Tan (2015) suggest that although consumers may have favourable pro-environmental attitudes, their perceptions towards 'being green' may influence their perceived readiness and thus their intention to engage in green consumption behaviour. The term 'being green' refers to engaging in environmentally friendly activities, including purchasing or using green products (Polonsky, 2011). Arli et al. (2015) suggested that 'being green' is yet to be perceived as a social norm in most countries, particularly emerging markets. When green social norms are relatively weak, consumers may experience only minimum or even no dissonance if there is a discrepancy between their attitudes and behaviour. As such, consumers' attitudes towards the environment might be inadequate to predict their behaviour.

Perceived readiness to be green is defined as 'a condition in which consumers perceive themselves as "ready" to engage in green consumption behaviour, such as buying green products' (Arli et al., 2018, p. 10). The scale for perceived readiness to be green was developed by Johnstone,

Yang, and Tan, (2014). It consists of three reversed-coded items: (1) *I do not have sufficient knowledge about environmental issues to make decisions about these types of products*; (2) *I do not have sufficient time to learn about environmentally friendly products*; and (3) *I have too many other responsibilities at the moment to think about environmentally friendly products.* Responses are measured on a 5-point Likert scale, ranging from 1 = 'Strongly disagree' to 5 = 'Strongly agree'.

In their earlier study on Indonesian consumers, Arli et al. (2015) found that consumers' perceived readiness to be green affects their intention to purchase green products. Whenever consumers perceive themselves as ready to be green, they are more likely to purchase green products.

In their subsequent research with a bigger sample (916 Indonesian students and non-students), Arli et al. (2018) reported that not only perceived readiness to be green positively influences consumers' intention to purchase green products but also it mediates the relationship between consumer attitudes towards green products and purchase intentions, perceived behavioural control and purchase intention, pro-environmental self-identity (i.e. whether consumers consider themselves to be pro-environment) and purchase intentions, as well as perceived sense of responsibility (i.e. what an individual perceives as their responsibility for environmental deterioration) and purchase intention.

These initial findings suggest that consumers' perceived readiness to be green plays an important role as one of the determinants of green product purchase intentions. Arli et al. (2015, 2018) argued that in countries where 'being green' is not yet considered as a social norm, engaging in responsible consumption behaviour is equivalent to 'behavioural change'. Therefore, an individual's readiness to change can serve as a proximal predictor of behavioural change. More importantly, Arli et al. (2018) suggest that favourable attitudes towards purchasing a green product may not translate into green product purchase intentions if consumers do not think that they are *ready* to be green. This may in part help to explain the attitude-behaviour gap in the responsible consumption context.

DISCUSSION

This chapter focuses on three key challenges to responsible consumption as identified from an intensive literature review: (1) consumers cannot be simply segmented into green and non-green consumers; (2) there is a gap between consumers' attitude towards and their actual responsible consumption behaviour; and (3) perceived readiness to be green may affect responsible consumption. How do the three studies (Study 1, Study 2, and a review of perceived readiness studies) examine these challenges? First, Study 1 shows that what is considered important by consumers may be different from what many managers or companies perceive. The CSR (corporate social responsibility) literature, for instance, has been predominantly focused on the managerial perspective (Aguinis & Glavas, 2012; Oberseder et al., 2013), where managers were reported to have a holistic view of social responsibility domains with regard to their stakeholders (Devinney et al., 2006; Oberseder et al., 2013, 2014). In contrast, 'most consumers cannot fully comprehend the overarching concept of CSR' (Oberseder et al., 2014, p. 111). As a result, consumers tend to approve and support CSR programs that are aligned with their interests (Morrison & Bridwell, 2011; Olander & Thogersen, 1995). Study 1 also suggests that the importance of social responsibility domains may be different between consumers in developed and developing countries. On the one hand, understanding which specific social responsibility areas were perceived to be important by consumers may help CSR managers create and implement more effective CSR initiatives. On the other hand, since consumers place different importance on different social responsibility domains, it may suggest that their perceived readiness to be green may be contextual (e.g. product/service dependent). For instance, the results of Study 1 indicate that the environmental domains (such as reducing energy consumption and disposing of waste correctly; see Table 12.1) were not perceived as being as important as the community domain (e.g. sourcing products and raw materials locally). In this context, consumers' perceived readiness to reduce their energy consumption might not be as high as their readiness to buy green products using local content materials. Therefore, Study 1 contributes to the relatively limited CnSR (consumer social responsibility) studies (Quazi et al., 2016; Vitell, 2015).

Second, using attitudes towards responsible consumption and responsible consumption behaviour as key variables, Study 2 proposes a typology of responsible consumption segments. The empirical study found support for the four identified segments (i.e. truly responsible, latent (potential) responsible, spurious responsible, and irresponsible segments). It addresses the attitude-behaviour gap issue by showing that there is only one segment (i.e. truly responsible) representing the consistent group of consumers who 'walk their talk' (Carrington et al., 2010). In the context of green household products (e.g. laundry detergent, dishwashing liquids, toilet paper rolls, and soaps) in Indonesia, the truly responsible segment represents between 58.6% and 61.8% of the consumers. The rest belongs to the other three segments. The findings have three important implications. First, responsible or sustainable consumption programs can be most effective if they are directed towards the right segment, that is, the truly responsible one. It would be interesting to extend this study into another context, for instance, examining why the 'pay-for-plastic bag' campaign failed in Indonesia. The reasons may include the wrong segment(s) being targeted or the largest segment for plastic bag users in Indonesia was possibly the irresponsible segment. The second implication is that research on responsible consumer consumption needs to integrate both attitudinal and behavioural measures (including using actual purchase/actions) to get a more comprehensive picture of the complex phenomenon. This can overcome the limitations of the purely attitude-based studies on responsible consumption. Another implication is that the proposed typology of responsible consumption segments may be further examined in different product and country contexts to investigate the attitude-behaviour gap. While most of the previous studies focus on the profiles of green consumers using demographic segmentation, the proposed typology provides a direct examination of the attitude-behaviour gap using the most relevant variables (i.e. attitudes towards responsible consumption and responsible consumption behaviour).

Third, previous studies reveal that perceived readiness to be green has a positive effect on green product purchase intentions and mediates the influence of consumer attitude towards green products and green product purchase intention (Arli et al., 2015, 2018). The findings suggest that perceived readiness to be green is a potential mediator explaining the attitude-behaviour gap. In other words, favourable attitudes towards responsible consumption may not translate into responsible consumption behaviour if consumers do not think that they are *ready* to be green (i.e. have sufficient knowledge about environmental issues, have sufficient time to learn about environmentally friendly products, and do not have too many other responsibilities at the moment to think about environmentally friendly products).

Concluding Remarks: Challenges in Creating Responsible Consumption in Emerging Markets

This chapter discusses the need for responsible consumption development in emerging markets. Through three studies, it highlights three major challenges in responsible consumption (i.e. (1) better understanding of consumer social responsibility, especially how consumers perceive different social responsibility domains; (2) targeting the 'right' responsible consumption segments; and (3) helping consumers to be ready to be green).

Furthermore, several other practical challenges in developing reasonable consumption in the context of emerging markets were also identified. The first challenge is how to inform, educate, and encourage consumers to be actively responsible. This needs more time and effort to deal with the 'potential responsible' and 'spurious responsible' segments.

Second, the responsible consumption issue involves how to 'normalize' green/responsible behaviours. It needs a consistent repositioning strategy to encourage the adoption of more responsible consumer practices, such as monitoring electricity consumption, recycling, taking own shopping bags to the shops, using energy-saving light bulbs, buying organic food, and using public transport whenever possible (Rettie, Burchell, & Barnham, 2014; Rettie, Burchell, & Riley, 2012). Not only might a normalization strategy increase an individual's readiness to be green, it may also attract more people to join the 'truly responsible' segment. Changing daily behaviour of individual consumers is the third challenge. This is particularly important when intervention strategies are not enough and identification/segmentation of consumers is not sufficient (McDonald et al., 2012). In other words, it is not easy to change a 'potential responsible' consumer, for instance, into a 'truly responsible' individual.

The fourth challenge is how to overcome barriers to be responsible consumers (Johnstone & Tan, 2015, p. 321): 'it is too hard to be green' (consumers' perceptions of external factors, such as marketers, government, and people who consumers live with, that make it difficult to adopt responsible consumption practices), 'green stigma' (less favourable perceptions towards green consumers and green messages), and 'green reservations' (consumers' ambivalence or uncertainty that greener consumption practices will make a difference to the environment). When these barriers can be overcome, consumers' "perceived readiness to be green" will increase. Last but not least, it needs an integrated effort of relevant parties, such as marketers, policy makers, consumers, religious leaders, and others, in creating a more responsible consumption. Consumer interests have to be incorporated in social responsibility initiatives or policies, because they play an important role in determining the successful implementation of such initiatives (Morrison & Bridwell, 2011; Olander & Thogersen, 1995; Vitell, 2015).

Despite the three studies in this chapter providing important insights into the challenges of responsible consumption in Indonesia, there are some limitations that may provide future research avenues. First, the three studies presented in this chapter involved different samples from different cities. It may be more comprehensive to examine the CnSR, responsible consumption segments, and perceived readiness to be green issues in one integrated study. Second, sustainable/responsible consumption is a complex issue. There are many other specific issues worth researching. For instance, further studies are needed to explore (1) investigating how marketing can help developing responsible consumers, especially in the context of bottom-of-the-pyramid, green, health-conscious, and the financially literate consumers (Giesler & Veresiu, 2014); (2) investigating barriers to responsible consumption behaviour in cross-cultural and multiple product category contexts; and (3) exploring other sustainable-related issues (e.g. voluntary simplicity, unethical behaviour of buying/using/committing to counterfeit products) in the emerging market contexts.

REFERENCES

Aguinis, H., & Glavas, A. (2012). What we know and don't know about corporate social responsibility: A review and research agenda. *Journal of Management, 38*, 932–968.

Ajzen, I. (1991). The theory of planned behavior. *Organizational Behavior and Human Decision Processes, 50*, 179–211.

Ajzen, I. (2002). *Constructing a TpB questionnaire: Conceptual and methodological considerations*. Retrieved October 14, 2016 from http://chuang.epage.au.edu.tw/ezfiles/168/1168/attach/20/pta_41176_7688352_57138.pdf

Ajzen, I. (2005). *Attitudes, personality, and behavior* (2nd ed.). New York: Open University Press.

Ajzen, I. (2011). The theory of planned behaviour: Reactions and reflections. *Psychology & Health, 26*, 1113–1127.

Ajzen, I., & Driver, B. L. (1992). Application of the theory of planned behavior to leisure choice. *Journal of Leisure Research, 24*, 207–224.

Arli, D., Tan, L. P., Tjiptono, F., & Yang, L. (2015, November–December). *Exploring consumers' readiness to be green in an emerging market*. Paper presented at the 2015 ANZMAC Conference: Innovation and Growth Strategies in Marketing, University of New South Wales, Sydney, Australia.

Arli, D., Tan, L. P., Tjiptono, F., & Yang, L. (2018). Exploring consumers' purchase intention toward green products in an emerging market: The role of consumers' perceived readiness. *International Journal of Consumer Studies, forthcoming*.

Baker, W. E., & Sinkula, J. M. (2005). Environmental marketing strategy and firm performance: Effects on new product performance and market share. *Journal of the Academy of Marketing Science, 33*, 461–475.

Barkham, P. (2016, October 3). Six billion plastic bags can't be wrong—So what do we tax next? *The Guardian.* Retrieved October 22, 2016 from https://www.theguardian.com/environment/shortcuts/2016/oct/03/six-billion-plastic-bags-cant-be-wrong-so-what-do-we-tax-next

Barnett, C., Cloke, P., Clarke, N., & Malpass, A. (2005). Consuming ethics: Articulating the subjects and spaces of ethical consumption. *Antipode, 37,* 23–45.

Bavani, M., & Wong, P. M. (2016, August 22). Billions of plastic bags still being used. *The Star.* Retrieved October 22, 2016 from http://www.thestar.com.my/metro/community/2016/08/22/billions-of-plastic-bags-still-being-used-six-years-have-gone-by-since-the-government-launched-the-n/

Belk, R. W. (1985). Issues in the intention behavior discrepancy. In J. N. Sheth (Ed.), *Research in consumer behavior* (Vol. I, pp. 1–34). Greenwich, CT: JAI Press.

Bhate, S., & Lawler, K. (1997). Environmentally friendly products: Factors that influence their adoption. *Technovation, 17,* 457–465.

Block, B. (2016, October 26). China reports 66-percent drop in plastic bag use. *Worldwatch.* Retrieved October 26, 2016 from http://www.worldwatch.org/node/6167

Bray, J., Johns, N., & Kilburn, D. (2011). An exploratory study in the factors impeding ethical consumption. *Journal of Business Ethics, 98,* 597–608.

Bredahl, L. (2001). Determinants of consumer attitudes and purchase intentions with regard to genetically modified food: Results of a cross-national survey. *Journal of Consumer Policy, 24,* 23–61.

Carrington, M., Neville, B., & Whitwell, G. (2010). Why ethical consumers don't walk their talk: Towards a framework for understanding the gap between the ethical purchase intentions and actual buying behaviour of ethically minded consumers. *Journal of Business Ethics, 97,* 139–158.

Caruana, R., & Chatzidakis, A. (2014). Consumer social responsibility (CnSR): Toward a multi-level, multi-agent conceptualization of the "other CSR". *Journal of Business Ethics, 121,* 577–592.

Chan, R., Wong, Y., & Leung, T. (2008). Applying ethical concepts to the study of 'green' consumer behavior: An analysis of Chinese consumers' intentions to bring their own shopping bags. *Journal of Business Ethics, 79,* 469–481.

Chatzidakis, A., Hibbert, S., Mittusis, D., & Smith, A. (2004). Virtue in consumption? *Journal of Marketing Management, 20,* 526–543.

Chow, L. (2016, August 5). Proof that charging customers for plastic bags reduces their use. *EcoWatch.* Retrieved October 22, 2016 from http://www.alternet.org/environment/england-proves-charging-consumers-plastic-bags-reduces-their-use

CIA. (2016). *The world factbook: Indonesia.* Retrieved May 25, 2016 from https://www.cia.gov/library/publications/resources/the-world-factbook/geos/id.html

322   F. TJIPTONO

Connelly, B. L., Ketchen, D. J., Jr., & Slater, S. F. (2011). Toward a "theoretical toolbox" for sustainability research in marketing. *Journal of the Academy of Marketing Science, 39*, 86–100.

Cronin, J. J., Smith, J. S., Gleim, M. R., Ramirez, E., & Martinez, J. D. (2011). Green marketing strategies: An examination of stakeholders and the opportunities they present. *Journal of the Academy of Marketing Science, 39*, 158–174.

Dans, E. (2015). *Volkswagen and the failure of corporate social responsibility.* Retrieved February 16, 2016 from http://www.forbes.com/sites/enrique-dans/2015/09/27/volkswagen-and-the-failure-ofcorporate-social-responsibility/#19b832356128

Devinney, T. M., Auger, P., Eckhardt, G., & Birtchnell, T. (2006, Fall). The other CSR: Making consumers socially responsible. *Stanford Social Innovation Review*, pp. 30–37.

Dick, A. S., & Basu, K. (1994). Customer loyalty: Toward an integrated conceptual framework. *Journal of the Academy of Marketing Science, 22*, 99–113.

Dickinson, R. A., & Carsky, M. L. (2005). The consumer as economic voter. In R. Harrison, D. Shaw, & T. Newholm (Eds.), *The ethical consumer* (pp. 25–36). London: Sage.

Eckhardt, G. M., Belk, R., & Devinney, T. M. (2010). Why don't consumers consume ethically? *Journal of Consumer Behaviour, 9*, 426–436.

Financial Times. (2014). Fortune 500 companies spend more than $15 bn on corporate social responsibilities. Retrieved January 14, 2015 from http://www.ft.com/cms/s/0/95239a6e-4fe0-11e4-a0a4-00144feab7de.html#axzz3xBGilLnl

Fishbein, M., & Ajzen, I. (1975). *Belief, attitude, intention and behavior: An introduction to theory and research.* Reading, MA: Addison-Wesley.

Fraj, E., & Martinez, E. (2007). Ecological consumer behaviour: An empirical analysis. *International Journal of Consumer Studies, 31*, 26–33.

Ganesan, S., George, M., Jap, S., Palmatier, R. W., & Weitz, B. (2009). Supply chain management and retailer performance: Emerging trends, issues, and implications for research and practice. *Journal of Retailing, 85*, 84–94.

Garland, R., & Gendall, P. (2004). Testing Dick and Basu's customer loyalty model. *Australasian Marketing Journal (AMJ), 12*(3), 81–87.

Giesler, M., & Veresiu, E. (2014). Creating the responsible consumer: Moralistic governance regimes and consumer subjectivity. *Journal of Consumer Research, 41*, 840–857.

Gleim, M. R., Smith, J. S., Andrews, D., & Cronin, J. J. (2013). Against the green: A multi-method examination of the barriers to green consumption. *Journal of Retailing, 89*, 44–61.

Grimmer, M., & Miles, M. P. (2017). With the best of intentions: A large sample test of the intention-behaviour gap in pro-environmental consumer behaviour. *International Journal of Consumer Studies, 41*, 2–10.

Gupta, S., & Ogden, D. T. (2009). To buy or not to buy? A social dilemma perspective on green buying. *Journal of Consumer Marketing, 26,* 376–391.

Handayani, P. (2016, October 22). View point: Plastic bags ban remains a dream in Indonesia. *The Jakarta Post.* Retrieved October 22, 2016 from http://www.thejakartapost.com/news/2016/10/08/view-point-plastic-bags-ban-remains-a-dream-indonesia.html

Hrubes, D., Ajzen, I., & Daigle, J. (2001). Predicting hunting intentions and behavior: An application of the theory of planned behavior. *Leisure Sciences, 23*(3), 165–178. Please refer to the discussion of the measure of behavior (p. 168).

Johnstone, M. L., & Tan, L. P. (2015). Exploring the gap between consumers' green rhetoric and purchasing behaviour. *Journal of Business Ethics, 132*(2), 311–328.

Johnstone, M.-L., Yang, L., & Tan, L. P. (2014, December). *The attitude-behavior gap: The development of a consumers' green perception scale.* Paper presented at the 2014 ANZMAC Conference, Griffith University, Brisbane, Australia.

Kollmuss, A., & Agyeman, J. (2002). Mind the gap: Why do people act environmentally and what are the barriers to pro-environmental behavior? *Environmental Education Research, 8,* 239–260.

Kotler, P., & Armstrong, G. (2014). *Principles of marketing* (15th ed.). Upper Saddle River, NJ: Pearson Education.

Laroche, M., Bergeron, J., & Barbaro-Forleo, G. (2001). Targeting consumers who are willing to pay more for environmentally friendly products. *Journal of Consumer Marketing, 18,* 503–520.

Lash, J., & Wellington, F. (2007). Competitive advantage on a warming planet. *Harvard Business Review, 85*(3), 94–102.

Lubin, D. A., & Esty, D. C. (2010). The sustainability imperative. *Harvard Business Review, 88,* 42–50.

Luchs, M., Naylor, R. W., Rose, R. L., Catlin, J. R., Gau, R., Kapitan, S., Mish, J., Ozanne, L., Phipps, M., Simpson, B., Subrahmanyan, S., & Weaver, T. (2011). Toward a sustainable marketplace: Expanding options and benefits for consumers. *Journal of Research for Consumers, 19,* 1–12.

Luo, X., & Bhattacharya, C. B. (2006). Corporate social responsibility, customer satisfaction and market value. *Journal of Marketing, 70,* 1–18.

Madden, T. J., Ellen, P. S., & Ajzen, I. (1992). A comparison of the theory of planned behavior and the theory of reasoned action. *Personality and Social Psychology Bulletin, 18,* 3–9.

Maignan, I., & Ferrell, O. C. (2001). Corporate citizenship as a marketing instrument: Concepts, evidence and research directions. *European Journal of Marketing, 35,* 457–484.

Maignan, I., Ferrell, O. C., & Hult, G. T. M. (1999). Corporate citizenship: Cultural antecedents and business benefits. *Journal of the Academy of Marketing Science, 27,* 455–469.

Manieri, T., Barnett, E. G., Valdero, T. R., Unipan, J. B., & Oskamp, S. (1997). Green buying: The influence of environmental concern on consumer behavior. *The Journal of Social Psychology, 137*, 189–204.

McCarty, J. A., & Shrum, L. J. (2001). The influence of individualism, collectivism, and locus of control on environmental beliefs and behavior. *Journal of Public Policy and Marketing, 20*, 93–104.

McDonagh, P., & Prothero, A. (2014). Sustainability marketing research: Past, present and future. *Journal of Marketing Management, 30*(11–12), 1186–1219.

McDonald, S., Oates, C. J., Alevizou, P. J., Young, C. W., & Hwang, K. (2012). Individual strategies for sustainable consumption. *Journal of Marketing Management, 28*, 445–468.

Menguc, B., & Ozanne, L. K. (2005). Challenges of the 'green imperative': A natural resource-based approach to the environmental orientation—Business performance relationship. *Journal of Business Research, 58*, 430–438.

Mittelstaedt, J. D., Shultz, C. J., Kilbourne, W. E., & Peterson, M. (2014). Sustainability as megatrend: Two schools of macromarketing thought. *Journal of Macromarketing, 34*, 253–264.

Mohr, L. A., Webb, D. J., & Harris, K. E. (2001). Do consumers expect companies to be socially responsible? The impact of corporate social responsibility on buying behavior. *Journal of Consumer Affairs, 35*, 45–72.

Moisander, J. (2007). Motivational complexity of green consumerism. *International Journal of Consumer Studies, 31*, 404–409.

Morley, K. (2016, July 30). Britain banishes plastic bags as 5p 'tax' sees usage plummet by 6 billion. *Telegraph*. Retrieved October 22, 2016 from http://www.telegraph.co.uk/news/2016/07/30/britain-banishes-plastic-bags-as-5p-tax-sees-usage-plummet-by-6/

Morrison, E., & Bridwell, L. (2011). Consumer social responsibility: The true corporate social responsibility. *Competition Forum, 9*(1), 144–149.

Muncy, J. A., & Vitell, S. J. (1992). Consumer ethics: An empirical investigation of the ethical beliefs of the final consumer. *Journal of Business Research, 24*(1), 297–312.

Newholm, T., & Shaw, D. (2007). Studying the ethical consumer: A review of research. *Journal of Consumer Behaviour, 6*, 253–270.

Oberseder, M., Schlegelmilch, B. B., & Gruber, V. (2011). Why don't consumers care about CSR?: A qualitative study exploring the role of CSR in consumption decisions. *Journal of Business Ethics, 104*(4), 449–460.

Oberseder, M., Schlegelmilch, B. B., & Murphy, P. E. (2013). CSR practices and consumer perceptions. *Journal of Business Research, 66*, 1839–1851.

Oberseder, M., Schlegelmilch, B. B., Murphy, P. E., & Gruber, V. (2014). Consumers' perceptions of corporate social responsibility: Scale development and validation. *Journal of Business Ethics, 124*, 101–115.

Olander, F., & Thogersen, J. (1995). Understanding consumer behavior as pre-requisite for environmental protection. *Journal of Consumer Policy, 18*(4), 345–385.

Peattie, K. (2001). Towards sustainability: The third age of green marketing. *The Marketing Review, 2,* 129–146.

Peattie, K., & Collins, A. (2009). Guest editorial: Perspectives on sustainable consumption. *International Journal of Consumer Studies, 33,* 107–112.

Pew Research Center. (2010). *Obama more popular abroad than at home, global image of U.S. continues to benefit.* Washington, DC: Pew Research Center.

Phipps, M., Ozanne, L. K., Luchs, M. G., Subrahmanyan, S., Kapitan, S., Catlin, J. R., Gau, R., Naylor, R. W., Rose, R. L., Simpson, B., & Weaver, T. (2013). Understanding the inherent complexity of sustainable consumption: A social cognitive framework. *Journal of Business Research, 66,* 1227–1234.

Pickett-Baker, J., & Ozaki, R. (2008). Pro-environmental products: Marketing influence on consumer purchase decision. *Journal of Consumer Marketing, 25,* 281–293.

Polonsky, M. J. (2011). Transformative green marketing: Impediments and opportunities. *Journal of Business Research, 64,* 1311–1319.

Porter, M., & van der Linde, C. (1995). Toward a new conception of the environment-competitiveness relationship. *The Journal of Economic Perspectives, 9,* 97–118.

Prothero, A., Dobscha, S., Freund, J., Kilbourne, W. E., Luchs, M. G., Ozanne, L. K., & Thøgersen, J. (2011). Sustainable consumption: Opportunities for consumer research and public policy. *Journal of Public Policy & Marketing, 30*(1), 31–38.

Pujari, D., Wright, G., & Peattie, K. (2003). Green and competitive: Influences on environmental new product development performance. *Journal of Business Research, 56,* 657–671.

Quazi, A., Amran, A., & Nejati, M. (2016). Conceptualizing and measuring consumer social responsibility: A neglected aspect of consumer research. *International Journal of Consumer Studies, 40,* 48–56.

Rettie, R., Burchell, K., & Barnham, C. (2014). Social normalisation: Using marketing to make green normal. *Journal of Consumer Behaviour, 13*(1), 9–17.

Rettie, R., Burchell, K., & Riley, D. (2012). Normalizing green behaviors: A new approach to sustainability marketing. *Journal of Marketing Management, 28*(3–4), 420–444.

Ribka, S. (2016, October 4). One step back in effort to curb plastic waste. *The Jakarta Post.* Retrieved October 22, 2016 from http://www.thejakartapost.com/news/2016/10/04/one-step-back-effort-curb-plastic-waste.html

Roberts, J. (1996). Green consumers in the 1990s: Profile and implications for advertising. *Journal of Business Research, 36,* 217–232.

Savitz, A. W., & Weber, K. (2006). *The triple bottom line: How today's best-run companies are achieving economic, social and environmental success—And how you can too.* New York: John Wiley, Inc.

Schwartz, S. H., & Howard, J. A. (1981). A normative decision-making model of altruism. In J. P. Rushton & R. M. Sorrentino (Eds.), *Altruism and helping behavior* (pp. 189–211). Hillsdale, NJ: Lawrence Erlbaum.

Sener, A., & Hazer, O. (2008). Values and sustainable consumption behavior of women: A Turkish sample. *Sustainable Development, 16*, 291–300.

Sheth, J. N., Sethia, N. K., & Srinivas, S. (2011). Mindful consumption: A customer-centric approach to sustainability. *Journal of the Academy of Marketing Science, 39*, 21–39.

Stern, P. C. (2000). Toward a coherent theory of environmentally significant behavior. *Journal of Social Issues, 56*(3), 407–424.

Stern, P. C., Dietz, T., Abel, T., Guagnano, G. A., & Kalof, L. (1999). A value-belief-norm theory of support for social movements: The case of environmental concern. *Human Ecology Review, 6*, 81–97.

Szmigin, I., Carrigan, M., & McEachern, M. G. (2009). The conscious consumer: Taking a flexible approach to ethical behavior. *International Journal of Consumer Studies, 33*(2), 224–231.

Tanner, C., & Kast, W. S. (2003). Promoting sustainable consumption: Determinants of green purchases by Swiss consumers. *Psychology & Marketing, 20*, 883–902.

Thatcher, A., & Yeow, P. H. P. (2016). Human factors for a sustainable future. *Applied Ergonomics, 57*, 1–7.

Tukker, A., Huppes, G., Guinée, J., Heijungs, R., de Koning, A., Van Oers, L., Suh, S., Geerken, T., Van Holderbeke, M., Jansen, B., & Nielsen, P. (2005). *Environmental impact of products (EIPRO): Analysis of the life cycle environmental impacts related to the total final consumption of the EU25.* Brussels, Germany: IPTS/ESTO, European Commission Joint Research Centre.

Ulusoy, E. (2016). Experiential responsible consumption. *Journal of Business Research, 69*, 284–297.

UNDP. (2016). *UNDP human development reports: Indonesia.* Retrieved June 6, 2016 from http://hdr.undp.org/en/countries/profiles/IDN

United Nations World Commission on Environment and Development. (1987). *Our common future.* Oxford, UK: Oxford University Press.

Valor, C., & Carrero, I. (2014). Viewing responsible consumption as a personal project. *Psychology & Marketing, 31*, 1110–1121.

Vitell, S. J. (2015). A case for consumer social responsibility (CnSR): Including a selected review of consumer ethics/social responsibility research. *Journal of Business Ethics, 130*, 767–774.

Webb, D. J., Mohr, L. A., & Harris, K. E. (2008). A re-examination of socially responsible consumption and its measurement. *Journal of Business Research, 61*, 91–98.

Wikipedia. (2016). *Semarang.* Retrieved October 11, 2016 from https://en.wikipedia.org/wiki/Semarang

Zelezny, L. C., Chua, P. P., & Aldrich, C. (2000). Elaborating on gender differences in environmentalism. *Journal of Social Issues, 56*, 443–457.

Zimmer, M. R., Stafford, T. F., & Stafford, M. R. (1994). Green issues: Dimensions of environmental concern. *Journal of Business Research, 30*(1), 63–74.

Zudianto, H. (2010). *Yogyakarta: Management of multiculturalism.* Retrieved September 5, 2015 from http://www.city.hamamatsu.shizuoka.jp/foreign/english/intercity_cooperation/pdf/congre_06.pdf

Promoting Green Technology Financing: Political Will and Information Asymmetries

Jothee Sinnakkannu and Ananda Samudhram

INTRODUCTION

Environmental pollution, degradation and the human factor have been identified as important research areas in green ergonomics (Thatcher, 2013). Hanson (2013) observes the broad convergence of the aims of the green agenda with that of ergonomics and the human factor, in examining the activities of humans that ultimately impact human welfare and well-being, wherein green ergonomics is defined as "ergonomics interventions that have a pro-nature focus" (Thatcher, 2013, p. 391).

Thatcher (2013) notes that green ergonomics embraces designs compelling changes in human behaviour. Such designs have emerged from disciplines that are not traditionally linked with ergonomics, including psychology, education and the biological sciences (e.g. Louv, 2005; Sterling, 2001; Wilson 1984). This chapter adds the discipline of finance to this list, exploring the effectiveness of financial incentives in the greening of Malaysian firms. In addition, it incorporates Thatcher and Yeow's (2016) system of systems approach, whereby a larger ecosystem is considered to be composed of smaller sub-systems. Individual firms comprise the sub-systems and incentives that reduce firm-level pollution

J. Sinnakkannu (✉) • A. Samudhram
School of Business, Monash University Malaysia, Bandar Sunway, Malaysia

© The Author(s) 2018 329
A. Thatcher, P. H. P. Yeow (eds.), *Ergonomics and Human Factors for a Sustainable Future*, https://doi.org/10.1007/978-981-10-8072-2_13

and will, in aggregate, lower the national (larger ecosystem) pollution levels. Samudhram, Siew, Sinnakkannu and Yeow (2016) apply the system of systems approach to corporate social responsibility (CSR) reporting, reasoning that holistic CSR reporting frameworks would induce firms to undertake meaningful CSR-related activities. The widespread adoption of activities that reflect CSR at the firm level, including attention to the health and well-being of the human factor, would in aggregate improve the human condition nationally. This chapter extends similar conceptualisations to green financing in Malaysia and other emerging economies. It seeks to enhance the greening of nations by formulating relevant policy-level recommendations that would promote the adoption of green processes at the firm level that in aggregate would help to green national economies.

Political will is essential for addressing global warming and climate change. Policy-level funding cuts (reflecting a lack of political will) would deter the implementation of activities that could check the negative impact of climate change (Kelhart, 2008). Sneed (2017) discusses potential implications of a lack of political support for the global climate change agenda during the Trump presidency. Barbi, Ferreira and Guo (2016) discuss the interplay between government will, in terms of establishing relevant policies for minimising China's greenhouse emissions, and practical reality, in terms of judiciously implementing relevant policies (including limited crackdowns on enterprises that pollute the land and the air) while seeking to achieve economic growth targets. In essence, government will, including the provision of relevant funding, would be able to effectively address activities that contribute to global warming.

Research examining the role of other factors, especially information asymmetries, which could deter the effectiveness of green technology policies driven by unambiguous political will, particularly in emerging economies, is scarce. The current chapter addresses this gap in the literature, by examining the case of the green technology financing scheme (GTFS) in Malaysia. The Malaysian government's national budget allocated 3.5 billion ringgit (amounting to almost a billion US dollars[1]) for enhancing the uptake of green technology loans by Malaysian firms, to green their production processes. Sixty per cent of the green technology loan amounts were guaranteed by the government (mitigating the non-repayment risks for the lending banks to some extent), with a 2% interest rate rebate to attract applicants. Surprisingly, a large portion of these budgeted amounts remained unspent. The study presented in this chapter addresses two overarching research questions. The first question is: why has the GTFS-based loan uptake rate been low? The second question asks: what should be done to improve these uptake rates?

This study found that firms were generally unaware of the overall benefits of the GTFS, which led to the low uptake rates. Thus, the dissemination of pertinent information to relevant stakeholders is important, in addition to government will, to ensure the effectiveness of policies designed to reduce a nation's carbon footprint.

Based on the Malaysian experience, this study formulates recommendations for effective national policies that could drive the widespread greening of national economies, especially in developing nations. This study follows up on Thatcher's (2013) views that multi-disciplinary research could advance the frontiers of green ergonomics research. In summary, it explores the role of green technology financing (from the finance discipline) in the greening of traditional economies (related to green ergonomics), which would help to reduce pollution and combat global warming and climate change and promote human health and well-being.

LITERATURE REVIEW

This section reviews the literature pertaining to green technology financing in general and Malaysia's green technology financing scheme specifically, in the "Green Technology Financing" and "Malaysia's Green Technology Financing Scheme" sections, respectively.

Green Technology Financing

The green economy improves human well-being and social equity, while significantly reducing environmental risks and ecological scarcities (United Nations, 2014). The active promotion of green growth would help to alleviate the harsh effects of climate change, including radical weather patterns and widespread economic damage (WEF, 2013). The investment required to realise the sustainable development goals that will address various environmental issues is estimated to average around 5–7 trillion US dollars per annum over 2015–2030, with developing nations alone requiring 3.3–4.3 trillion US dollars. Based on current investment patterns, an estimated annual investment gap of USD 2.5 trillion has to be bridged to finance sustainable development in developing countries (WEF, 2013).

Eskelson, Antal, Fidanza, Leclercq and Rosca (2016) and De Serres, Murtin, and Nicoletti (2010) indicate that one of the main barriers for the development of green business is access to finance. Crespi, Ghisetti and Quatraro (2015) offer that market and incentive-based instruments can be used by policymakers to promote green economies, along with command and control regulations, voluntary agreements and information- and education-based instruments. Such market-based incentives include emissions trading, environmental taxes and charges, deposit-refund systems, subsidies and compensation mechanisms and green purchasing. Banks could also play an important role in the development of green economies (IFC, 2013).

Market-based instruments employ incentives to combat environmental issues in a manner that aligns the self-interest of pertinent organisations with the policymakers' aims (Stavins, 2003) and encompass a wide range from subsidies to compensation strategies (EEA, 2005). Loans with low or no interest charges could also be used to promote environmentally friendly economic activities (Crespi et al., 2015). Pertinent financial incentives, that are also important for promoting the greening of the resource-constrained SME sector (Pimenova & Van der Vorst, 2004), include subsidies, grants, soft loans and tax concessions (Bradford & Fraser, 2008; Clement & Hansen, 2003; Mir & Feitelson, 2007).

In contrast to much of the published research, that tends to explore the effectiveness of various green policies for combating climate change in developed nations, this chapter examines the implementation of pertinent policies in an emerging economy, namely, Malaysia. Generally, in cash strapped developing and emerging economies, there is a perception that the stringent application and enforcement of policies to combat pollution may have negative economic repercussions (e.g. Barbi et al., 2016). As such, the commitment to combat global warming at international conventions could be viewed as lip service, lacking real political will, which in turn could result in limited effectiveness in meeting targets that are meant to limit environmental pollution, such as greenhouse gas emissions. In the case of Malaysia, however, billions of ringgit (nearly one billion US dollars) was set aside in the national budgets since 2009, in the form of interest rate subsidies as well as green loans backed by government guarantees, to green Malaysian firms and industries. There was a surprisingly low uptake of these budgetary allocations. The current study, financed by Malaysian Ministry of Science, Technology and Innovation

(MOSTI), examines the reasons for this low uptake. It finds that strong political will, as indicated by the billions of ringgit allocated to mitigate environmental degradation and pollution, alone is insufficient to effectively combat global warming. This study identifies various mitigating factors, that have received little attention in the extant literature, that need to be considered and cogently addressed by policymakers, in order to effectively combat global warming especially in developing and emerging economies. The next section details the Malaysian setting and discusses Malaysia's green policies.

Malaysia's Green Technology Financing Scheme

From 2001 to 2010, Malaysia's average annual CO_2 (carbon dioxide) emission levels have exceeded the world's and upper-income nations' averages (as per Fig. 13.1). In contrast to the average CO_2 emission levels of OECD member nations, which have trended downwards, the data from Malaysia indicates an unhealthy rising trend. Malaysia is committed to reducing the nation's carbon footprint. Various green policies, including

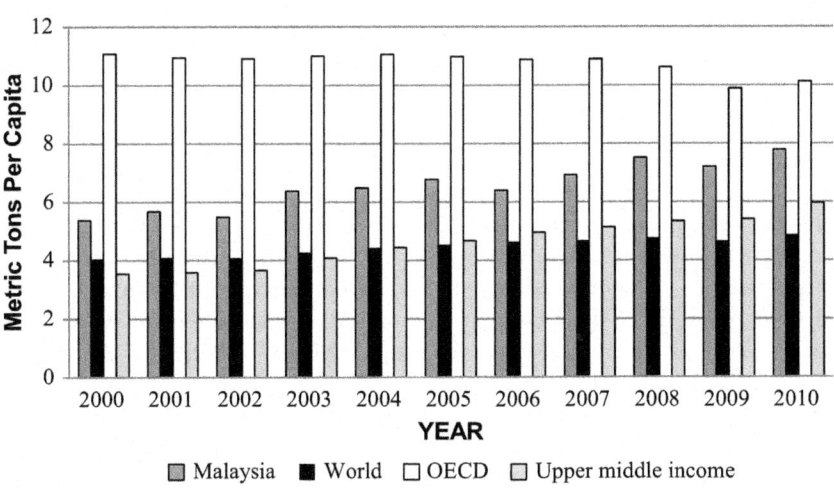

Fig. 13.1 CO_2 emissions: 2000–2010. Malaysia, compared with averages of the world, OECD and upper middle-income nations (Data Source: World Bank DataBank)

interest rate subsidies for green loans, have been established with the aim of achieving this national objective (KeTTHA, 2015).

The anticipated green economic growth would also provide long-term benefits. Malaysia's strong economic growth in the 1990s, driven by high-volume, low-cost production strategies, began to slow down in the new millennium. Past economic strategies transformed Malaysia from a poor country to an upper middle-income nation. However, due to the current high labour costs and the levelling off of labour productivity in the manufacturing sector, new strategies are required to propel future economic growth (Tenth Malaysia Plan, 2010; Tu, Ho, Chau, & Yu, 2016). Green technology loans would promote green economic growth, which would in turn support the transformation of Malaysia into a high value-added, high-income economy by 2020 (GTFBC, 2013a). Various green projects are expected to generate a total Gross National Income (GNI) of RM53 billion by 2020 (GTFBC, 2013b).

APEC (2014, p. iv) notes that a "critical component for sustainable low-carbon energy promotion is the financial framework as renewable energy investment can be expensive. To help mitigate this challenge, Malaysia has several financial incentives and strategies in place to encourage investment in renewable energy applications".

Malaysia's national Green Technology Policy, launched in 2009, reflects the policymakers' political will and commitment to the development of green technology that will ensure sustainable development and the conservation of the environment for future generations (ibid.).

In 2010, RM1.5 billion (approximately USD450 million) was allocated to the Green Technology Funding Scheme, GTFS (WTO, 2014), to support soft loans for the adoption of green technology by Malaysian firms. In 2013, an additional RM2 billion (approximately USD600 million) budgetary allocation was announced (GTFBC, 2013a). The GTFS is designed to encourage Malaysian firms to pursue green economic activities in general. In particular, this scheme would motivate Malaysia's small- and medium-scale enterprises (SMEs), which have limited resources, to go green. The Malaysian government provided a 2% interest rate subsidy, in addition to guaranteeing 60% of the GTFS-based loan through the Credit Guarantee Corporation (CGC). The GTFS was restructured in 2013, wherein the GTFS-based loan applicants would get either a 30% green loan guarantee or a 2% off the total annual interest rate (APEC, 2014). This restructuring would allow more firms to participate in the GTFS.

Table 13.1   GTFS-based loan eligibility and general criteria

Project eligibility	Must be certified by GreenTech. Malaysia		
Project criteria	Must be located within Malaysia, utilising local and imported technology		
Eligibility	Legally registered Malaysian-owned companies (at least 70%) in all economic sectors		
Firm type	*Producers*	*Users*	*Incentives*
Financing size	Maximum RM50 mil	Maximum RM10 mil	2%
Financing tenure	Up to 15 years	Up to 10 years	Rebate on interest rate charged by financial institutions, with 60% loan
Government guarantee	60% of the loan amount by CGC		guarantee. As of 2013, either a 2% interest rebate
Audit	Every 6 months over the life of the project		or 30% loan guarantee
Relevant agencies	Green Technology Corporation (GreenTech) Credit Guarantee Corporation (CGC)		

Adapted: KeTTHA (2011)

The GTFS projects a reduction of CO_2 emissions by 2.67 metric tons per year and the creation of 3018 green jobs (KeTTHA, 2015). Table 13.1 summarises the eligibility criteria and key details of GTFS-based loans.

Before applying for the GTFS-based loans, prospective GTFS-based loan applicants must submit their proposed projects to the Malaysian Green Technology Corporation (GreenTech Malaysia). GreenTech Malaysia validates the positive green impacts of the proposed projects and issues green project certificates. This green project certificate is then submitted, together with a completed GTFS-based loan application, to pertinent financial institutions. The banks review the GTFS-based loan applications and provide loans to the successful applicants. As of September 2012, 209 projects had been issued green certificates. However, only 67 GTFS-based loans were approved, comprising an approximate aggregate loan amount of RM800 million (GTFBC, 2013b). Although the GTFS-based loan disbursement picked up over time (ibid.), a large portion of the GTFS-based loan allocation remained unused. As of 16 January 2014, RM1.24 billion had been approved, to 97 applications, and the undisbursed fund amounted to RM2.24 billion (APEC 2014). The banks had rejected many of the GTFS-based loan applications because they were perceived to be too risky, suggesting that the Malaysian financial institutions could be "unfamiliar with financing green projects, while the applicants may be new businesses and do not meet the credit requirements" (APEC, 2014, p. 53). With the restructuring of the GTFS in 2013, wherein either a 30% loan guarantee or a 2% interest rate rebate was offered, the financial institutions could regard the GTFS loans to be more risky, which could further impede the uptake of the GTFS (ibid.). In addition, a limited understanding of the scheme (Tu et al., 2016) may have also contributed to the low uptake rates. These developments indicate a need to study the low uptake of the GTFS in detail, which could lead to informed approaches for lifting the uptake rates. The pertinent study is discussed in the following section.

EMPIRICAL STUDY

Methodology

The empirical study presented herein followed up on an earlier (Phase 1) exploratory investigation that involved interviews with the regulators (GreenTech Malaysia), the participating financial institutions (i.e. banks that would disburse the GTFS-based loans) and the GTFS-based loan applicants. These interviews indicated an expectations gap, whereby GTFS-based loan applicants were under the impression that just meeting all of the criteria in Table 13.1 (particularly the project certifications from GreenTech) was sufficient for successful GTFS-based loan applications. The bankers, however, also consider the long-term financial viability of the new business models entailed by the adoption of the green technologies. These financial considerations are important since the banks assume a large portion of the GTFS-based loan non-repayment risk.

The empirical part of the study (Phase 2) explores these information asymmetries further by examining various details requested in the GTFS-based loan application forms. Structured questionnaires were provided to all GTFS-based loans applicants listed in the GTFS database, who were situated in the Klang Valley. The Klang Valley was selected because it includes major commercial and industrial hubs and houses a large portion of the GTFS-based loan applicants. The current study analyses the first set of incoming responses from these applicants, consisting of 12 successful and 17 unsuccessful loan applicants. This study focuses on only the energy producers, who comprise 69% of the approved GTFS loan applicants, from 2009 to 2014 (Tu et al., 2016). There could be some additions to these early results, when all of the responses have been collected and analysed at the end of the study. However, the current analysis pro-vides important insights into the information that influences the GTFS-based loan approvals.

Research Questions and Hypothesis Development

The literature indicates that GTFS-based loan applications could be rejected due to a perception that the underlying projects carry high risks (APEC, 2014). Bankers could mitigate these risks by giving approvals to firms that indicate an ability to pay back the GTFS loans, by examining various key characteristics of the applicants. Such characteristics include the applicant's paid-up capital, years of project experience, project costs, amount of loan applied for and the extent to which the loan covers the project cost (loan/project cost). Research questions (RQs) 1–5, detailed below, are based on this assumed risk-averse orientation of bankers in approving GTFS-based loans.

Paid-Up Capital
Firms with higher paid-up capital would have more resources to pay back loans, compared with firms that have lower paid-up capital, ceteris paribus. Thus, firms with higher paid-up capital are likely to have lower non-payment risk than firms with relatively lower paid-up capital. As such, risk-averse bankers would approve GTFS-based loans to firms that have a higher paid-up capital, to reduce the banks' non-payment risk exposure. RQ 1 and H1 reflect this reasoning:

RQ1: Is the paid-up capital for firms that obtained the GTFS-based loan approval higher than that of the unsuccessful GTFS-based loan applicants?
H1: The mean paid-up capital for firms that obtained GTFS-based loan approvals was higher than the mean paid-up capital for the GTFS-based loan applications that were rejected.

Years of Experience in the Project
The literature indicates that new businesses may not "meet the credit requirements" (APEC, 2014, p. 53). Risk-averse bankers would regard new businesses to bear greater GTFS-based loan non-payment risk, in comparison to older establishments that have more years of experience, ceteris paribus. Therefore, bankers would provide GTFS-based loan

approvals to firms that have more experience, ceteris paribus, where a greater number of years in the project serve as a proxy for more experience.

RQ2: Is the mean number of years on the project greater for approved GTFS-based loan applicants, compared to rejected applicants?

H2: The number of years on the project for firms that obtained GTFS-based loan approvals is higher than that for the rejected applicants.

Project Cost

The literature indicates that the perceived risks associated with GTFS-based loans could serve as a barrier to the provision of approvals to GTFS-based loan applicants, which in turn would dampen the GTFS-based loan uptake rates (APEC, 2014). Risk-averse bankers would provide GTFS loan approvals to applicants that submit lower project costs, because non-repayment of the lower project costs would result in a lower risk exposure for the banks, ceteris paribus.

RQ3: Is the mean project cost lower for approved loan applications, compared with rejected applications?

H3: The mean project cost for firms that obtained GTFS-based loan approvals is lower than the mean project costs for the rejected applicants.

Amount of GTFS-Based Loan Applied For

The literature indicates that risks are an important consideration in the GTFS-based loan approvals by financial institutions (APEC, 2014). Bankers could reduce the loan non-payment risk exposure by approving projects that apply for lower GTFS-based loan amounts. RQ4 considers that bankers would approve GTFS-based loan applications that require lower loan amounts, since the lower loan quantum would limit the bank's overall loan non-repayment risk exposure, ceteris paribus. An added consideration is that when the GTFS-based loan applicant's requested loan amount comprises only a small portion of the overall project costs, wherein the applicant would also rely on other sources of finance for undertaking the project, bankers would approve the GTFS-based loan. The availability of additional funding that supplements the bank loan would provide an assurance of the financial viability of the underlying green project, especially when bankers are "unfamiliar with green financing projects" (APEC,

2014, p. 53). In line with this reasoning, RQ5 considers that bankers will approve GTFS-based loans that comprise a smaller proportion of the overall green project cost.

RQ4: Is the average GTFS-based loan amount applied for lower for approved loan applicants, compared to the rejected applicants?

H4: The mean GTFS-based loan amount applied for is lower for firms that obtained GTFS-based loan approvals than the mean GTFS-based loan amount requested by the rejected applicants.

RQ5: Is the average GTFS-based loan applied for/project cost ratio lower for approved applicants, compared to rejected applicants?

H5: The mean GTFS-based loan applied for/project cost ratio is lower for firms that obtained GTFS-based loan approvals than that for the rejected applicants.

Energy Categorisation and Source of Technology

Furthermore, bankers appear to be unfamiliar with the risks associated with green technology projects (APEC, 2014). The green energy production technologies within the solar and electricity categorisations tend to be better known than the other energy categorisations. Risk-averse bankers would prefer to approve loans to the energy categorisations that they are familiar with, ceteris paribus. As such, RQ6 and H6 consider whether the GTFS-based loan approvals favour projects within certain energy categorisations.

RQ6: Is the energy categorisation of applicants a significant factor in getting GFTS-based loans approvals?

H6: There is a significant difference between the approval rates for GTFS-based loans for the different energy categorisations.

The green technology that underlies the project proposed by the GTFS-based loan applicant could be totally home-grown in Malaysia, or be sourced or imported from abroad. Alternatively, it could be a mixture of home-grown and externally sourced technology. Risk-averse bankers would approve projects based on imported technology, perceiving imported technology as being more viable than home-grown innovations and combinations of home-grown and imported technology. The reliance on the imported label would mitigate the limited familiarity with the GTFS-based loan schemes (APEC, 2014). RQ7 and H7 are based on this reasoning.

RQ7: Is the source of the underlying technology a significant factor in getting GFTS-based loans approvals?
H7: There is a significant difference between the approval rates for GTFS-based loans for the different energy source categorisations.

Information Asymmetries
The interviews conducted in the earlier part of this study indicate that the GTFS-based loan applicants do not always provide all of the information needed by the bankers to make informed loan application decisions. Considering that bankers would require specific documentation for making informed loan approval judgements, this comment indicates an asymmetry between the information that the bankers appear to look for and the actual information supplied by the GTFS-based applicants, which is consistent with the general lack of key information regarding the GTFS (Tu et al., 2016). An examination of the reasons that were given for the rejection of the GTFS-based loan application would shed light on the factors that have led to low approval rates, and the subsequent low uptake, of GTFS-based loans. These reasons would be able to provide insights into the specific asymmetries, which can then be addressed through policies specifically tailored to overcome these asymmetries.

RQ8: What were the reasons given for the rejection of the GTFS loan application?

In comparison to the first seven questions, RQ8 is open ended and is applicable only to the rejected GTFS-based loan applications.

Sample and Analyses

All of the GTFS-based loan applicants who were located in the Klang Valley were contacted and asked if they were willing to participate in the survey. The researchers visited the respondents who were willing to participate and distributed questionnaires which asked for the following information:

1. Paid-up capital of the firm.
2. Years of experience in the proposed GFTS-based project.
3. Total cost of the GTFS-based project.

4. Amount of GTFS-based loan applied for.
5. Energy categorisation of the project (e.g. solar, electricity, wind, etc.).
6. Source of the underlying green technology (home-grown, imported or combined).
7. Reasons given by the bank for the rejection of the GTFS-based loan application (open ended and applicable to only the unsuccessful loan applicants).

The responses from 29 surveys (12 approved and 17 rejected GTFS-based loan applications), representing the set that was available at the time of this writing, were analysed. T-tests were used to compare the relevant means, and contingency tables were used to test the distribution of relevant categorisations, within the sample of applicants who were successful and unsuccessful in securing GTFS-based loans. The next section details the first (interviews) and second (surveys) phases of this study and provides summary statistics and an analysis of the empirical data obtained from the surveys.

Interviews and Surveys: Data, Analyses and Discussions

Tu et al. (2016) indicate that there is limited understanding of the GTFS in general, as such potential participants perhaps do not show interest in this scheme because they do not really know how it would benefit them. In addition, APEC (2014, p. 41) lists insufficient "institutional measures to meet informational.... needs" amongst the reasons for the low penetration of renewable energy projects in Malaysia. Thus, there is a general awareness that various information asymmetries exist, wherein policymakers and regulators' understanding regarding the benefits of green financing schemes does not seem to have been passed to potential beneficiaries amongst the wider business community. Yet, there is little in-depth research-based information regarding the specific nature of these asymmetries. Such detailed information would help policymakers to identify precise areas of information asymmetry, which would in turn help in the formulation of approaches to disseminate relevant information in a manner that targets and effectively overcomes this barrier to widespread adoption of green technologies.

Face-to-face interviews were undertaken to explore the information asymmetries indicated in Tu et al. (2016) and APEC (2014) in depth, specifically in relation to the GTFS. Information obtained from these interviews, together with that garnered from the literature review regarding the GTFS in Malaysia, was used to develop pertinent survey questions. The results from the analysis of the empirical survey data were then used to develop policy-level recommendations for boosting the GTFS uptake rates.

First Phase: Open-Ended Interviews with Regulators, Applicants and Bankers

The face-to-face interviews were conducted with the regulators (viz. GreenTech), GTFS-based loan applicants and bankers. The regulators were represented by senior managers of GreenTech. Two GTFS-based loan applicants and two banks that participated in the GTFS were contacted for interviews. The interviewees were selected randomly, from the pool of all GTFS-based loan applicants and participating banks, in the Klang Valley. The limited geographical reach was necessitated by resource constrains, particularly to contain travel, room and board costs.

The interviewees were asked to discuss their views on why the uptake of GTFS loans was low. The comments garnered from these interviews are summarised in Table 13.2. The key information asymmetries and information gaps identified via these interviews are described in Fig. 13.2.

Second Phase: Structured Interviews That Examine These Information Asymmetries in Detail

The GTFS-based loan applications were surveyed in the second phase. The surveyed material included both successful and unsuccessful applications. The relevant data and analysis are presented in Tables 13.2, 13.3, 13.4 and 13.5, within the corresponding subsections.

Subsection 3.3.2.1 presents the summary statistics, in Tables 13.3 (metric data) and 13.4 (t-tests). The pertinent analysis is provided in subsection 3.3.2.2, in Tables 13.5 (chi-square tests for categorical data) and 13.6 (summary of findings). The responses to the open-ended question, regarding supporting documentation, are summarised in subsection 3.3.2.3, in Fig. 13.3.

Table 13.2 Key findings from interviews

Regulators	Banks	Applicants
Twofold objective: projects must demonstrate 1. Scientific validity 2. Financial sustainability	Certification only ensures scientific validity	Loan applications were rejected, although a lot of time, effort and money were spent on getting green project certifications
Scientific validity—via certification (explicit) Financial viability (implicit, not explicitly stated)	CGC[a] takes up only 60% of non-payment risk. Banks take up the remaining 40%. As of 2013, banks take up 70% of the loan non-repayment risk. Therefore, banks look for evidence of financial viability of the projects to safeguard against non-payment risk	Actual criteria for loan approvals were not clear—some companies with low capitalisation and new companies with little project experience appear to be given loan approvals. Approved loan amounts fall below the applied for amounts
Certification board's role: *Only* ensures scientific validity, not financial viability[b]	Many applicants *do not* focus on documentation regarding the project's financial viability, which leads to low approval rates	There seems to be no point in applying for GTFS loans because there is very little chance of getting approvals despite fulfilling all of the *stated* criteria

[a]*CGC* Credit Guarantee Corporation
[b]Recently, the regulators decided to include a banker in the team that assesses projects submitted for certification. Nevertheless, the objective of the certification remains as an assessment of the scientific validity of the submitted projects, rather than financial viability. As such, the conclusions of this study are applicable to the current setting

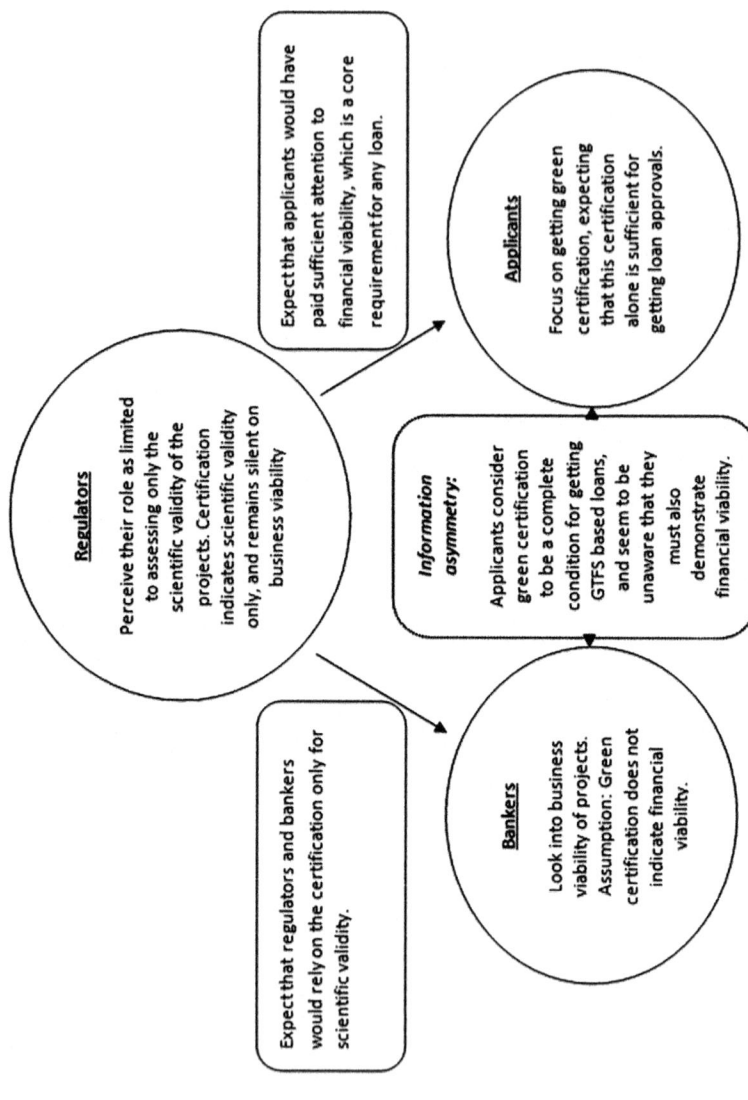

Fig. 13.2 Information asymmetries and expectations gaps

Summary Statistics

Table 13.3 Summary statistics: paid-up capital, prior experience, costs and amount of loan applied for in RM (millions) and as percentage of project cost

Summary	Paid-up capital (RM millions)			Years of project experience			Project costs (RM millions)			GTFS loan applied for (RM millions)			Loan/project cost (%)		
Statistics	O	A	R	O	A	R	O	A	R	O	A	R	O	A	R
Mean	19.83	10.76	26.24	9.84	10.23	9.59	19.61	28.26	13.50	14.31	19.90	10.69	81.55	80.31	82.35
Median	10.00	5.50	10.00	10.00	10.00	10.00	10.93	12.80	10.00	10.00	11.50	10.00	80.00	87.79	80.00
Minimum	0.20	0.20	1.00	0.5	0.5	3	2.00	2.75	2.00	2.00	3.14	2.00	47.14	47.14	62.50
Maximum	100.0	50.00	100.00	25	25	20	100.0	100.0	30.50	50.00	50.00	25.75	100	100	100
Count	29	12	17	28	11	17	29	12	17	28	11	17	28	11	17

Key: *O* overall, *A* approved, *R* rejected

Summary statistics: energy producer category, source of technology

Applicant's	Energy producer category				Technology source			
status	Solar	Electricity	Others	Total	Home-grown	Combined	Imported	Total
Overall	9	18	2	29	7	21	1	29
Approved	4	6	2	12	3	8	1	12
Rejected	5	12	0	17	4	13	0	17

Data Analysis

Table 13.4   Tests of significance: means of approved verses rejected applications

Means tested	t-statistics	p-values (one tailed)	Conclusions
Paid-up capital	1.7275**	0.0485	Approved loan applications have a significantly lower mean paid-up capital than the rejected applications
Project experience	0.2406	0.4063	No significant difference between the mean numbers of years of experience in the project between approved and rejected loan applications
Project costs	1.5530*	0.0732	Mean project costs are significantly higher for approved applications than rejected applications
Loan amount (applied for)	1.6594*	0.0615	Mean amount applied for in approved applications is significantly higher than the amount of GTFS loan applied for in the rejected applications
Loan/project cost	0.26973	0.39565	No significant difference between the mean percentage of the project cost covered by the approved and rejected loan applications

Note: Based on the means presented in Table 13.3

Significant at *10% or **5% levels

Table 13.5 χ^2-tests: impact of energy producer and technology source categorisations on GTFS loan approvals

Categorisation[a]	χ^2-statistics	p-values (one tailed)	Conclusions
Energy producers: *Solar, electricity, others*	3.3486	0.1874	No significant difference between the energy producer categorisations in the approved and rejected applications
Technology source: *Home-grown, combined, imported*	1.5163	0.4685	No significant difference in the technology source categorisations between approved and rejected applications

[a]Based on categorisations presented in Table 13.4

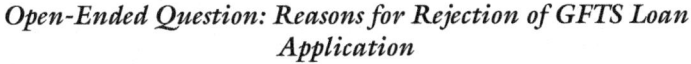

Open-Ended Question: Reasons for Rejection of GFTS Loan Application

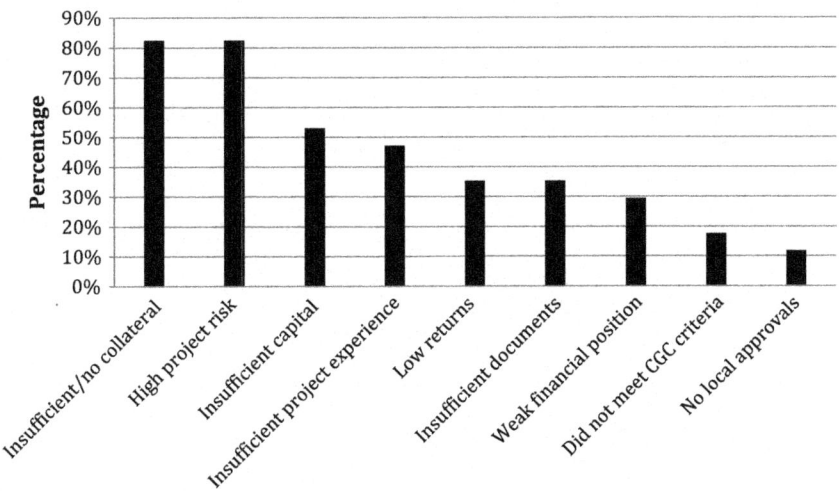

Fig. 13.3   Analysis of reasons for rejection of GTFS-based loan applications

DISCUSSION OF THE FINDINGS

In summary, this study finds that GTFS loan approval rates are higher for firms with lower paid-up capital, relatively higher project costs and relatively higher amounts of loan applied for. There is no significant difference between the accepted and rejected applications based on energy production categorisations, the mean number of years of experience in the project, source of technology and percentage of project cost covered by the loan amount applied for.

Therefore, factors such as lower paid-up capital, relatively higher project costs and relatively higher amounts of loans applied for, which could be indicative of higher risk, did not disfavour the loan applications. Several other factors that could mitigate the riskiness of the projects, such as relatively larger number of years of experience in the project, also do not appear to be favoured in the loan approval process. These empirical findings support the comments given by the GTFS loan applicants in the interviews, wherein they indicated that some companies with low capitalisation and new companies with little project experience appear to be given

Table 13.6 Summary of findings

RQ	Findings
1	**Approved loan applications have a significantly lower paid-up capital than the rejected applicants. *H1 is rejected*
2	There is no significant difference between the mean numbers of years of experience in the project between approved and rejected loan applications. *H2 is rejected*
3	*Mean project costs are significantly higher for approved applications than rejected applications. *H3 is rejected*
4	*The mean amount applied for in approvals is significantly higher than the amount of GTFS loan applied for in the rejected applications, at the 10% level of confidence. *H4 is rejected*
5	There is no significant difference between the mean percentage of the project cost covered by the loan applications in the approved and rejected samples. *H5 is rejected*
6	There is no significant difference between the energy production categorisations in the approved and rejected applications. *H6 is rejected*
7	There is no significant difference in the technology source (as local, imported and combined) between approved and rejected GTFS loans. *H7 is rejected*
8	The most common reasons for rejections were insufficient collateral, high project risk and insufficient capital

Levels of significance: *10%, **5%

loan approvals (Table 13.2). These empirical findings are generally counter-intuitive to the idea that banks tend to reject many GTFS-based loan applications because the green projects are perceived to be too risky, wherein risk mitigation factors are based on conventional perceptions, such as relatively higher paid-up capital and extensive project experience.

RQ8 examines the specific rationale of the banks, for the rejection of GTFS-based loan applications. The most common reasons were insufficient collateral and high project risk, with just over half of the rejections being due to insufficient capital, while about 40% of the rejections were due to insufficient project experience. These findings are consistent with the observations that the perceived riskiness of green projects (APEC, 2014) and support the idea that green developers (in this case, indicating GTFS loan applicants) "need to come up with a proper financing model with a proper cash flow to give confidence to the lenders when they evaluate the projects" (GTFBC, 2013b, p. 8). In essence, the reasons for rejections are generally risk related, and insufficient measures by applicants to mitigate this risk, such as providing sufficient collateral.

Thus, a general insight emerges wherein the GTFS-based loan applicants' perception of risk mitigation factors, such as higher paid-up capital and extensive years of project experience, does not align with the bankers' perception of risk mitigation factors, such as pertinent cash flow projections that would indicate the financial viability of the green project. Interestingly, the need to document the long-term financial viability of the GTFS-based green projects is not explicitly stated, emphasised or publicised in the current guidelines.

POLICY RECOMMENDATIONS AND FUTURE WORK

These findings are counter-intuitive to expectations that loan approvals would be based on the perceived riskiness of the applicants. Applicants with relatively lower paid-up capital, applications for relatively higher loan amounts and relatively higher project costs, that could be indicative of higher risk, appear to have a higher chance of approval. There is no significant difference in the percentage of the project cost covered, between the applicants who received approvals and the applications that were rejected.

Recommendation 1

Policymakers should highlight in informational brochures, briefings and workshops for potential GTFS-based loan applicants that sometimes loans for higher project costs may be approved. This move would encourage response from potential applicants who wish to undertake large, financially viable, projects. The disbursement of GTFS-based loans to large, financially viable projects will help to lift the GTFS-based uptake rates appreciably.

Recommendation 2

The informational material must also highlight the common reasons for GTFS-based loan application rejections, such as insufficient collateral, high project risk and insufficient expertise. Potential applicants could then become aware of these reasons and ensure that measures to reduce the risk for the lenders, such as sufficient collateral and robust cash flow studies, are in place prior to applying for the GTFS-based loans. Information regarding documentation that demonstrates the project's financial viability should be provided in information sessions on the GTFS. This information will guide prospective GTFS-based loan applicants in the preparation of documentation that will showcase the financial viability of the project, thus improving chances of getting approvals. As a result, the approval rates for GTFS-based loans will improve, lifting the uptake rates for these loans.

Recommendation 3

The banks could also prepare special instructional booklets that provide guidelines for GTFS-based loan applications, indicating what the bankers look for in approving GTFS-based loans. Such actions, by the bankers, would help to bridge the missing link in the information set currently available to GTFS-based loan applicants, wherein the applicants seem to get the impression that just fulfilling GreenTech Malaysia's criteria for the green technology certification is sufficient for obtaining the GTFS-based loan approvals. Complementing the policymakers' efforts, the bankers must explicitly declare that the demonstration of financial viability, in addition to green certification, is important for obtaining GTFS-based loan approvals. These instructional booklets should be handed to each GTFS-based loan applicant together with the GTFS-based loan application forms, and should provide clear examples of key reports, such as projected cash flow statements, that will help to demonstrate the financial viability that the banks are looking for.

The information referred to recommendations 1–3 can be released in the instructional booklets, videos and talks to SME groups that are eligible for the GTFS-based loans, by both the policymakers and banks.

Future studies could undertake interviews and surveys of current and future loan applicants and bank officers involved in the GTFS approval process, to get a deeper insight into the extent to which loan applicants and loan approvers consider the various factors identified in research questions 1–7.

LIMITATIONS

This research incorporates the responses from an early phase of a larger ongoing study. When additional responses are incorporated, as the study progresses, some of the findings might become modified. However, the findings of this study are consistent with the indications of a gap between the loan applicants' perceptions and the loan approvers' expectations regarding the key factors that characterise successful loan applications. As such, this limitation is unlikely to impact the final conclusion of the overall study significantly.

Conclusions

This study examines the reasons for low uptake of GTFS loans in Malaysia, which are meant to drive the greening of the Malaysian economy. There appears to be an expectations gap between the loan applicants and loan approvers regarding the key factors that are important for GTFS-based loan approvals. In particular information asymmetries, wherein bankers look for the financial viability of the proposed green projects while the GTFS-based applicants focus on the scientific validly, without being aware of the need to document the financial viability as well, have led to the low GTFS-based loan approvals. The dissemination of the findings of this study to GTFS loan applicants and bankers is recommended, to overcome this expectations gap between the bankers and the GTFS-based loan applicants. The education of potential GTFS-based applicants through seminars, talks to relevant associations, brochures, books and videos would be able to complement and help to boost the effectiveness of the market-based incentives provided by the GTFS.

While political will is viewed as a key independent variable in addressing global pollution and environmental degradation, additional confounding variables must also be considered to boost the effectiveness of policies designed to combat climate change. This study offers that even when there is strong political will, as evidenced by the billions of ringgit (approximately a billion US dollars) allocated in the Malaysian national budget for the GTFS initiative, moderating factors such as information asymmetries could dampen the effectiveness of the policies that address global warming. As such, policies in this area must explicitly incorporate information dissemination mechanisms, such as seminars, workshops and public talks, to educate relevant stakeholders and potential participants regarding key factors that will make the policies work at maximum potential (such as the need to address the financial sustainability of the green projects). Such information dissemination mechanisms will help policymakers to successfully combat global warming, which will in turn help to contain climate change-related health impacts on the general populace. These educational instruments would also be useful for policymakers in other emerging and developing economies, wherein relatively low financial literacy rates could dampen the effectiveness of policies designed to develop environmentally and financially sustainable green economies.

Acknowledgement The authors thank the Malaysian Ministry of Science, Technology and Innovation for a large E-Science grant that financed this research.

NOTE

1. APEC (2014) considers that RM10 million is equivalent to approximately three million US dollars, reflecting the general currency exchange rates at the time that these policies were declared. Based on this exchange rate, the budgeted allocation of RM3.5 billion, for the Malaysian green technology scheme, amounts to approximately one billion US dollars.

REFERENCES

APEC. (2014). *Peer review on low carbon energy policies in Malaysia.* Final report, Asia-Pacific Economic Cooperation. https://www.wto.org/english/tratop_e/tpr_e/s292_e.pdf. Accessed 13 Mar 2017.

Barbi, F., Ferreira, L., & Guo, S. (2016). Climate change challenges and China's response: Mitigation and governance. *Journal of Chinese Governance, 1,* 324–339.

Bradford, J., & Fraser, E. D. G. (2008). Local authorities, climate change and small and medium enterprises: Identifying effective policy instruments to reduce energy use and carbon emissions. *Corporate Social Responsibility and Environmental Management, 15,* 156–172.

Clement, K., & Hansen, M. (2003). Financial incentives to improve environmental performance: A review of Nordic public sector support for SMEs European environment. *The Journal of European Environmental Policy, 13,* 34–47.

Crespi, F., Ghisetti, C., & Quatraro, F. (2015). *Taxonomy of implemented policy instruments to foster the production of green technologies and improve environmental and economic performance.* Working paper no. 90, WWWforEurope. Retrieved March 6, 2017, from http://www.foreurope.eu/fileadmin/documents/pdf/Workingpapers/WWWforEurope_WPS_no090_MS216_partI.pdf

De Serres, A., Murtin, F., & Nicoletti, G. (2010). *A framework for assessing green growth policies.* OECD economics department working paper series, Working paper no. 774. OECD Publishing, Paris.

EEA. (2005). *Market-based instruments for environmental policy in Europe.* Technical report no 8/2005, European Environment Agency, Copenhagen, Denmark.

Eskelson, D., Antal, I., Fidanza, B., Leclercq, M., & Rosca, A. (2016). *Green business models and the green finance landscape.* GREEN-WIN project 642018 RIA. Retrieved March 7, 2017, from http://green-win-project.eu/sites/default/files/D4.1-green_investment_landscape-V2.pdf

GTFBC. (2013a, January). Green financing, the new frontier. Address of the Deputy Governor, Malaysian Central Bank. Green technology financing bankers' conference 2012. *The Banker's Journal Malaysia, 140*, 3–6.

GTFBC. (2013b, January). Green financing, the new frontier. Address of the Secretary General for the Ministry of Energy, Green Technology and Water, Malaysia. Green technology financing bankers' conference 2012. *The Banker's Journal Malaysia, 140*, 6–8.

Hanson, M. (2013). Green ergonomics. Challenges and opportunities. *Ergonomics, 56*, 399–408.

IFC. (2013). *Mobilizing public and private funds for inclusive green growth investment in developing countries*. A stocktaking report prepared for the G20 development working group, International Finance Corporation: World Bank Group. Retrieved March 7, 2017, from http://www.ifc.org/wps/wcm/connect/topics_ext_content/ifc_external_corporate_site/cb_home/publications/publication_mobilizinggreeninvestments/ http://www-iam.nies.go.jp/aim/event_meeting/2015_cop21_japan2/file/03_malaysia.pdf

Kelhart, M. D. (2008). The sound of silence at the Environmental protection agency. *Bioscience, 58*, 924.

KeTTHA. (2011). *Green technology future opportunities in Malaysia*. Vienna Spring Dialogue 2011, Malaysian Green Technology Corporation.

KeTTHA. (2015). *Implementation of green technology policy in Malaysia*. Ministry of Energy, Green Technology & Water Malaysia. Retrieved March 13, 2017, from http://www-iam.nies.go.jp/aim/event_meeting/2015_cop21_japan2/file/03_malaysia.pdf

Louv, R. (2005). *Last child in the woods: Saving our children from nature-deficit disorder*. Chapel Hill, NJ: Algonquin Books.

Mir, D. F., & Feitelson, E. (2007). Factors affecting environmental behavior in micro-enterprises: Laundry and motor vehicle repair firms in Jerusalem. *International Small Business Journal, 25*, 383–415.

Pimenova, P., & Van der Vorst, R. (2004). The role of support programmes and policies in improving SMEs environmental performance in developed and transition economies. *Journal of Cleaner Production, 12*, 549–559.

Samudhram, A., Siew, E.-G., Sinnakkannu, J., & Yeow, H. P. (2016). Towards a new paradigm: Activity level balanced sustainability reporting. *Applied Ergonomics, 57*, 94–104.

Sneed, A. (2017). *Trump day 1: Global warming's fate*. Retrieved March 4, 2017, from https://www.scientificamerican.com/article/trump-day-1-global-warmings-fate/

Stavins, R. N. (2003). Experience with market-based environmental policy instruments handbook of environmental economics. In K. G. Mäler & J. R. Vincent (Eds.), *Handbook of environmental economics* (Vol. 1, pp. 355–435). Amsterdam, The Netherlands: Elsevier.

Sterling, S. (2001). *Sustainable education: Re-visioning learning and change.* Bristol, UK: Green Books.

Tenth Malaysia Plan. (2010). *The tenth Malaysia plan: 2011–2015.* Putrajaya, Malaysia: The Economic Planning Unit, Prime Minister's Department.

Thatcher, A. (2013). Green ergonomics: Definition and scope. *Ergonomics, 56,* 389–398.

Thatcher, A., & Yeow, P. H. P. (2016). A sustainable system of systems approach: A new HFE paradigm. *Ergonomics, 59,* 167–178.

Tu, F., Ho, C. S., Chau, L. W., & Yu, X. (2016). *Promoting urban sustainability through green technology in Malaysia.* MIT-UTM Malaysia Sustainable Cities Program. Retrieved March 11, 2017, from https://malaysiacities.mit.edu/paperTu

United Nations. (2014). *United Nations environment assembly fact sheet.* Retrieved March 5, 2017, from http://www.unepfi.org/fileadmin/events/2014/nairobi/unea_symposium_factsheet.pdf

WEF. (2013). *The Green Investment report the ways and means to unlock private finance for green growth.* A report of the green growth action alliance, World Economic Forum. Retrieved March 7, 2017, from http://www3.weforum.org/docs/WEF_GreenInvestment_Report_2013.pdf

Wilson, E. O. (1984). *Biophilia.* Cambridge, MA: Harvard University Press.

WTO. (2014). *Trade policy review: Malaysia.* World Trade Organisation.

CHAPTER 14

Lives We Have Reason to Value

Dave Moore

INTRODUCTION

The twinned aims of HFE are generally explained to be enhancing system performance whilst also optimising human wellbeing. This simple explanation serves us well, but the term 'wellbeing' will mean different things in different contexts. For the purposes of this chapter, I will be treating 'wellbeing' in accordance with the position International Labour Organization (ILO) promoted in the 1980s: 'work concerns man as a whole: not just muscle and nerves, but intelligence, capabilities, feelings and aspirations' (Clerc, 1985, p. 24). Of these I will be focussing most on aspirations. We as HFE professionals are still in business presumably because clients/funders recognise the merit of these aims, and we get good enough results in both areas for them to come back and ask for more. How we measure success in these aims is a little fuzzy at times, but overall there is agreement when dealing with industry, the military and consumers of specific products and services.

In these settings we have frameworks of pre-existing metrics to refer to, particularly in system performance, and even where a triple (Elkington,

D. Moore (✉)
Department of Built Environment Engineering, Faculty of Design and Creative Technologies, Auckland University of Technology, Auckland, New Zealand

© The Author(s) 2018
A. Thatcher, P. H. P. Yeow (eds.), *Ergonomics and Human Factors for a Sustainable Future*, https://doi.org/10.1007/978-981-10-8072-2_14

355

2004)—economic, social and environmental—or quadruple bottom line reporting system is in use. Across New Zealand the fourth element is predominantly understood to be culture, as in the example of the New Zealand Bus Co (http://www.nzbus.co.nz/sustainability).

These reporting systems whilst appearing forward-thinking may not be individually well-thought-through enough. Some researchers (Bebbington, Higgins, & Frame, 2009) found New Zealand organisations moving to 'sustainability reporting' simply as an accepted part of a differentiation strategy, but without rational explanations for the detail of their schemes. For corporate contributions to sustainable futures, 'off the shelf' reporting packages are inadequate; organisations need specific measures linked to specific aims.

The bulk of practitioner work gets focussed for us by the client. They have us investigate within limited sub-systems, at best at the Child-Sibling level (Wilson, 2014), and only very rarely with meaningful reference to a Parent System. Cycles are also generally short enough for us to evaluate our success within the timescale of a project, or at least within the tenure of whoever is paying for our services. This justifies the investment in HFE, and gives the scope for iterative improvements (the next few contracts). Slow-onset occupational health issues are notable exceptions to this having longer cycle times—which in part explains why we as societies do such a poor job regarding health as opposed to safety. Our effectiveness cannot be proven quickly enough, return on investment is unclear and so contracts to act in this area are fewer.

I see two fundamental departures from HFE business-as-usual in the proposal that we evolve to engage proactively in designing for a sustainable future. First, the reference population is not a workplace team, army squad or self-selected body of consumers that are all in the market for a small ride-on lawnmower, for example. The unit of interest is human society as a whole, and the ecology of a biosphere of which it is an integral part. So there is an issue of scale and the cutting across of multiple physical, administrative, linguistic, professional, social and cultural boundaries that this species has purposefully constructed. Generally in the populations we are paid to engage with, a workable degree of alignment can also be assumed. But at a regional or national level—whilst there may be broad agreement about what success looks like—the ideas on *how* success will be achieved can differ greatly dependent, not least, upon political ideology. Interim measures of success based on milestones will therefore differ too, and so acceptable, short-term success measures for improving system performance and human wellbeing will not be the same. We need new approaches to measuring success for this kind of work. As Wilkin (2010) suggested, we also need to acknowledge the professional values we hold and specify these—our work is not value-free.

The second fundamental departure from business-as-usual that I see is the need to work with people we haven't worked with before. Multidisciplinary and interdisciplinary teamwork is familiar, and some would argue essential, to a systems approach. Given the need for new approaches, I would suggest though, that simply bringing together the disciplines is not enough for the challenges discussed in this book. Transdisciplinary exercises open the possibility for new ways of visualising, relating and acting. For example, it is predicted that here in New Zealand climate change will lead to a greater number of extreme weather events. The economy is heavily dependent upon the export of primary produce, and seismic activity is constant—periodically disastrous—as in Christchurch and more recently Kaikoura. The natural environment therefore has a very direct bearing on our ability in this country to build capital in its various forms. Contextual factors (Tappin, Vitalis, & Bentley, 2016) are those influences that are highly significant but over which the domestic sector has no control—such as the local weather and overseas market fluctuations. The latter were also, in turn, being influenced by the weather globally.

So climate change is predicted to increase the volatility of growing cycle conditions, with multiple knock-on impacts. One of these impacts known already (Tappin, Bentley, & Vitalis, 2008) to HFE professionals could well be an increase in musculoskeletal disorders amongst people working in the meat processing sector. In drought conditions farmers who graze animals on pasture grass will send more stock to the abattoirs to avoid having to buy feed. The surge in stock arriving has been shown by Tappin et al. (2008) to lead to unsustainable work practices including very early starts as part of extended working hours. The links between such work organisational pressures and MSDs are well established.

The work reported by Tappin et al. (2016) used a participatory approach at a whole-of-industry level in an attempt to understand how the impact of contextual factors such as undesirable weather patterns could be mitigated at sector, company, and plant level. This was the first time that such an approach had been employed rigorously by HFE professionals in New Zealand. However, it looked at just one aspect of the many issues that the ripple effects of climate change will raise. If we were to look holistically at the impacts of climate change in an entire region, with the full diversity of enterprises—as opposed to just following one line of impact down to one type of cost (MSD amongst meat workers)—then the variety of specialisms needed around the table, and potential new relationships to explore, increases exponentially.

This chapter looks firstly at the questions of alignment to shared goals and how success in designing for a sustainable future may be measured. It then uses the examples of two more recent transdisciplinary exercises— one small and industry-based (Gaskin, Edwin, Moore, & Guard, 2015) and a far larger regional adaptive governance project, with terms of reference extending for the next 100 years (Moore & Barnard, 2012).

MEASURING SYSTEM PERFORMANCE WHERE THE SYSTEM IS IMMENSE

This section looks at potential shared goals in transdisciplinary work aimed at designing for a more sustainable future.

The Need for Shared Goals

In his seminal work on the history of HFE, David Meister (1999, p. 143) assumed the system concept to be the foundation of HFE. He noted that whilst we didn't invent it, nor always work day-to-day with that wider perspective, it provides our conceptual structure. His definition of the system aligns with that of most others in that it includes a specified goal; a joint purpose. For a practitioner engaged in industry work, it is an important standpoint. Without boundaries or agreed measures of success, we may never get all the data collected, the final report accepted nor our consultancy invoice settled.

This book addresses an altogether higher level of question though. As a species, what is our agreed aim regarding the ideal state of our physical environment and our relationship with it? Many would argue that universally we want a better world for our children, and so we want to improve, not just sustain. So there is a balance to be drawn between returning the planet to a pristine pre-human condition, and destroying, as we have been, the systems that support us for reasons that make sense only in the short-term.

Social and Economic Goals and Decent Work

The physical environment is perhaps the most straightforward piece in the sustainability jigsaw. More problematic is socioeconomic sustainability, and the setting of aspirational targets. Any global initiative that seeks as a starting point, universal common goals, in any degree of detail, will struggle.

Not only due to the obvious differences in cultural and religious/spiritual values, but also because of disagreement about what desirable short to mid-term political and commercial interventions would look like. Simple long-term aims that benefit us are more palatable, and understandable to voters and employees, than immediate measures that may incur personal cost. It is now more than a year since the UK voted in a referendum to leave the EU, but the vast interactive network of implications for the employment market alone is still emerging into public consciousness. Doubtless the vast majority want a healthy, prosperous country; but the definitions of healthy and prosperous will differ, as do the ideas about how to achieve that state.

The ILO Decent Work campaign has established a very modest set of goals as a result, too modest in fact for any practical use in any given country. The over-simplification renders the list of little use.

Equality of Opportunity and Social Mobility

The most difficult aspect of all, I suggest, regards sustainability of standards of living, or ideally, upward social mobility. If we all want better for our children than we ourselves had, then unless our children are also more talented, harder working, and luckier than us then that cannot be expected. The other alternative is that we, as nations, choose to (or continue to choose to) suppress peer competition actively through social and economic mechanisms that support hereditary privilege. The argument against restricting the development of individual human capital though— their health, skills, experiences and connectedness—is an old one that more recently has been backed by longitudinal research.

Thomas Paine, in his highly influential pamphlet *Common Sense* (1776) argued for an American independence from Britain and one without monarchy. 'One of the strongest natural proofs of the folly of hereditary right in kings, is, that nature disapproves it' (p. 40). Even more controversially he went beyond this rejection of absolute power to also condemn the inherited social advantage of those circling within the wider orbit of the throne.

> Most wise men, in their private sentiments, have ever treated hereditary right with contempt; yet it is one of those evils, which when once established is not easily removed; many submit from fear, others from superstition, and the more powerful part shares with the king the plunder of the rest. (p. 41)

Paine saw a limiting effect on overall human development through the disadvantage by design of large sections of the population describing it as 'an insult and an imposition on posterity' (p. 40). It has been suggested that the consequences of such impositions will also be unaffordable in coming years, even if democracies saw them as desirable and morally defensible. Citing multiple measures as examples, Pickett and Wilkinson (2010) argue that the indirect costs of inequality across societies are now shown to be unsustainable. Their substantial work covering 20 of the wealthy nations shows that these costs are borne not just by the poor but also the rich, and this message has subsequently prompted individual countries to take active steps towards reducing social inequality ('Closing the Gap').

> ...politics should now be about social relations and how we can develop harmonious and sustainable societies (The Equality Trust).

Human Development Indices as Multiple and Evolving Goals

The assumption that a single shared vision, a joint purpose, has been attained (or is at least attainable) where it is actually not will operate as a barrier to development. At a company level, the *Our Vision* statement in its frame on the Board Room wall is not something that any sane employee would die for, and it is understood that most staff would not be in work tomorrow if they won the National Lottery. Personal and corporate goals co-exist and compromise is reached. This is a very familiar territory for the HFE community. Beyond even work-life balance though, at the outer (society/biosphere) layers of our concentric rings models, how much agreement and alignment on ultimate aims can we expect as a species?

Even where there is a shared sense of a final goal, such as your country winning a world cup, ideas on the intermediate goals and related strategies will not be shared. So the immediate goals and milestones remain contentious.

In the context of designing for a sustainable future agreeing on an end point goal, say 100 years into the future is probably the least contentious. In the human development field the problems arise in agreeing on the intermediary gauges of success, and these have been a long-term focus for researchers such as Martha Nussbaum (2011). Inevitably though perhaps, given the complexity of the task, critics such as Alkire (2005) have still pointed to a perceived lack of cross-cultural influence in the goals that have evolved. Others have taken a different approach altogether. Amartya Sen, one of those behind the influential Human Development Index (http://hdr.undp.org/en) instead sees the aim of human development as enhancing individual effective freedom, the ability of people to live the lives 'that they have reason to value'.

On the face of it this sounds uncontroversial, but acceptance becomes a bigger ask when we think through the implications. We are being asked to help people achieve change, whether or not we personally agree with their values, or the worldview that these belong to. This points to an altogether different approach with less to do with what we want to achieve as a species, and more to do with how we collectively make decisions— irrespective of the outcomes. The process and refinement of democracy are ultimately more important than any immediate outcomes.

HFE PROFESSIONALS IN TRANSDISCIPLINARY EXERCISES

Project 1. Multiple Value Cost-Benefit Studies: An Example from the Fishing Sector

In the 2011–12 year a New Zealand commercial fishing company experienced a new-staff turnover of more than 50 percent. This cost the company an estimated NZ\$3.5 million. Prompted in part by this experience and the high level of injuries, a series of studies were commissioned. The most pressing question posed initially was that of cost/benefit. How much do we get back for putting a certain amount (of effort) in? The intermediary questions related to how success was measured closer to the tasks that earned the profits on board the vessels. So, the study moved from simply how the chief financial officer alone might define success in the annual report to how all the people in the various jobs might define it. If changes needed to be made it mattered how intervention success might be gauged and who potentially would walk away within months of being trained.

Figures 14.1 and 14.2 (multiple cost-benefit analysis [MCBA]) show the various values described by the crew and a matrix of these values as prioritised by the different crew jobs. The Skipper and First Mate run the vessel and will probably be at sea all their lives. This is their career and pay is not the primary concern, nor lifestyle that they are able to achieve on land between voyages. Many spend their free time also out at sea, catching

What success looks like

The MCBA firstly asked how success would be measured by various crew members

Fig. 14.1 What success on board looks like—elements identified by the different levels of staff on a commercial fishing vessel

fish for fun. The Deckhands and Factory Workers that bring in and process the fish below decks are the lowest paid jobs and most likely these workers will be part of the 50 percent churn. It is a job and about getting the most money without sacrificing too much shore life.

This was a small study (Gaskin et al., 2015) and the detailed findings are less important than the principle. Building whatever understanding we can about what will allow an intervention to be embraced or rejected by the different groups on board is clearly essential. Ideally we do this before the interventions are designed, and certainly before the implementation design is finalised. Do the people on this vessel share an overriding joint purpose? Theoretically yes (company survival), in practice I would say no (as demonstrated in Fig. 14.2). System efficiency only matters at lower levels because it gets them home quicker otherwise they derive no other direct benefit, unlike the people running the operation. If this company fails, then the fish and market demand will remain and another company will take over, and it too will need staff at lower levels. From an HFE perspective the purposes surely are multiple, as reflected in the matrix. They are also evolving because the personal circumstances of each individual are also changing (e.g. the arrival of a new baby, relationship problems on shore, a wedding to find money for, etc.).

New Zealand is a legally bi-cultural, multicultural migrant nation. Here in Auckland 60 percent of the workforce were born overseas. The crew in the study above did not have such a wide variety of ethnicities as commonly

SKIPPER & 1ST MATE Shift hours: 12 on/off	X	X	X	X	X	X	X	X				X			X
2nd Mate - Medic Shift hours: 12 on/off	X			X	X	X				X	X	X			
ENGINEER Shift hours: 12 on/off				X		X	X		X		X	X			
FACTORY MANAGER & SUPERVISOR Shift hours: 12 on/off	X				X	X			X		X				
DECKHANDS Shift hours: 6 on/off							X	X	X		X	X			X
FACTORY WORKER Shift hours: 6 on/off						X					X	X		X	
TOTAL	1	3	1	1	2	3	2	3	2	1	5	4	1	1	2

Fig. 14.2 Matrix of the values identified as being of highest personal priority by each crew level

found in this country, but the principle and method described here potentially accommodate the values of sub-populations that would not be guessed at by the company otherwise.

Project 2. Weaving the Korowai: East Cape of New Zealand

Introduction
The origins of this project are reported in Moore and Barnard (2012), a rare exercise in that it integrated physical and social impact studies in one approach. Since this work, a 100-year memorandum of understanding has been signed by the New Zealand government indicating a commitment to work with the Regional Council, Iwi (tribe), and other bodies to restore the health of the Waiapu river catchment damaged substantially by erosion since European settlement and the ensuing deforestation. The author is part of an International Advisory Board formed by SCION (Forest Research Inc. NZ), a Crown Research Institute. A pivotal learning point from the early work in the Waiapu catchment area was that decisions that made sense at the time from a dominant western European worldview had led directly and indirectly to the damage that needed repairing; and that the full set of solutions required would be found in a more broadly informed inclusive worldview.

Worldviews and Timeframes

> He said that what men do not understand is that what the dead have quit is itself no world but is also only the picture of the world in men's hearts. He said that the world cannot be quit for it is eternal in whatever form as are all things within it (McCarthy, 1994, p. 413).

The relationship between the living, the dead and the land is a common area for complications and clashes in large studies and projects, and it goes far beyond day-to-day issues on-site such as disturbing human remains. Funder reporting timeframes and corporate memories have become shorter with time, but increasing awareness of the importance of consulting properly with all stakeholders and knowledge holders demands the opposite. Exercises in New Zealand with Iwi, such as the work in the Waiapu catchment involving land use, are now commonly looking a century ahead or longer. Designing for end-users whose parents are not yet conceived is an unusual brief by contemporary western standards, but for these projects and HFE professionals in transdisciplinary teams concerned with sustainability, it is highly appropriate. Our normal western HFE consultation process and time allocations will be inadequate for this requirement. A major part of transdisciplinary work is therefore building understanding about how good decisions are made in all the subpopulations within the system and designing projects that facilitate these.

Interventions and the Wider System

There has been a thin and generally poorly funded thread of activity in HFE that has sought local solutions for local problems, very much in line with the approach of Sen (1999). Historically these solutions have been at a micro level, considering tools and tasks, often in subsistence settings (O'Neill & Moore, 2017), and in some cases with apparent disregard or naivety regarding the wider aspirations and challenges of the peoples concerned. The small scale of the projects can be blamed for this. But certainly, in New Zealand at least, there has also been a lack of interest in systems thinking generally—and/or conflicting political wills in projects at a transdisciplinary scale (Jollands & Harmsworth, 2007). To ignore this wider picture requires the abandoning of a holistic approach and hence also the '… and Wellbeing' twin aims of HFE.

The 'dust belt' disasters in the USA between the World Wars led to massive social dislocation, with predictable consequences in the receiving environment. Unsustainable farming methods on the former open grasslands in Oklahoma and adjacent states put whole districts of families on the road to California, where supply of labour outstripping demand prompted abuses. The experiences set out in the John Steinbeck 1939 classic *Grapes of Wrath* capture the nature and scale of the indirect costs of unsustainability, triggering—in this case—mass migration from the Dustbowl regions to the west coast. Social polarisation and conflict in the receiving environment (farms of California) being examples of these knock-on costs.

> If you who own the things people must have could understand this, you might preserve yourself. If you could separate causes from results, if you could know that Paine, Marx, Jefferson, Lenin were results, not causes, you might survive. But that you cannot know. For the quality of owning freezes you forever into "I", and cuts you off forever from the "we". (Chap. 14)

A superior but more expensive hand tool only improves the life of the small farmer or contractor if it means they can earn the same amount more quickly, with less effort or by being able to employ from a wider pool of people. If everyone has one and market forces dictate, then tonnage rates will adjust, and they will almost certainly earn the same as they did, but now with the extra burden of having to produce more. All those contractors or small farmers will also have to have more capital tied up in tools. The system(s) and limitations within which rural people operate cannot be ignored if HFE is to live up to its billing of embracing a systems approach. Where the micro-economic impact goes unreported, a reductionist fascination with load carriage biomechanics does HFE a disservice.

Building Capital in All the Forms That They Value

At present (mid-2017) the Weaving the Korowai project is still at the level of investigating the aspirational economic mix and resultant range of jobs and employment patterns. A significant element in this is the degree of financial seeding capital that will be required of prospective owners or people seeking training, in a region where the average household income is NZ$11 000 a year, placing them in the poorest 1 percent of people in the country (Statistics New Zealand, 2016).

The Sustainable Livelihoods Framework (SLF) (Ostrom, 2009) is one of a number of tools that can be used to explore the structure and strength of communities. This social anthropology device was used in the formative stages of the Weaving the Korowai project to facilitate discussions on both social impact and social aspiration. Similar to the multiple value (cost) benefit analysis (MVCA aka MCBA) approach described earlier, it allowed the researchers to learn more about how security is built by specific groups and individuals and not just in the form of savings in the bank, income from jobs or how much would be realised if they cashed up sold the house. In addition to financial capital, the analysis captured:

- Natural capital (the resources of their environment including fish stocks, healthy soil, clean air)
- Physical capital (tools and equipment including the infrastructure to get produce to market)
- Social capital (family, networks, friendships, support groups, trust, reciprocal understanding and precedents)
- Human capital (skills including literacy and communication, awareness including specific cultural knowledge, labour potential, spiritual strength, being respected, holding self-respect)

In specific times or situations the forms of capital we have reason to value most will shift. The predominant patterns of value may also differ between those who associate most strongly with European cultures and those who align more with indigenous worldviews. For example, people who introduce themselves as being *of* a particular piece of land, rather than *from* it. Formally we see indigenous knowledge, values and worldviews being increasingly sought, but far less often incorporated in policy and programme design. Sustainability for such people is surely more favoured where these proven models of capital are understood and, where appropriate for their future, reinforced.

THE HFE PROFESSIONAL'S ROLE

In the transdisciplinary Weaving the Korowai project, I have been aware of being introduced by my capabilities—rather than any familiar professional descriptor. Unsettling at first for some (including me), it is probably common, and probably essential. Unlike multi- or interdisciplinary teams, the aim is for actors to not be limited by the stances of their original backgrounds. A distinction is this freedom to reconstruct or develop the required theoretical frameworks from new. From discussions with others in this team, I learned that the question 'what actually can I contribute here?' was one most on our minds. It wasn't only me struggling to understand what aspects of my training and experience could be applied but also how and when.

Erik Hollnagel (2017) argues that HFE '... as a practical solution should be based on a small number of simple principles with a strong empirical foundation' (p. 45). He proposes five simple principles—the Nitty Gritty (of HFE)—which I have found to provide a useful framework in assessing and categorising my work.

1. His first principle concerns accepting work As Done (not As Imagined), with the attendant trade-offs and workarounds. In the context of an entire region and its people, for me the relevance here is in building an understanding about what is already working and why. The East Cape area has a history of interventions with disproportionate side effects that have resulted in a net loss. Approaches such as Appreciative Enquiry that bring together the system into one room to catalogue assets and chart existing successes are a crucial starting point, especially in communities with little room to absorb loss.

2. Principle two, his Minimum Action rule, aligns closely with the first principle. An observation from this work is that on a day-to-day basis the people are looking for a better version of what they already

have and trust. Minimising action in the form of reduced laws and regulations is also popular from a political perspective at present. Ministerial directives in New Zealand require that anyone proposing an addition to our rules also suggests how to repeal two others at the same time.

3. Form following function (and vice versa) is described as the third principle and is perhaps harder to immediately extrapolate to a community level. The original use of the expression 'Form Follows Function (FFF)' emerged in the 1930s to describe the aesthetic design approach of the Modernist architects. Groups such as Bauhaus in Germany sought to eschew classic practices, for example, the use of classic Greek and Roman styling and the practice of hiding the more mundane or ostensibly unattractive parts of a building. Joints, hydrants and elevators were to be celebrated—not covered up with decoration or camouflaged as nymphs.

Therefore, in a simple HFE design context, the utility of FFF is clear; for example, airports can helpfully display where everything is, and what everything does for first-time users by use of open space, good sight lines and consistent placement and circulation.

The vice versa rider refers to the affordances quality. The designer can make a wall mounted heater that is very clearly a wall mounted heater, but if by dint of location and height it also affords a useful seat, then it will be used as one. One form but two functions, one of which may lead to damage or injury through (in hindsight) foreseeable misuse by people forced to queue alongside it—as is common in airports.

At a transdisciplinary complex system level, the principle still applies. Intuitive design with faithful affordances works because it draws upon the existing experiential language of the population. Examples might include communication design, the library of references used when describing success or failure and decision-making processes. In the case of the latter, Form Following Function could be clearly recognised in a comparison of two different ways of applying adaptive governance to resource allocation. Where the function is universal participation, the wider range of methods for conducting informed dialogues (including the pre-requisite addressing of information asymmetry) will be expressed.

As for the broken airport heaters—honest, iterative development through simulation, trialling and so on will minimise unwanted affordances.

4. What You Look For Is What You See (WYLFIWYS) is principle four. Of the five principles this is a sign that I suspect should be nailed to the wall in the meeting room for any transdisciplinary endeavour. It applies in several ways. Individually, we arrive with preconceptions, plus years of single profession immersion—guaranteeing misunderstandings and probably prejudices. Most professionals don't feel limited by their 'original professional stances'—but elevated. Transdisciplinary processes involve systematic questioning which can be disturbing for some who may feel that their professional worth is under threat. Collectively I would suspect WYLFIWYS applies in transdisciplinary teams as it would for any other group. The usual biases will be at play when one option involves unavoidable conflict/cost/extra work/discomfort and so on and another option doesn't.

5. The fifth and final simple principle that Hollnagel (2017) suggests is Show What's Going On. At face value this would seem to be the most universally understandable of the HFE principles to bring to a transdisciplinary team. However, in long-term projects, this may not be true or will be very difficult. In short-cycle exercises in industry, results of interventions can be almost immediate, so the sense of action/reaction is relatively quick and easy to establish. In the transdisciplinary macro-project world, the most significant interventions can be changes in national policy or radical changes in land use of tens of thousands of hectares involving multiple private, corporate and tribal owners scattered around the world. The task of showing what's going on is no less important but requires, at a minimum, the breaking down of action and reaction into sub-cycles and a skilled communication team. The challenges are akin to control-display issues for large vessels that respond slowly, but with potential lags of years not minutes.

A Concluding Reflection

Lange-Morales, Thatcher and García-Acosta (2014) have proposed six values for the HFE discipline: (1) respect for human rights, (2) respect for the Earth, (3) appreciation of complexity, (4) respect for diversity, (5) respect for transparency and openness and (6) respect for ethical decision-making. The experience of the author reported in this chapter supports a seventh value: 'HFE respects democracy, as the best way we have so far found to assist people live the lives that they have reason to value'.

Acknowledgments

- Tim Barnard, SCION Research and the Advisory Board of the Weaving the Korowai Project. The Ministry of Business, Innovation and Employment: funders of the Weaving the Korowai Project 2016–2019 phase.
- Members of the New Zealand Fishing Safety Forum, and the Centre of Occupational Health and Safety Research at Auckland University of Technology: funders of the multi-value cost-benefit study.

REFERENCES

Alkire, S. (2005). *Valuing freedoms: Sen's capability approach and poverty reduction.* Oxford, NY: Oxford University Press.

Bebbington, J., Higgins, C., & Frame, R. (2009). Initiating sustainable development reporting: Evidence from New Zealand. *Accounting, Auditing & Accountability Journal, 22,* 588–625.

Clerc, J. M. (Ed.). (1985). *Introduction to working conditions and environment.* Geneva, Switzerland: International Labour Organization.

Closing the Gap. *The spirit level for New Zealand.* http://www.closingthegap.org.nz/the-spirit-level-for-new-zealand/. Accessed 24 Aug 2017.

Elkington, J. (2004). Enter the triple bottom line. In A. Henriques & J. Richardson (Eds.), *The triple bottom line: Does it all add up?* London, UK: Earthscan.

(The) Equality Trust. Slides of key graphs and tables from The Spirit Level (2009). Downloaded 24 Aug 2017. https://www.equalitytrust.org.uk/resources/the-spiritlevel

Gaskin, H., Edwin, M., Moore, D., & Guard, D. (2015). Multi-value cost-benefit analysis in the NZ fishing sector. In G. Lindgaard & D. Moore (Eds.), *Proceedings of the 19th Triennial International Ergonomics Association Congress,* Melbourne, 2015. http://ergonomics.uq.edu.au/iea/proceedings/Index.html

Hollnagel, E. (2017). The nitty-gritty of human factors. In S. Shorrock & C. Williams (Eds.), *Human factors and ergonomics in practice improving system performance & human wellbeing in the real world.* Boca Raton, FL: CRC Press.

Jollands, N., & Harmsworth, G. (2007). Participation of indigenous groups in sustainable development monitoring: Rationale and example from New Zealand. *Ecological Economics, 62,* 716–726.

Lange-Morales, K., Thatcher, A., & García-Acosta, G. (2014). Towards a sustainable world through human factors and ergonomics: It is all about values. *Ergonomics, 57*(11), 1603–1615.

McCarthy, C. (1994). *The crossing.* (page 413). London: Picador.

Meister, D. (1999). *The history of human factors and ergonomics*. Mahwah, NJ: Lawrence Erlbaum Associates.

Moore, D., & Barnard, T. (2012). With eloquence and humanity? Human factors/ergonomics in sustainable human development. *Human Factors: The Journal of the Human Factors and Ergonomics Society, 54*(6), 940–951. https://doi.org/10.1177/0018720812468483. Special Edition of Keynote Addresses from the 18 IEA Congress, Recife, Brazil. HFES, Santa Monica, CA, USA.

Nussbaum, M. (2011). *Creating capabilities: The human development approach*. Cambridge, MA: Harvard University Press.

O'Neill, D., & Moore, D. (2017). The evolving realities of HF/E practice in agriculture. In S. Shorrock & C. Williams (Eds.), *Human factors and ergonomics in practice improving system performance & human wellbeing in the real world*. Boca Raton, FL: CRC Press.

Ostrom, E. (2009). A general framework for analysing sustainability of social-ecological systems. *Science, 325*, 419–422.

Paine, T. (1776, February 14). *Common sense; Addressed to the inhabitants of America* (W&T Bradford's 3rd edition). Philadelphia. Downloaded 24 Aug 2017 from Project Gutenberg. https://www.gutenberg.org/ebooks/147

Pickett, K., & Wilkinson, R. (2010). *The spirit level: Why equality is better for everyone*. London: Penguin.

Sen, A. (1999). *Development as freedom*. Oxford, NY: Oxford University Press.

Statistics New Zealand. (2016). *Data in calculator*. http://www.stuff.co.nz/business/money/80229052/interactive-see-how-your-household-income-compares

Steinbeck, J. (1939). *Grapes of wrath*. New York: Viking.

Tappin, D., Bentley, T., & Vitalis, T. (2008). The role of contextual factors for musculoskeletal disorders in the New Zealand meat processing industry. *Ergonomics, 51*, 1576–1593.

Tappin, D., Vitalis, T., & Bentley, T. (2016). The application of an industry level participatory ergonomics approach in developing MSD interventions. *Applied Ergonomics, 52*, 151–159.

Wilkin, P. (2010). The ideology of ergonomics. *Theoretical Issues in Ergonomics Science, 11*, 230–244.

Wilson, J. (2014). Fundamentals of systems ergonomics/human factors. *Applied Ergonomics, 45*, 5–13.

CHAPTER 15

Ergonomics and Human Factors for a Sustainable Future: Suggestions for a Way Forward

Andrew Thatcher and Paul H. P. Yeow

INTRODUCTION

In this book we have looked at the ways in which human factors and ergonomics (HFE) can contribute to a more sustainable future. As Nickerson (1992) noted, in essence the HFE aim with regard to sustainability can be summarised as facilitating behaviour change. This can be achieved through using a number of HFE strategies including design using behaviour-shaping constraints (Vicente, 1998), the design of feedback mechanisms or the provision of information (Drury, 2008, 2014; Martin, Legg, & Brown, 2013; Vicente, 1998), and the design of decision-support systems (Drury, 2008, 2014). More specifically, the suggestions for the domains where HFE can contribute to sustainability are through designing for reduced/durable/recycled resource use (Hanson, 2013; Thatcher, 2013),

A. Thatcher (✉)
Psychology Department, University of the Witwatersrand,
Johannesburg, South Africa

P. H. P. Yeow
School of Business, Monash University Malaysia, Bandar Sunway, Malaysia

© The Author(s) 2018   373
A. Thatcher, P. H. P. Yeow (eds.), *Ergonomics and Human Factors for a Sustainable Future*, https://doi.org/10.1007/978-981-10-8072-2_15

the design of jobs to support work within the green economy (Hanson, 2013; Thatcher, 2013), the design to support corporate sustainability (Steimle & Zink, 2006; Zink, Steimle, & Fischer, 2008), and the design of disaster management services (Hanson, 2013; Moore & Barnard, 2012).

Numerous authors have identified a range of different places where HFE interventions would be most relevant. This list is extensive and can only be summarised here. Within the domain of design for reduced/durable/recycled resource use are products and systems that include the efficient use of energy, water, food, land, materials, transportation, and cities, and the reduction of various types of waste (Hanson, 2013; Martin et al., 2013; Moray, 1995; Nickerson, 1992; Radjiyev, Qiu, Xiong, & Nam 2015; Thatcher, 2013). In the domain of job design, the emphasis is on ensuring wellbeing, health, and safety (see Docherty, Forslin, & Shani, 2002) across a wide array of sectors including recycling, renewable energy installations, organic farming, and work in extreme climatic environments (Hanson, 2013). The corporate sustainability domain overlaps partially with the job design domain with suggestions for improving health and safety as well as wellbeing. However, suggestions in this domain also include designing appropriate corporate social responsibility initiatives, considering the design of organisations across geographical space, and ensuring sustainable economic success (Steimle & Zink, 2006; Zink, 2014; Zink, Steimle, & Fischer, 2008). Suggested work in the disaster management domain includes designing appropriate security systems to prevent violence and terrorism (Moray, 1995), healthcare and emergency services to cope with natural and humanitarian disasters (Moore & Barnard, 2012; Steimle & Zink, 2006), and the design of other complex systems to avert disasters (Steimle & Zink, 2006).

Drury (2014) emphasises that HFE's role might play out at four levels. At the most basic level, HFE should be involved with trying to assist behaviour change at the personal level, regardless of the context (i.e. at home, at work, at play, etc.). HFE interventions that occur at this level are primarily about changing consumption, waste reduction, and lifestyle choices and behaviours. At the next level, HFE should be involved in behaviour change at the work level. This would involve influencing groups to change behaviours to reduce waste and optimise the efficient use of resources. At the third level, HFE should be involved with changing behaviour at the general public level. For Drury (2014), behaviour change at this level would primarily be through designing feedback systems to optimise efficient use of resources. At the broadest level, HFE should be involved with design of systems to support decision-making behaviour at the government or policy level.

Having considered the great potential that HFE has for contributing solutions to these sustainability challenges and having reviewed what we has already been conducted, we now turn our attention to what we consider to be the future goals for our discipline. Since this book is about HFE's role in enabling a sustainable future, we express these goals in terms of where we are now ("from") and what skills and ideas we still need to develop ("to"). We have identified five goals that we believe emerge naturally from the work presented in this book. These themes are (1) from specialised, to multidisciplinary, to transdisciplinary; (2) from systems HFE to complexity HFE; (3) from positivism to value-laden science; (4) from mitigation to adaptation; and (5) from general to local solutions. We acknowledge that these goals are strongly influenced by Moray's (1995) assessment of what our discipline needs to do to meet the global challenges facing humanity published more than 20 years ago. We discuss each of these goals in more detail in the following sections.

Goal 1: From Specialised to Multidisciplinary, Interdisciplinary, and Transdisciplinary

As we mentioned in Chap. 1, one of the features of this book is the wide range of disciplines that have contributed to compiling this collection. As Moray (1995) noted, the problems that emerge from sustainability require expert input from many different disciplinary perspectives. The HFE discipline itself is naturally adept at drawing knowledge and expertise from many different disciplines including an understanding of physiology, anatomy, biomechanics, psychology, sociotechnical systems, and design theory. However, with the challenges that emerge from sustainability, it will be necessary to engage more broadly with the social sciences such as sociology, political science, anthropology, philosophy, human resources, and the management sciences. Chapters 2, 3, 4, 8, 11, 12 and 13 do this to some extent. Also, given the damage that we are currently inflicting on our natural environment, it will also be necessary to engage with the ecological and biological sciences. The work contained in Thatcher and Yeow (2016) and in Richardson et al. (2017) goes some way to making connections between HFE and the ecological sciences. This work is also included in Chaps. 2 and 7 of this book.

This is not the first time that the necessity for diverse disciplinarity within HFE has been raised. Wisner (1985) called for more engagement between HFE and anthropology; Moray (2000) called for greater connections between HFE, anthropology, and politics; Boudeau, Wilkin, and Dekker (2014) called for greater engagement between ergonomics and politics, while Wilkin (2010) called for a closer look at the philosophy of HFE. We are sure that these types of debates and discussions will make many people within the HFE discipline feel decidedly uncomfortable. For some, the discomfort is felt because these proposals call for people within the HFE discipline to further share and dilute their specialised expertise. For others, these proposals may feel as if the HFE discipline is spreading itself too thin. Following Wilson (2014) we would argue that it is our understanding of systems that include humans that makes the HFE discipline distinct. But in order to meet this self-appointed mandate in the context of sustainability challenges means, we will also need to understand how the human systems interact in ever-larger groupings (e.g. at sociological, anthropological, and political levels). In addition, sustainability means understanding something about how ecological systems function and how our behaviour and interactions with these life-supporting systems can support or destroy them.

However, Lang et al. (2012) and Stokols, Misra, Runnerstrom, and Hipp (2009) have argued that the challenges presented by sustainability require disciplines to move beyond a multidisciplinary approach towards an interdisciplinary approach, or even a transdisciplinary approach. An interdisciplinary approach involves a level of cooperation in order to achieve a synthesis between different theories and methods. A transdisciplinary approach requires not just cooperation and synthesis but an integration of disciplinary knowledge and methods to create new, unified theoretical frameworks not limited by their original disciplinary stances. Fiore, Phillips, and Sellers (2014) have given a useful overview of transdisciplinary research and the central role such an approach might play in integrating the HFE discipline with other disciplines attempting to address sustainability challenges. In particular, it could be argued that the HFE discipline might contribute a unique blend of knowledge related to design, human physiology, human anatomy, and human behaviour and other aspects of human psychology.

There are very few published studies that demonstrate the types of roles that an HFE practitioner can play in assisting transdisciplinary teams to address sustainability challenges. One such example is the work of Moore and Barnard (2012). In this work they report on the role of the HFE practitioner in supporting the activities of the Sustainable Livelihoods Framework that also involved specialists from social anthropology, economics, planning, physical sciences, and representatives of community. The underlying goals of the project were to ensure that the deeply impoverished communities in the study area could develop sufficient social, natural, economic, and cultural capital to survive and thrive into the future (Moore & Barnard, 2012). As Moore and Barnard (2012) concluded, the role of the HFE specialist in this transdisciplinary team was to "build [an] understanding about the characteristics of people, including not only their physical and cognitive capabilities and limitations, but also the unique sets of aspirations, knowledge, and skills that they have reason to value" (p. 948). It should also be acknowledged that this understanding could not be achieved without also understanding something about the ecological, financial, and political constraints encountered by these communities.

In this book there are several chapters that extend on the interdisciplinary and transdisciplinary perspective of HFE. From a transdisciplinary perspective, there are a number of chapters that attempt to create new theoretical and methodological approaches through the integration across disciplines. Chapter 2 integrates ecological science theories with HFE theories. Chapter 3 integrates human resources and corporate social responsibility theories with HFE theories. Chapter 4 integrates marketing, information systems, and environmental science theories with HFE theories. Chapter 10 integrates management science theories with HFE theories. From an interdisciplinary perspective, Chap. 11 attempts to merge economics with environmental science and green ergonomics, while Chap. 13 looks at how political science theories can be used to address HFE issues. We would argue that this is a good start, but more work needs to be done in this area.

Goal 2: From Systems HFE to Complexity HFE

Several authors have noted that HFE is a systems discipline (Carayon, 2006; Dul et al., 2012; Wilson, 2014; Zink, 2014). The systems that the HFE discipline is interested in understanding are those that include humans and traditionally have spanned several levels of complexity from "simple" human-tool or human-task systems to more complex sociotechnical systems. However, as Siemieniuch, Sinclair, and Henshaw (2015) have observed, an even deeper understanding of complex systems is required to address the sustainability challenges. The theoretical models that have been developed within the HFE discipline so far each draw our attention to the need to embrace an understanding of complex systems (García-Acosta, Pinilla, Larrahondo, & Morales, 2014; Steimle & Zink, 2006; Thatcher, 2013; Thatcher & Yeow, 2016; Zink, 2014). Dekker, Hancock, and Wilkin (2013) went further by specifically outlining the qualities of complexity that require our understanding. These qualities include the need to understand local relationships, dynamic interactions, fuzzy boundaries, and emergent properties (Dekker et al., 2013). A detailed discussion of each of these concepts is beyond the scope of this chapter, but interested readers may wish to start with Dekker et al. (2013) and continue their reading with Cilliers (1998) and Norberg and Cumming (2008).

Each of the chapters in this book addresses the issue of complexity within systems in some way. We don't go through all the chapters in detail here but instead highlight a few examples. Chapter 2 tackles the issue of complexity in HFE systems-of-systems through trying to find ways to navigate through complex networks of interacting HFE systems. Chapter 3 demonstrates that understanding the relatively simple concept of "decent work" requires an understanding of the more complex issues of global supply chains, child labour, slave labour, and organisational ethics. Chapter 10 examines the issue of global supply chains more closely and the implications that this has for how we model work systems, by looking at the interrelated impacts of outsourcing and digitisation and how this complicates our understanding of the global production of work. Chapter 11 looks at the complex socio-ecological relationships associated with keeping a river clean. In this chapter, the authors consider how HFE might be used to understand and support the interrelationships between the various stakeholders that use a fresh water source, including the organisations that use water for production, the farmers who use water for agriculture, the communities that draw water for cleaning and consumption, and the government agency that regulate water use.

In support of what Zink and Fischer suggest in Chap. 10, we believe it will be necessary for complex systems theory to be introduced into the curricula of HFE educational programmes if the HFE discipline is going to make a difference to sustainability problems. The complexity that requires the attention from HFE stretches across time (such as product lifecycle ergonomics (Zink, 2014) or the sustainable system-of-systems perspective (Thatcher & Yeow, 2016)) and place (such as supply chain ergonomics (Hasle & Jensen, 2012)). Walker et al. (2010) make several cogent arguments as to why complex systems thinking should be an important factor to considering understanding HFE systems, not least would be because humans are the source of much of the complexity (Bar-Yam, 2002). However, Salmon, Walker, Read, Goode, and Stanton (2017) have questioned whether HFE has the existing evaluation tools to deal with this level of complexity. Currently our way of modelling HFE issues in complex systems is based on accident analysis methods such as Accimap (Rasmussen, 1997), the Systems Theoretic Accident Model and Process (STAMP) (Leveson, 2004), and the Human Factors Analysis and Classification System (HFACS) (Shappell & Wiegmann, 2012), or systems analysis methods such as Event Analysis of Systemic Teamwork (EAST) (Walker et al., 2006), the Functional Resonance Analysis Method (FRAM (Hollnagel, 2012), or Cognitive Work Analysis (CWA) (Vicente, 1999). However, these methods typically assess individuals and teams as the unit of analysis, rather than entire hierarchies of systems. One recent method to emerge is the Cognitive Work Analysis Design Toolkit (CWA-DT). The CWA-DT combines the traditional CWA approach with a participatory approach (Read, Salmon, Lenné, & Jenkins, 2015). This method shows promise because it has a transdisciplinary focus. While these methods may be useful for modelling sociotechnical systems, sustainability issues actually require the modelling of socio-ecological-technological systems. Further developments are therefore clearly required to integrate complexity thinking with HFE.

GOAL 3: FROM VALUE-FREE TO VALUE-LADEN SCIENCE

Wilkin (2010) argued that HFE likes to think of itself as an objective science that is, by implication, value-free. In this value-free conceptualisation, HFE sees itself as a discipline where "reliable knowledge is based on facts about the world that can be measured and verified through observation" (Wilkin, 2010, p. 234). However, this way of thinking within HFE assumes that interactions and behaviour take place within a closed system that is largely predictable. Arguably, very few HFE systems can truly be described as closed systems. As was shown in Chap. 2, in HFE the biological system (i.e. the human) interacts with social systems, embraced within various levels of political systems, financial systems, and ecological systems. While behaviour relevant for HFE may be measurable and observable at a localised micro level in a laboratory, HFE outcomes are far more difficult to reliably predict in the field. An example of how values are important in making informed decisions with more sustainable outcomes is to consider the case of alternative vehicle fuels. One of the options as an alternative energy source to fossil fuels in the vehicle industry is biofuels. It can be shown scientifically, both in a laboratory and in field testing, that biofuelled vehicles emit fewer greenhouse gases than fossil-fuelled vehicles (Pacala & Socolow, 2004). At face-value then, there may be important local health and wellbeing benefits for urban populations where these vehicles operate. However, a value-laden approach invites us to consider the values of the entire system, not just the scientific benefits of biofuels over fossil fuels. There are now numerous studies that suggest that there may be significant negative effects for human health and wellbeing from changing land use (Fargione, Hill, Tilman, Polasky, & Hawthorne, 2008). In particular, land that was previously being used to plant crops to feed people was being used to plant crops that were harvested for biofuels, causing food availability crises in some regions and rising food prices globally. Even more concerning, was the clearing of additional land (usually forested) to reap the benefits of additional income from biofuels. Clearing efforts have significantly increased the amount of carbon in the atmosphere through burning and by removing the carbon sinks such as trees (Fargione et al., 2008). In addition, Melillo et al. (2009) have noted that biofuel production results in increased greenhouse gas emissions from nitrous oxide due to increased fertilizer used to stimulate biofuel crop growth. The net effect of moving to biofuels is therefore likely to be reduced human health and wellbeing over a far greater area. The need for HFE to embrace this complexity has already been addressed in Goal 2 above. What is important to note from Wilkin's (2010) argument is that the predominant paradigm within HFE is that it assumes that it is value-free, but it is in fact a discipline that is actually value-laden, but that the values are not actually specified.

What values should HFE choose? Wilkin (2010) argues that the studies HFE chooses to conduct, the funding HFE chooses to seek, and the industries that support HFE initiatives all determine the values of the discipline. For Hancock and Drury (2011), HFE research and practice primarily aims to address the quality of life for the people who were the specified subject of HFE investigations. Here, the benefit is for the people funding the investigations and the relatively few direct recipients of those investigations. In addition, the values that drive this exercise are largely those of financial stability and the quality of work-life of a few. In particular, Hancock and Drury (2011) noted that the primary funders of HFE work (at least in the USA) were the military and large corporations. In fact, Moray (1995) referred to this traditional HFE role as supporting the "world of western liberal capitalism" (p. 1691), by which he meant the goal of HFE was to make the workplace more tolerable and effective/productive for workers in industrialised economies. These observations suggest that HFE already has an unstated set of values and that they benefit the few, rather than the many. Moray (1995) argued that the HFE discipline needs a clearly articulated (and presumably also an actively debated) set of values to guide the questions we should ask and the solutions that we seek.

What would these values look like? Dekker et al. (2013) considered values for the HFE discipline specifically for sustainability concerns. The values that Dekker et al. (2013) identified for HFE in a sustainability context were embracing complexity and emergence (i.e. Goal 2 as articulated here). More specifically, embracing complexity referred to a need to understand how local interactions have global consequences and to understand how interactions change over time. Embracing emergence meant anticipating that there could be unforeseeable consequences. Dekker et al. (2013) concluded that there should be further discussion about the appropriate values for HFE in the context of sustainability challenges. The only study that has clearly set out to define values for the HFE discipline is Lange-Morales, Thatcher, and García-Acosta (2014). Lange-Morales et al. (2014) accepted this challenge and developed a set of six values for HFE. These values are (1) respect for human rights, (2) respect for the Earth, (3) appreciation of complexity, (4) respect for diversity, (5) respect for transparency and openness, and (6) respect for ethical decision-making. Appreciation of complexity is also noted as Goal 2, and respect for diversity is partly captured by Goal 1 and Goal 5 in this chapter. There is yet to be a robust debate as to whether these are appropriate values for the HFE discipline. In this chapter we have already discussed the need to deepen our understanding of complexity. Next we will discuss the need to respect diversity.

GOAL 4: FROM MITIGATION TO ADAPTATION

Incropera (2016) recommended two concurrent paths towards addressing sustainability challenges: mitigation and adaptation. The first path of mitigation involves reducing the rate of resource consumption per person to levels that are ecologically sustainable. This involves thinking about how our behaviour, products, and systems might be modified to reduce our current rate of impact on limited resources. From a human factors and ergonomics (HFE) perspective, this means designing products and systems that are more efficient and effective in utilising non-renewable resources or by finding ways to change our behaviour to adopt renewable resources or to reduce wastage of resource. Most of the examples presented and reviewed in this book portray various attempts at mitigation (i.e. reducing our current impact to forestall the chances of disaster in the future). Chapters 4, 5, 6, 9, 11, 12 and 13 each give examples of HFE work that addresses mitigation approaches. There are now numerous HFE examples of empirical work looking at interface design to ensure efficient use of resources (Durugbo, 2013; Fang & Sun, 2016; Harvey, Thorpe, & Fairchild, 2013; Katzeff, Nyblom, Tunheden, & Torstensson, 2012; Kobus et al., 2013; Revell & Stanton, 2016; Sauer, Wiese, & Rüttinger, 2002, 2003, 2004), design to understand and encourage the sustainable use of sustainable products (Cocron et al., 2013; Franke, Arend, McIlroy, & Stanton, 2016; Fréjus & Guibourdenche, 2012; Lee & Kang, 2013; Stanton et al., 2013; Stedmon, Winslow, & Langley, 2013; Young, Birrell, & Stanton, 2011), and the integration of employee wellbeing and effectiveness with sustainability initiatives (Bolis, Brunoro, & Sznelwar, 2016; Thatcher & Milner, 2014).

However, at the current rate of world population growth (Van den Bergh & Rietveld, 2004), it is likely that mitigation will be insufficient to stave off future disaster. Radical changes are required in human behaviour, possibly involving population control, to prevent the collapse of human-supporting ecosystems. In the absence of such radical behaviour changes, the second concurrent path that is required is adaptation. This means creating resilient products and systems that will be able to cope with the inevitable changes to the planet's ecosystems. Of special interest to HFE are products and systems that will allow humans to adapt to these changes. Some of these changes have already started to occur and therefore a concurrent strategy is already required. This book does not consider the adaptation requirements in any depth. Climate change is going to result in significant changes to the environments in which people need to perform work. For example, rising temperatures in most parts of the world will affect the physical wellbeing of people who need to perform physical work tasks (Kjellstrom, Gabrysch, Lemke, & Dear, 2009). This means an HFE examination of the tasks that can be performed or the design of tools and equipment that will allow the tasks to be performed under the modified conditions. Rising sea levels, for those people living near the sea, will affect where people live, the work that they will be able to perform, and the interconnections with other people (either moving people closer together if land becomes scarce, or separating people if islands start to form). Changing rainfall patterns and temperatures will affect which crops can be grown and which livestock can be farmed, significantly impacting on farming and food availability. What is needed is resilient socio-ecological-technical systems.

GOAL 5: FROM GENERAL TO LOCAL SOLUTIONS

Using biological systems as a basis, Fiksel (2003) identified a number of key properties that could be transferred to the design of engineered systems to make them more resilient. Key among these properties is diversity. For Fiksel (2003) diversity refers to whether the (engineered) system contains multiple forms or allows for multiple behaviours. More forms and behaviours give the system a greater chance to recover from unusual disturbances and hence support sustainability. Lange-Morales et al. (2014) incorporated respect for diversity as one of the core values of HFE for sustainability. Diversity within the HFE discipline is often operationalised as cross-cultural design, but Lange-Morales et al. (2014) have suggested that we need to go further and understand the diversity of place (i.e. the geographical and cultural setting) and ecological diversity (i.e. our interactions with other biological entities). As a consequence of global variability, Moray (1995) argued that few HFE solutions are truly universal.

Lange-Morales et al. (2014) suggested that one of the ways to respect diversity and to foster variability is to encourage local HFE solutions for local HFE problems. Not only does this increase diversity but it is also a way of distributing and building HFE expertise and providing local employment. In addition, these types of indigenous HFE solutions are more likely to be accepted by local users as is commonly found in participatory HFE approaches (Imada, 1991; Martin et al., 2013). People who have to live and work with the consequences of HFE interventions are more likely to accept those interventions if they feel some ownership of the intervention or the evaluation process. Wisner's (1985) anthropotechnology approach takes a similar stance, warning of the dangers of simply transferring technology globally without due consideration of the cross-cultural, anthropological, geographical, and managerial implications.

There are now numerous parts of the world where a combination of colonial work practices and ill-considered technology transfers have left a complex array of working environments that seldom take due consideration of indigenous systems or cultures. In addition, since a large proportion of work worldwide actually takes place in the informal economy (Benjamin, Beegle, Recanatini, & Santini, 2014) where traditional HFE approaches seldom reach, HFE needs to re-think how it is to grow and make a difference. Moving from global to local solutions is an important way to bridge this gap.

CONCLUSIONS

In Chap. 1, we laid out the case for sustainability. In that discussion we demonstrated how humans are already a clear and present danger to the planet and the ecosystems that support human habitation. The problems are severe and are only likely to become more critical in the coming decades. The human influence on the planet is now so significant that geologists have argued that we have entered the Anthropocene age (Steffen, Grinevald, Crutzen, & McNeill, 2011). We also made it clear that these challenges are anthropogenic and therefore HFE as a discipline is well placed to make a significant contribution to addressing these challenges. As we also acknowledged in Chap. 1, this book does not pretend to address all the challenges raised by sustainability that have a clear link back to HFE. We do believe though that this book makes a significant start. In particular, the chapters in this book indicate that there has now been a great deal of work on reducing various resource use and waste production. Evidence for these types of HFE interventions can be found in Chaps. 4, 5, 6, 7, 9, 11, 12 and 13. This work represents interventions at the personal level (Chap. 4), at the work level (Chaps. 7 and 9), at the public level (Chaps. 5 and 6), and at the government/policy level (Chaps. 11, 12 and 13). This book also contains two chapters on the design of sustainable work systems (Chaps. 3 and 10) and a chapter on corporate sustainability (Chap. 8). The two chapters on sustainable work systems and the chapter on corporate sustainability are each at the work level.

From this overview it is easy to see that there are two obvious gaps in our knowledge. The most glaring omission is work that seeks to develop systems resilient to natural and humanitarian disasters. Moore and Barnard (2012) have published some work in this regard as have Meshkati, Tabibzadeh, Farshid, Rahimi, and Alhanaee (2016). As we create and build more complex, dangerous systems in close proximity to communities, the risk for a major crisis increases dramatically as evidenced by recent disasters at Fukushima, Deepwater Horizon, the tsunamis in Japan and Indonesia, and flooding from Tropical Storm Sandy around New York. There are concerns that these events are a portent of what is still to come. The second omission is with regard to what HFE can do to influence behaviour at the personal level. Since HFE is primarily concerned with work contexts, it is not surprising that much of our effort has gone towards understanding what we can do to address sustainability challenges at the local, public, and regional level because this is where financial incentives can be more readily realised. However, the HFE interventions themselves might not be sustainable unless it is people themselves that change their behaviours. In part, Chaps. 4, 5 and 6 address or review research that is aimed at addressing behaviour change at the personal level, but clearly more work is needed from the HFE discipline in this regard.

Finally, we would like to suggest that one of the limiting factors in connecting sustainability and HFE is the current definition of HFE. The International Ergonomics Association's website gives the following definition, approved at the IEA Congress in 2000:

> Ergonomics (or human factors) is the scientific discipline concerned with the understanding of interactions among humans and other elements of a system, and the profession that applies theory, principles, data and methods to design in order to optimize human well-being and overall system performance.

We would argue that this definition implies that the systems of interest to HFE are closed systems with linear relationships between humans and the other components of the system. We feel that this is not the most up-to-date view of the types of systems with which many HFE researchers and practitioners actually engage, with many more systems now requiring a more complex, systemic understanding. This would imply the need to consider an expanded definition in order to include these types of systems. Wilson (2014) and Walker et al. (2017) have already challenged HFE to think beyond linear systems to embrace the complexities of system ergonomics. In this book, we embrace the emerging notions of systems ergonomics and invite HFE to extend systems thinking to include the wicked problems (Murphy, 2012) associated with sustainability challenges.

References

Bar-Yam, Y. (2002). *Complexity rising: From human beings to human civilization, a complexity profile.* Oxford, UK: UNESCO Publishers.

Benjamin, N., Beegle, K., Recanatini, F., & Santini, M. (2014). *Informal economy and the World Bank.* Policy research working paper 6888.

Bolis, I., Brunoro, C. M., & Sznelwar, L. I. (2016). Work for sustainability: Case studies of Brazilian companies. *Applied Ergonomics, 57,* 72–79.

Boudeau, C., Wilkin, P., & Dekker, S. W. (2014). Ergonomics as authoritarian or libertarian: Learning from Colin Ward's politics of design. *The Design Journal, 17,* 91–114.

Carayon, P. (2006). Human factors of complex sociotechnical systems. *Applied Ergonomics, 37,* 525–535.

Cilliers, P. (1998). *Complexity and postmodernism: Understanding complex systems.* London: Routledge.

Cocron, P., Bühler, F., Franke, T., Neumann, I., Dielmann, B., & Krems, J. F. (2013). Energy recapture through deceleration – Regenerative braking in electric vehicles from a user perspective. *Ergonomics, 56,* 1203–1215.

Dekker, S. W., Hancock, P. A., & Wilkin, P. (2013). Ergonomics and sustainability: Towards an embrace of complexity and emergence. *Ergonomics, 56,* 357–364.

Docherty, P., Forslin, J., & Shani, A. B. (Eds.). (2002). *Creating sustainable work systems: Emerging perspectives and practice.* London: Routledge.

Drury, C. G. (2008). The future of work in a sustainable society. In K. J. Zink (Ed.), *Corporate sustainability as a challenge for comprehensive management* (pp. 199–214).

Drury, C. G. (2014). Can HF/E professionals contribute to global climate change solutions? *Ergonomics in Design, 22,* 30–33.

Dul, J., Bruder, R., Buckle, P., Carayon, P., Falzon, P., Marras, W. S., et al. (2012). A strategy for human factors/ergonomics: Developing the discipline and profession. *Ergonomics, 55,* 377–395.

Durugbo, C. (2013). Improving information recognition and performance of recycling chimneys. *Ergonomics, 56,* 409–421.

Fang, Y. M., & Sun, M. S. (2016). Applying eco-visualisations of different interface formats to evoke sustainable behaviours towards household water saving. *Behaviour & Information Technology, 35,* 748–757.

Fargione, J., Hill, J., Tilman, D., Polasky, S., & Hawthorne, P. (2008). Land clearing and the biofuel carbon debt. *Science, 319*(5867), 1235–1238.

Fiksel, J. (2003). Designing resilient, sustainable systems. *Environmental Science & Technology, 37,* 5330–5339.

Fiore, S. M., Phillips, E., & Sellers, B. C. (2014). A transdisciplinary perspective on hedonomic sustainability design. *Ergonomics in Design, 22,* 22–29.

Franke, T., Arend, M. G., McIlroy, R. C., & Stanton, N. A. (2016). Ecodriving in hybrid electric vehicles – Exploring challenges for user-energy interaction. *Applied Ergonomics, 55,* 33–45.

Fréjus, M., & Guibourdenche, J. (2012). Analysing domestic activity to reduce household energy consumption. *Work, 41*(Supplement 1), 539–548.

García-Acosta, G., Pinilla, M. H. S., Larrahondo, P. A. R., & Morales, K. L. (2014). Ergoecology: Fundamentals of a new multidisciplinary field. *Theoretical Issues in Ergonomics Science, 15,* 111–133.

Hancock, P. A., & Drury, C. G. (2011). Does human factors/ergonomics contribute to the quality of life? *Theoretical Issues in Ergonomics Science, 12,* 416–426.

Hanson, M. A. (2013). Green ergonomics: Challenges and opportunities. *Ergonomics, 56,* 399–408.

Harvey, J., Thorpe, N., & Fairchild, R. (2013). Attitudes towards and perceptions of eco driving and the role of feedback systems. *Ergonomics, 56,* 507–521.

Hasle, P., & Jensen, P. L. (2012). Ergonomics and sustainability – Challenges from global supply chains. *Work, 41*(Supplement 1), 3906–3913.

Hollnagel, E. (2012). *FRAM: The functional resonance analysis method: Modelling complex socio-technical systems.* Burlington, VT: Ashgate.

Imada, A. S. (1991). The rationale of participatory ergonomics. In K. Noro & A. S. Imada (Eds.), *Participatory ergonomics* (pp. 30–49). London: Taylor & Francis.

Incropera, F. P. (2016). *Climate change: A wicked problem.* New York: Cambridge University Press.

Katzeff, C., Nyblom, Å., Tunheden, S., & Torstensson, C. (2012). User-centred design and evaluation of EnergyCoach – An interactive energy service for households. *Behaviour & Information Technology, 31,* 305–324.

Kjellstrom, T., Gabrysch, S., Lemke, B., & Dear, K. (2009). The 'Hothaps' programme for assessing climate change impacts on occupational health and productivity: An invitation to carry out field studies. *Global Health Action, 2,* 2082.

Kobus, C. B., Mugge, R., & Schoormans, J. P. (2013). Washing when the sun is shining! How users interact with a household energy management system. *Ergonomics, 56,* 451–462.

Lang, D. J., Wiek, A., Bergmann, M., Stauffacher, M., Martens, P., Moll, P., et al. (2012). Transdisciplinary research in sustainability science: Practice, principles, and challenges. *Sustainability Science, 7,* 25–43.

Lange-Morales, K., Thatcher, A., & García-Acosta, G. (2014). Towards a sustainable world through human factors and ergonomics: It is all about values. *Ergonomics, 57,* 1603–1615.

Lee, S. Y., & Kang, M. (2013). Innovation characteristics and intention to adopt sustainable facilities management practices. *Ergonomics, 56,* 480–491.

Leveson, N. G. (2004). A new accident model for engineering safer systems. *Safety Science, 42,* 237–270.

Martin, K. K., Legg, S. S., & Brown, C. C. (2013). Designing for sustainability: Ergonomics – Carpe diem. *Ergonomics, 56,* 365–388.

Melillo, J. M., Reilly, J. M., Kicklighter, D. W., Gurgel, A. C., Cronin, T. W., Paltsev, S., et al. (2009). Indirect emissions from biofuels: How important? *Science, 326,* 1397–1399.

Meshkati, N., Tabibzadeh, M., Farshid, A., Rahimi, M., & Alhanaee, G. (2016). People-technology-ecosystem integration: A framework to ensure regional interoperability for safety, sustainability, and resilience of interdependent energy, water, and seafood sources in the (Persian) Gulf. *Human Factors, 58,* 43–57.

Moore, D., & Barnard, T. (2012). With eloquence and humanity? Human factors/ergonomics in sustainable human development. *Human Factors, 54,* 940–951.

Moray, N. (1995). Ergonomics and the global problems of the twenty-first century. *Ergonomics, 38,* 1691–1707.

Moray, N. (2000). Culture, politics and ergonomics. *Ergonomics, 43,* 858–868.

Murphy, R. (2012). Sustainability: A wicked problem. *Sociologica, 6,* 1–23.

Nickerson, R. S. (1992). What does human factors research have to do with environmental management? *Proceedings of the Human Factors and Ergonomics Society Annual Meeting, 36,* 636–639.

Norberg, J., & Cumming, G. S. (Eds.). (2008). *Complexity theory for a sustainable future.* New York: Columbia University Press.

Pacala, S., & Socolow, R. (2004). Stabilization wedges: Solving the climate problem for the next 50 years with current technologies. *Science, 305,* 968–972.

Radjiyev, A., Qiu, H., Xiong, S., & Nam, K. (2015). Ergonomics and sustainable development in the past two decades (1992–2011): Research trends and how ergonomics can contribute to sustainable development. *Applied Ergonomics, 46,* 67–75.

Rasmussen, J. (1997). Risk management in a dynamic society: A modelling problem. *Safety Science, 27,* 183–213.

Read, G. J., Salmon, P. M., Lenné, M. G., & Jenkins, D. P. (2015). Designing a ticket to ride with the Cognitive Work Analysis Design Toolkit. *Ergonomics, 58,* 1266–1286.

Revell, K. M., & Stanton, N. A. (2016). Mind the gap – Deriving a compatible user mental model of the home heating system to encourage sustainable behaviour. *Applied Ergonomics, 57,* 48–61.

Richardson, M., Maspero, M., Golightly, D., Sheffield, D., Staples, V., & Lumber, R. (2017). Nature: A new paradigm for well-being and ergonomics. *Ergonomics, 60,* 292–305.

Salmon, P. M., Walker, G. H., Read, G. J. M., Goode, N., & Stanton, N. A. (2017). Fitting methods to paradigms: Are ergonomics methods fit for systems thinking? *Ergonomics, 60,* 194–205.

Sauer, J., Wiese, B. S., & Rüttinger, B. (2002). Improving ecological performance of electrical consumer products: The role of design-based measures and user variables. *Applied Ergonomics, 33,* 297–307.

Sauer, J., Wiese, B. S., & Rüttinger, B. (2003). Designing low-complexity electrical consumer products for ecological use. *Applied Ergonomics, 34,* 521–531.

Sauer, J., Wiese, B. S., & Rüttinger, B. (2004). Ecological performance of electrical consumer products: The influence of automation and information-based measures. *Applied Ergonomics, 35,* 37–47.

Shappell, S. A., & Wiegmann, D. A. (2012). *A human error approach to aviation accident analysis: The human factors analysis and classification system.* Burlington, VT: Ashgate.

Siemieniuch, C. E., Sinclair, M. A., & de Henshaw, M. J. C. (2015). Global drivers, sustainable manufacturing and systems ergonomics. *Applied Ergonomics, 51,* 104–119.

Stanton, N. A., McIlroy, R. C., Harvey, C., Blainey, S., Hickford, A., Preston, J. M., et al. (2013). Following the cognitive work analysis train of thought: Exploring the constraints of modal shift to rail transport. *Ergonomics, 56*(3), 522–540.

Stedmon, A. W., Winslow, R., & Langley, A. (2013). Micro-generation schemes: User behaviours and attitudes towards energy consumption. *Ergonomics, 56,* 440–450.

Steffen, W., Grinevald, J., Crutzen, P., & McNeill, J. (2011). The Anthropocene: Conceptual and historical perspectives. *Philosophical Transactions of the Royal Society of London A: Mathematical, Physical and Engineering Sciences, 369,* 842–867.

Steimle, U., & Zink, K. J. (2006). Sustainable development and human factors. In W. Karwowski (Ed.), *International encyclopedia of ergonomics and human factors* (2nd ed.). London: Taylor & Francis.

Stokols, D., Misra, S., Runnerstrom, M. G., & Hipp, J. A. (2009). Psychology in an age of ecological crisis: From personal angst to collective action. *American Psychologist, 64,* 181–193.

Thatcher, A. (2013). Green ergonomics: Definition and scope. *Ergonomics, 56,* 389–398.

Thatcher, A., & Milner, K. (2014). Changes in productivity, psychological wellbeing and physical wellbeing from working in a 'green' building. *Work, 49,* 381–393.

Thatcher, A., & Yeow, P. H. (2016). A sustainable system of systems approach: A new HFE paradigm. *Ergonomics, 59,* 167–178.

Van den Bergh, J. C. J. M., & Rietveld, P. (2004). Reconsidering the limits to world population: Meta-analysis and meta-prediction. *BioScience, 54,* 195–204.

Vicente, K. J. (1998). Human factors and global problems: A systems approach. *Systems Engineering, 1,* 57–69.

Vicente, K. J. (1999). *Cognitive work analysis: Toward safe, productive, and healthy computer-based work.* Mahwah, NJ: Lawrence Erlbaum Associates.

Walker, G. H., Gibson, H., Stanton, N. A., Baber, C., Salmon, P., & Green, D. (2006). Event analysis of systemic teamwork (EAST): A novel integration of ergonomics methods to analyse C4i activity. *Ergonomics, 49*, 1345–1369.

Walker, G. H., Salmon, P. M., Bedinger, M., & Stanton, N. A. (2017). Quantum ergonomics: Shifting the paradigm of the systems agenda. *Ergonomics, 60*, 157–166.

Walker, G. H., Stanton, N. A., Salmon, P. M., Jenkins, D. P., & Rafferty, L. (2010). Translating concepts of complexity to the field of ergonomics. *Ergonomics, 53*, 1175–1186.

Wilkin, P. (2010). The ideology of ergonomics. *Theoretical Issues in Ergonomics Science, 11*, 230–244.

Wilson, J. R. (2014). Fundamentals of systems ergonomics/human factors. *Applied Ergonomics, 45*, 5–13.

Wisner, A. (1985). Ergonomics in industrially developing countries. *Ergonomics, 28*(8), 1213–1224.

Young, M. S., Birrell, S. A., & Stanton, N. A. (2011). Safe driving in a green world: A review of driver performance benchmarks and technologies to support 'smart' driving. *Applied Ergonomics, 42*, 533–539.

Zink, K. J. (2014). Designing sustainable work systems: The need for a systems approach. *Applied Ergonomics, 45*, 126–132.

Zink, K. J., Steimle, U., & Fischer, K. (2008). Human factors, business excellence and corporate sustainability: Differing perspectives, joint objectives. In K. J. Zink (Ed.), *Corporate sustainability as a challenge for comprehensive management* (pp. 3–18). Heidelberg, Germany: Physica-Verlag.

Index[1]

A

Acceptance, 47, 61, 83, 86, 136–140,
146, 149, 305, 361
Activity theory, 13, 15
Adaptation
adaptive control, 140–141
adaptive cycles, 38
Adoption, 10, 13, 15, 81, 122,
135–151, 211, 272, 290, 303,
319, 330, 334, 336, 341
Asymmetries
information, 24, 255, 329–351, 368
legislative, 2–4
resource, 2, 4, 24
Attention restoration theory
(ART), 170
Australia, 9, 136, 221, 227, 228, 304

B

Bank loans for green technology, 338
Battery electric vehicles (BEVs), 13,
15, 136–144, 146, 148, 149

Biophilia
biophilia hypothesis,
162, 163, 177
biophilic design, 15, 161–182
Brazil, 9, 216

C

Charging
behaviour, 144, 145, 150
style, 144–146, 150
Climate change, 2–4, 10, 23, 29,
78, 87, 100, 102, 135, 161,
301, 306, 330–333, 351,
357, 382
Collective, 57, 86, 99–102, 196, 200,
206, 277, 279, 281
Collectivist, *see* Collective
Colour, 175–176, 228, 288
Comfort
acoustic, 217, 228
thermal, 224, 225, 228
visual, 179, 228

[1]Note: Page numbers followed by 'n' refer to notes.

© The Author(s) 2018
A. Thatcher, P. H. P. Yeow (eds.), *Ergonomics and Human Factors for a
Sustainable Future*, https://doi.org/10.1007/978-981-10-8072-2

Communication, 2, 64, 231, 246, 251, 254, 255, 283, 284, 291, 313, 366, 368, 369

Complex systems, *see* System

Complexity, 7, 11, 12, 16, 24, 26, 29–31, 35, 36, 42, 65, 66, 113, 118, 122–124, 129, 148, 191, 206, 215, 277, 360, 369, 375, 377–381, 385

Consumer social responsibility (CnSR), 300, 301, 304, 306–315, 317, 318, 320

Consumption
energy, 78, 80, 111–113, 115, 123–125, 141, 146, 148, 150, 177, 226, 317
resources, 213, 226, 381

Corporate social initiatives, 61, 299, 374

Corporate social responsibility (CSR), 60, 61, 206, 245, 262, 263n1, 299, 300, 304, 307, 310, 313, 317, 330, 377

Crowd work, 244, 246, 247, 253, 255, 256, 259

D

Daylight, 172, 173, 176–177, 217, 228

Decent work
agenda, 48–51, 62
blue collar work, 244–247, 250–258, 261–263
decent work and sustainability, 51–52, 244, 255, 256, 259
definition, 13, 47–69, 251
and the International Labour Organization (ILO), 48–51, 62, 66, 68, 245, 248–249, 252, 254, 261, 355, 359

the enterprise, 54, 56, 59–62, 64, 65
the individual, 11, 48, 49, 55–57, 62–65
knowledge work, 244, 247, 250–258, 263
precarious work, 67
the society, 49, 55–59, 61, 64–66

Decision-making, 35, 68, 69, 88, 112, 280, 284, 291, 368, 369, 375

Degradation, 2–6, 172, 280, 286, 329, 333, 351

Disease, 2, 3, 49, 57, 169, 196, 197, 287, 288

Diversity, 24, 39, 144, 163, 167, 202, 357, 369, 381, 383

E

Eco-driving, *see* Transport

Ecology, 6, 25, 28, 86, 127, 138, 167, 197, 213, 220, 224, 227, 243, 248, 250, 251, 259–261, 277, 279, 299, 302, 331, 356, 375–377, 379, 383

Eco-mobility, *see* Transport

Electric vehicles (EVs), 10, 15, 135–151

Energy consumption
energy efficiency, 111, 112, 115, 119, 122, 126, 136, 146–150, 212, 213, 218, 219, 225, 234
energy flows, 150

Energy management, *see* System

Environmental knowledge, 99, 182, 286, 316, 318

Ergoecology, 7, 10, 47

Ethics
responsibility, 302
values (*see* Values)

Exploitation, 4, 27, 50, 66, 67, 197

F

Financing
fish pollution (*see* Pollution)
Green Technology Financing
Scheme (GTFS), 329–351
information asymmetry
(*see* Asymmetries)

G

Germany, 136, 253, 368
Global supply chain, *see* Supply chain
Global value creation, *see* Value
creation
Global warming, *see* Climate change
Governance, 60, 253, 262, 272, 281,
284, 290, 358, 368
polycentric, 279–284, 288, 291
Green building
Building Research Establishment
Environmental Assessment
Method (BREEAM),
221, 224, 225
Comprehensive Assessment System
for Building Environmental
Efficiency (CASBEE), 221,
225–226
construction, 211–215, 217,
219–221, 223, 224, 226, 228,
231, 233, 234, 276
definition, 212–214
DGNB Label, 221, 228–229
ergonomics credits, 222–225,
228, 230
GreenStar, 221, 227–228
indoor environmental quality, 28, 33,
171, 174, 225, 226, 229
Leadership in Energy and
Environmental Design (LEED),
171, 172, 221–223, 225
rating systems, 171, 172, 220–222,
224–227, 229–233

SBTool, 226–227
WELL, 222, 223, 229–230
Green ergonomics, 10, 25, 47, 78,
139, 141, 144, 151, 162,
171–172, 191, 216, 218, 220,
233, 271–291, 329, 331, 377
Greenhouse gases (GHG), 2, 3, 29,
39, 271, 332, 380
Green technology, 330, 334, 336,
339, 341, 350
Green Technology Financing Scheme
(GTFS), 329–351

H

Health, 3, 4, 6, 33, 49, 54, 57, 59–62,
116, 162, 163, 165, 166,
168–171, 173, 177, 178, 181,
193, 194, 196–199, 201–206,
215–217, 219, 221–225, 228,
229, 231, 233, 256, 271–273,
275, 277, 279, 284–288, 290,
304, 330, 331, 351, 356, 359,
364, 374, 380
Home heating systems, 10, 13, 15,
115, 118, 120, 122, 126
Human development, 6, 196, 248,
360–361
Human resource management
(HRM), 16, 54, 57, 252, 262

I

Indonesia, 40, 81, 301,
305, 308, 310, 314,
317–319, 385
Institutions, 57, 271–291, 335,
336, 338
Interdisciplinary, 41, 215, 357, 367,
375–377
Interface design, 116, 117, 122,
141, 382

L
Life-cycle
  natural, 163, 167
  social, 260, 261

M
Maintenance, 33, 59, 116, 178, 213,
    214, 219, 220, 247, 256,
    259–261, 280
  building maintenance, 33, 212, 220,
    221, 226, 227, 230, 232
Malaysia, 40, 81, 86, 99, 100, 102,
    273, 274, 283–286, 290, 291,
    305, 330, 332–334, 339, 341,
    342, 351
Mental models, 113–121, 123–129,
    148, 149, 204
Meso-ergonomics, 11, 13
Micro-ergonomics, 11–13
Mitigation, 100, 102, 213, 348, 349,
    375, 381–383
  interventions, 6, 12, 13, 26, 33–36,
    38–40, 57, 69, 80, 81, 103,
    139, 144, 149, 162, 169, 173,
    174, 182, 218, 220, 230, 250,
    252, 253, 256, 262, 263, 279,
    290, 319, 329, 359, 361, 362,
    365–367, 369, 374, 383–385

N
Natural environment, *see* System,
    ecosystem
Natural resource use, 271–291
Nature connection, 165–171
Networks
  Network Theory, 31, 33, 35–39, 41
  social networks, 36, 38, 81
New Zealand, 221, 227, 356, 357,
    362, 364–368

O
Occupant wellbeing, *see* Wellbeing
Organisations, 7, 13, 28, 30, 35, 37,
    39, 100, 253, 281, 283, 290,
    332, 356, 378
  design, 12, 15, 374

P
Participatory
  approaches, 15, 251, 357, 379
  ergonomics, 41, 258
Plants, 173, 174, 177–182, 220,
    260, 274, 276, 284, 285, 290,
    357, 380
Politics, 13, 41, 67, 193,
    360, 376
  political view, 4
Pollution
  fish, 284–288
  sewage, 274, 288
  waste, 4, 40, 274, 305
  water, 272, 273, 279, 283–286,
    288, 290, 291
Product design, 245, 256
Productivity, 52, 53, 56, 69, 80,
    161, 162, 173, 174, 177,
    178, 180, 205, 214–216,
    223, 232, 233, 246, 248,
    251, 252, 334
Pro-environmental behaviour,
    83, 162, 168
Psychodynamic, 13, 15, 191–206
Psychological wellbeing, *see* Wellbeing

Q
Quality of life, 50, 58, 202, 203, 215,
    224, 248, 380
  See also Quality of work-life
Quality of work-life, 381

R

Range anxiety, 140, 142, 143
  See also Range stress
Range stress, 15, 140–144, 149
Readiness, 261, 300, 301, 306, 307,
  315–320
Recycling, 15, 31, 81, 83, 85, 87,
  151, 256, 258–261, 319, 374
Regulators, 336, 341
Renewable resources, 54, 144, 171, 382
Resilience, 10, 12, 169, 214
  resilient system, 382–384
Responsible consumption
  emerging market, 15, 299–320
  responsible computer consumption,
    77–103
  responsible resource consumption,
    213, 381
Restoration
  mental, 170
  physical, 180

S

Self-efficacy, 83, 86–87, 94–96,
  100, 102
Social mobility, 359–360
Soft Systems Methodology, 33–35
South Africa, 219, 221, 227, 228
Stress
  psychological stress, 179, 205
  stress recovery theory (SRT), 170
  work stress, 180
Supply chain, 8, 10, 12, 13, 15, 30,
  48, 191, 203, 243–263, 378, 379
  supply chain ergonomics, 8, 10, 12,
    15, 30, 245, 379
Sustainability
  corporate, 61, 191–206, 374, 384
  definition, 5, 10, 52, 195
  Human Factors and Sustainable
    Development (HFSD), 8, 9

sustainability challenges, 7, 24,
  42, 135, 136, 375, 376, 378,
  381, 385
sustainable development, 5–6, 9, 10,
  47, 48, 65, 67, 191
sustainable work, 51, 384
  (*see also* Decent work)
WCED definition, 5
Sustainable, 150, 171
  appliances, 14, 111–129
  buildings, 172, 191–206, 211–234
    (*see also* Green building)
  consumption, 15, 191, 300, 303,
    305, 311, 318, 320
  devices, 10, 14, 111–129, 191
  marketing, 299–305, 310
  organisations, 12, 13, 15, 32,
    191–206
  Sustainable Development Goals,
    243, 263, 331
  technology, 7, 13, 114, 118,
    124–126, 260
  transport, 135–136
    (*see also* Eco-mobility)
  work, 8, 10, 11, 14, 25, 48, 51, 54,
    59, 61, 62, 65–68, 194, 196,
    198, 199, 201, 205, 249, 256,
    258, 260–263, 384
System
  agricultural system, 276–277
  boundary, 12, 15, 23–42
  complex system, 7, 24, 25, 27, 30,
    38, 40, 113, 120, 125,
    191–206, 214, 233, 368, 374,
    378, 379
  control system, 128, 226,
    227, 232
  decision-support system, 373
  definition, 358
  ecosystem, 2–6, 40, 48, 78, 213,
    245, 273, 276, 291, 329, 330,
    382, 384

System (*cont.*)
  energy management system (EMS),
    80, 127
  natural system, 11, 25–27, 78, 102,
    164, 172, 277, 279, 290
  performance, 16, 79, 80, 230, 233,
    272, 355, 356, 358–361, 385
  social system, 5, 129, 379
  sociotechnical system, 11, 375,
    377, 379
System-of-systems
  definition, 385
  sustainable system-of-systems
    (SSoS), 8, 10, 11, 15,
    23–42, 378

T
Theory of Planned Behaviour (TPB),
    80, 82–86, 94, 96–99, 102, 137,
    139, 305
Thermal comfort, 224, 225, 228
Tragedy of the commons, 57, 58, 277,
    279, 283
Transdisciplinary, 41, 51, 357, 358,
    361–369, 375–377, 379
Transport
  eco-driving, 87, 136, 141, 146–150
  public, 10, 27, 319
Triple bottom line (TBL)
  economic capital, 28
  natural capital, 28
  social capital, 28

U
United Kingdom (UK), 40, 79,
    111–114, 118, 120, 125, 221,
    304, 359
United Nations Environment
  Programme (UNEP), 260
United States of America (USA), 39,
    63, 79, 113, 136, 221, 276, 304,
    365, 381

V
Value creation, 243–263
Values
  ethics, 15
  human rights, 369, 381
  transparency, 40, 369, 381
Views of nature, 178–179, 217

W
Water pollution, *see* Pollution
Water quality, *see* Pollution
Wellbeing
  definition, 355
  mental, 167–169, 203
  physical, 167–169, 382
  psychological, 168, 203
Wicked problems, 2, 23, 42, 385
Workplace layout, 26, 27, 29, 31–33,
    231–233
World Health Organisation (WHO),
    262, 275

Printed by Printforce, the Netherlands